ANIMAL MODELS of COGNITIVE IMPAIRMENT

FRONTIERS IN NEUROSCIENCE

Series Editors
Sidney A. Simon, Ph.D.
Miguel A.L. Nicolelis, M.D., Ph.D.

Published Titles

Apoptosis in Neurobiology
Yusuf A. Hannun, M.D., Professor of Biomedical Research and Chairman/Department of Biochemistry and Molecular Biology, Medical University of South Carolina
Rose-Mary Boustany, M.D., tenured Associate Professor of Pediatrics and Neurobiology, Duke University Medical Center

Methods for Neural Ensemble Recordings
Miguel A.L. Nicolelis, M.D., Ph.D., Professor of Neurobiology and Biomedical Engineering, Duke University Medical Center

Methods of Behavioral Analysis in Neuroscience
Jerry J. Buccafusco, Ph.D., Alzheimer's Research Center, Professor of Pharmacology and Toxicology, Professor of Psychiatry and Health Behavior, Medical College of Georgia

Neural Prostheses for Restoration of Sensory and Motor Function
John K. Chapin, Ph.D., Professor of Physiology and Pharmacology, State University of New York Health Science Center
Karen A. Moxon, Ph.D., Assistant Professor/School of Biomedical Engineering, Science, and Health Systems, Drexel University

Computational Neuroscience: Realistic Modeling for Experimentalists
Eric DeSchutter, M.D., Ph.D., Professor/Department of Medicine, University of Antwerp

Methods in Pain Research
Lawrence Kruger, Ph.D., Professor of Neurobiology (Emeritus), UCLA School of Medicine and Brain Research Institute

Motor Neurobiology of the Spinal Cord
Timothy C. Cope, Ph.D., Professor of Physiology, Emory University School of Medicine

Nicotinic Receptors in the Nervous System
Edward D. Levin, Ph.D., Associate Professor/Department of Psychiatry and Pharmacology and Molecular Cancer Biology and Department of Psychiatry and Behavioral Sciences, Duke University School of Medicine

Methods in Genomic Neuroscience
Helmin R. Chin, Ph.D., Genetics Research Branch, NIMH, NIH
Steven O. Moldin, Ph.D, Genetics Research Branch, NIMH, NIH

Methods in Chemosensory Research
Sidney A. Simon, Ph.D., Professor of Neurobiology, Biomedical Engineering, and Anesthesiology, Duke University
Miguel A.L. Nicolelis, M.D., Ph.D., Professor of Neurobiology and Biomedical Engineering, Duke University

The Somatosensory System: Deciphering the Brain's Own Body Image
Randall J. Nelson, Ph.D., Professor of Anatomy and Neurobiology, University of Tennessee Health Sciences Center

The Superior Colliculus: New Approaches for Studying Sensorimotor Integration
William C. Hall, Ph.D., Department of Neuroscience, Duke University
Adonis Moschovakis, Ph.D., Institute of Applied and Computational Mathematics, Crete

New Concepts in Cerebral Ischemia
Rick C. S. Lin, Ph.D., Professor of Anatomy, University of Mississippi Medical Center

DNA Arrays: Technologies and Experimental Strategies
Elena Grigorenko, Ph.D., Technology Development Group, Millennium Pharmaceuticals

Methods for Alcohol-Related Neuroscience Research
Yuan Liu, Ph.D., National Institute of Neurological Disorders and Stroke, National Institutes of Health
David M. Lovinger, Ph.D., Laboratory of Integrative Neuroscience, NIAAA

***In Vivo* Optical Imaging of Brain Function**
Ron Frostig, Ph.D., Associate Professor/Department of Psychobiology, University of California, Irvine

Primate Audition: Behavior and Neurobiology
Asif A. Ghazanfar, Ph.D., Primate Cognitive Neuroscience Lab, Harvard University

Methods in Drug Abuse Research: Cellular and Circuit Level Analyses
Dr. Barry D. Waterhouse, Ph.D., MCP-Hahnemann University

Functional and Neural Mechanisms of Interval Timing
Warren H. Meck, Ph.D., Professor of Psychology, Duke University

Biomedical Imaging in Experimental Neuroscience
Nick Van Bruggen, Ph.D., Department of Neuroscience Genentech, Inc., South San Francisco
Timothy P.L. Roberts, Ph.D., Associate Professor, University of Toronto

The Primate Visual System
John H. Kaas, Department of Psychology, Vanderbilt University
Christine Collins, Department of Psychology, Vanderbilt University

Neurosteroid Effects in the Central Nervous System
Sheryl S. Smith, Ph.D., Department of Physiology, SUNY Health Science Center

Modern Neurosurgery: Clinical Translation of Neuroscience Advances
Dennis A. Turner, Department of Surgery, Division of Neurosurgery, Duke University Medical Center

Sleep: Circuits and Functions
Pierre-Hervé Luoou, Université Claude Bernard Lyon I, Lyon, France

Methods in Insect Sensory Neuroscience
Thomas A. Christensen, Arizona Research Laboratories, Division of Neurobiology, University of Arizona, Tucson, AZ

Motor Cortex in Voluntary Movements
Alexa Riehle, INCM-CNRS, Marseille, France
Eilon Vaadia, The Hebrew University, Jeruselum, Israel

Neural Plasticity in Adult Somatic Sensory-Motor Systems
Ford F. Ebner, Vanderbilit University, Nashville, TN

Advances in Vagal Afferent Neurobiology
Bradley J. Undem, Johns Hopkins Asthma Center, Baltimore, MD
Daniel Weinreich, University of Maryland, Baltimore, MD

The Dynamic Synapse: Molecular Methods in Ionotropic Receptor Biology
Josef T. Kittler, University College, London
Stephen J. Moss, University of Pennsylvania

Animal Models of Cognitive Impairment
Edward D. Levin, Duke University Medical Center, Durham, NC
Jerry J. Buccafusco, Medical College of Georgia, Augusta, GA

ANIMAL MODELS of COGNITIVE IMPAIRMENT

Edited by

Edward D. Levin

Duke University Medical Center
Durham, NC

Jerry J. Buccafusco

Medical College of Georgia
Augusta, GA

CRC Press
Taylor & Francis Group
Boca Raton London New York

CRC Press is an imprint of the
Taylor & Francis Group, an **informa** business
A TAYLOR & FRANCIS BOOK

CRC Press
Taylor & Francis Group
6000 Broken Sound Parkway NW, Suite 300
Boca Raton, FL 33487-2742

First issued in paperback 2019

ISBN-13: 978-0-8493-2834-3 (hbk)
ISBN-13: 978-0-367-39067-9 (pbk)
Library of Congress Card Number 2006040945

Library of Congress Cataloging-in-Publication Data

Animal models of cognitive impairment / edited by Edward D. Levin, Jerry J Buccafusco.
      p. cm. -- (Frontiers in neuroscience)
  ISBN 0-8493-2834-9 (alk. paper)
  1. Cognition disorders--Animal models. I. Levin, Edward D. II. Buccafusco, Jerry J. III. Frontiers in neruoscience (Boca Raton, Fla.)

RC553.C64A55 2006
616.8--dc22   2006040945

**Visit the Taylor & Francis Web site at
http://www.taylorandfrancis.com**

**and the CRC Press Web site at
http://www.crcpress.com**

Preface

Jerry J. Buccafusco

In recent years new approaches to drug discovery have provided a multitude of potential therapeutic targets, particularly for disorders of the central nervous system. High-throughput screening of vast libraries of peptide and nonpeptide compounds has provided an unprecedented number of drug entities that require evaluation for efficacy towards these newly established therapeutic targets. Protein-protein, and ligand-protein binding formats have been relegated to high capacity reaction plates and biochips. With this level of efficiency, the pharmaceutical industry should look forward to new heights of success in terms of the preclinical development of new and exciting drugable entities. Yet despite this optimism, it remains difficult to translate these findings from ELISA wells or chip spots directly to the clinic. This is particularly true for centrally acting compounds. Drug development strategies for the treatment of CNS disorders and neurodegenerative diseases appear to defy the older strategy of high selectivity and high potency. This strategy is being slowly replaced by the recognition that for many disease entities there potentially exist a wide variety of therapeutic targets. A drug molecule that targets two or more synergistically interacting targets could prove more effective than a compound highly selective and highly potent at one binding site. Evidence for this approach can be readily appreciated from the success enjoyed by more recently utilized "atypical" antipsychotic drugs in the treatment of schizophrenia. This class of agents is typified by poor receptor selectivity and moderate binding affinities. Similar situations occur with certain antidepressant and anti-Parkinson's drugs.

The multitarget model is perhaps even more pertinent to cognition-enhancing agents. New drugs developed to enhance memory and cognition have been targeted to a wide variety of neurotransmitter and neurohormone receptors. This laboratory has had the good fortune to be able to study a large number of these drugs in essentially the same nonhuman primate model (see Chapter 13). It is perhaps not so surprising that a function as complex and well integrated as cognition would utilize numerous neurotransmitter and hormonal pathways. Also, since neurodegenerative diseases often require both symptomatic treatment and disease modification, an efficient therapeutic approach would be to package both in the same molecule. Another aspect of cognitive pharmacology is the ability of many drugs that enhance cognition to induce a pharmacodynamic action that outlasts the presence of the drug in blood or brain compartments (see Chapter 13). It is not yet possible to predict whether, or to what degree, a compound will exert this protracted beneficial action. The implication is that knowledge of the pharmacodynamic profile of a new drug will be important in determining appropriate dosing regimens in the clinical setting.

The only means by which to predict the utility of a new potential cognition-enhancing drug is to study it in one or more of the relevant animal models. Though this component of the drug discovery pipeline represents the most restrictive pipe, it is perhaps one of the most important. Whether a new drug is discovered in academic or industrial settings, the compound will require significant animal testing prior to clinical study. The costs associated with clinical trials are so significant that both safety and good efficacy need to be validated in animal models prior to continued study of the drug in humans. The purpose of this book is to acquaint the reader with some of the most utilized of the animal models for cognition impairment. From mice to rats to monkeys these models have been used successfully in the context of drug discovery. Some have also provided new insights into disease processes.

About the Editors

Edward D. Levin, Ph.D., is a Professor of Psychiatry and Behavioral Sciences at Duke University Medical Center. He has appointments in the Departments of Psychiatry, Pharmacology, and Psychology and in the Nicholas School of the Environment at Duke University. He earned a B.A. in Psychology at the University of Rochester, NY, in 1976; he then earned an M.S. in Psychology in 1982 and a Ph.D. in Environmental Toxicology in 1984 at the University of Wisconsin. He was an NIH-sponsored postdoctoral fellow in the Psychology Department at the University of California at Los Angeles and was a visiting scientist at Uppsala University in Sweden. He has conducted research and taught at Duke University since 1989. Dr. Levin's research interests concern behavioral neuroscience, with emphasis on behavioral pharmacology and toxicology. He is particularly concerned with drug and toxicant effects on neuroplasticity, including applications to cognitive function, addiction, and sensorimotor modulation.

Jerry J. Buccafusco, Ph.D., is director of the Alzheimer's Research Center, in the Department of Pharmacology and Toxicology of the Medical College of Georgia. He holds the rank of Professor of Pharmacology and Toxicology and Professor of Psychiatry and Health Behavior. He holds a joint appointment as Research Pharmacologist and Director of the Neuropharmacology Laboratory at the Department of Veterans Affairs Medical Center. Dr. Buccafusco also is President and CEO (and founder, established March 1, 2000) of Prime Behavior Testing Laboratories, Inc. (Evans, GA), a contract research company for the preclinical evaluation of cognition-enhancing therapeutic agents.

Dr. Buccafusco was trained classically as a chemist, receiving an M.S. degree in inorganic chemistry from Canisius College in 1973. His pharmacological training was initiated at the University of Medicine and Dentistry of New Jersey where he received a Ph.D. degree in 1978. His doctoral thesis concerned the role of central cholinergic neurons in mediating a hypertensive state in rats. Part of this work included the measurement of several components of hypothalamically mediated escape behavior in this model. His postdoctoral experience included two years at the Roche Institute of Molecular Biology under the direction of Dr. Sydney Spector. In 1979 he joined the Department of Pharmacology and Toxicology of the Medical College of Georgia. In 1989 Dr. Buccafusco helped found and became the director of the Medical College of Georgia, Alzheimer's Research Center. The center hosts several core facilities, including the Animal Behavior Center, which houses over 50 young and aged macaque monkeys who participate in cognitive research studies.

Contributors

Kelly M. Banna
Department of Psychology
Auburn University
Auburn, Alabama

Anna I. Baranova
School of Medicine
Commonwealth University
Richmond, Virginia

Jerry J. Buccafusco, Ph.D.
Department of Pharmacology and
  Toxicology
Medical College of Georgia
Augusta, Georgia
Veterans Administration Medical Center
Augusta, GA,
Prime Behavior Testing Laboratories
Evans, Georgia

Daniel T. Cerutti, Ph.D.
Department of Psychological and Brain
  Sciences
Duke University
Durham, North Carolina

Michael W. Decker, Ph.D.
Neuroscience Research
Abbott Laboratories
Abbott Park, Illinois

Wendy D. Donlin
Department of Psychology
Auburn University
Auburn, Alabama
Department of Psychiatry and
  Behavioral Sciences
Johns Hopkins School of Medicine
Baltimore, Maryland

Robert J. Hamm, Ph.D.
School of Medicine
Commonwealth University
Richmond, Virginia

Edward D. Levin, Ph.D.
Department of Psychiatry and
  Behavioral Sciences
Duke University Medical Center
Durham, North Carolina

Dave Morgan, Ph.D.
Department of Pharmacology
University of South Florida
Tampa, Florida

M. Christopher Newland, Ph.D.
Department of Psychology
Auburn University
Auburn, Alabama

Elliott M. Paletz
Department of Psychology
University of Wisconsin
Madison, Wisconsin

Merle G. Paule, Ph.D.
Behavioral Toxicology Laboratory,
  Division of Neurotoxicology
National Center for Toxicology
  Research/FDA
Jefferson, Arkansas

Amir H. Rezvani, Ph.D.
Department of Psychiatry and
  Behavioral Sciences
Duke University Medical Center
Durham, North Carolina

Deborah C. Rice, Ph.D.
Maine Center for Disease Control and
  Prevention
Woolwich, Maine

Ramona Marie Rodriguiz, Ph.D.
Department of Psychiatry and
  Behavioral Sciences
Duke University Medical Center
Durham, North Carolina

Cindy S. Roegge, Ph.D.
Department of Psychiatry and
  Behavioral Sciences
Duke University Medical Center
Durham, North Carolina

Helen J.K. Sable, Ph.D.
Department of Veterinary Biosciences,
  College of Veterinary Medicine
University of Illinois
Urbana, Illinois

Susan L. Schantz, Ph.D.
Department of Veterinary Biosciences,
  College of Veterinary Medicine
University of Illinois
Urbana, Illinois

Jay S. Schneider, Ph.D.
Department of Pathology, Anatomy and
  Cell Biology
Thomas Jefferson University
Philadelphia, Pennsylvania

Alvin V. Terry, Jr., Ph.D.
College of Pharmacy
University of Georgia
Augusta, Georgia
Clinical Pharmacy
Medical College of Georgia
Augusta, Georgia

William C. Wetsel, Ph.D.
Department of Psychiatry and
  Behavioral Sciences
Duke University Medical Center
Durham, North Carolina

Mark D. Whiting, Ph.D.
School of Medicine
Commonwealth University
Richmond, Virginia

Lu Zhang, Ph.D.
Pfizer Global Research and
  Development
Pfizer, Inc.
New London, Connecticut

Contents

List of Illustrations

Dedication

To Regina, Chris, and Marty
Jerry Buccafusco

To Risa, Holly, and Laura
Ed Levin

1 Introduction

Edward D. Levin
Duke University Medical Center

Jerry J. Buccafusco
Medical College of Georgia

Animal models of cognitive impairment are critically important for determining the neural bases of learning, memory, and attention. These cognitive functions are the result of complex interactions of a variety of neural systems and thus cannot be well studied by simple *in vitro* models. Animal models of cognitive impairment are critical for determining the neural basis of cognitive function as well as for testing the efficacy of potential therapeutic drugs and the neurocognitive toxicity of environmental contaminants and drugs of abuse. A variety of models have used classic monkey, rat, and mouse models. Newer, nonmammalian complementary models with fish, flies, and flatworms are being developed. These will play an important role in both high-throughput screening of potential toxic or therapeutic compounds and in the determination of the neuromolecular bases of cognitive function.

Pharmacological models are the most commonly used models of cognitive dysfunction. They are key for determining how selected neurotransmitter-receptor systems are involved in various aspects of cognitive function such as learning, memory, and attention. Pharmacological models are also key for testing the possible use of cognition-enhancing drugs for potential treatment of cognitive impairments such as those seen in Alzheimer's disease and other aging-related syndromes, attention deficit hyperactivity disorder (ADHD), Parkinson's disease, and schizophrenia. Acetylcholine is the best-characterized transmitter system in the basis of cognitive function. Both muscarinic and nicotinic cholinergic receptors have been found to be critically involved. Glutamate systems, particularly those using NMDA (N-methyl-D-aspartate) receptors, have also extensively been shown to be involved in cognition. Drugs of abuse can also produce syndromes of cognitive impairment. The best example is ethanol, which impairs a variety of cognitive functions. All of these areas are covered in detail in Section I of this book (Chapters 2–5).

Applications of animal models of cognitive impairment have been quite successful in the realm of neurobehavioral toxicology. Lead is the prime example. Persisting cognitive impairments after developmental exposure to lead have been quite well modeled in monkeys and rodents. Other metals, notably mercury, have also been well studied with animal models. The same approach can also be taken with nonmetal toxicants. Notably, polychlorinated biphenyls (PCBs) have been shown to cause persistent cognitive effects in monkey as well as rodent models. Neurotoxicants can be used in quite specific ways as probes of the involvement of

1

particular brain systems in the basis of cognitive function. MPTP (1-methyl-4-phenyl-1,2,3,6-tetrahydropyridine) is an excellent example of how this approach has worked both for the discovery of the importance of a particular neural system (midbrain dopamine neurons) in cognitive function and as a forum for the development of new drug treatments for a disease (e.g., Parkinson's disease) involving that system.

Mice are becoming increasingly valuable in efforts to determine the molecular bases of cognitive function. In addition, genetically manipulated mice are increasingly being used in the development of models for new drug development. For these uses it is important to devise a valid, reliable, and quick battery of tests to determine cognitive function in mice. Application of transgenic mice to problems presented by amyloid deposition with aging is an especially promising forum for the development of new treatments for Alzheimer's disease and other aging-related cognitive impairments. Pharmacological models have shown that acetylcholine plays key roles in the neural bases of cognitive functions. Cholinergic-receptor knockout mice are being used to good effect in determining the role of various aspects of the cholinergic systems in cognitive function.

Animal models can quite well simulate specific syndromes of cognitive impairment where the inciting faction is well known. Prime examples of this approach include studies of aging and neurotrauma. Aging studies, especially with long-lived species such as monkeys, readily demonstrate aging-induced cognitive impairment and serve as a fine basis for developing new treatments and novel drugs. Neurotrauma causes cognitive impairment in animal models in quite similar ways as in humans. The specific mechanisms underlying such impairment and the therapeutic treatments for it can be well studied in animal models.

New nonmammalian models of cognitive impairment are being developed. These models are sometimes called alternative models but are better termed complementary models, because they are best used not in place of mammalian models but to complement them. Mammalian and nonmammalian models each have their own sets of advantages and disadvantages, which can be used in a mutually complementary fashion in a strategy of research advancement. Mammals have a high degree of neuroanatomic similarity to humans, but they are generally expensive and time-consuming models to use. Nonmammalian models can be very useful in high-throughput studies for identifying potential toxicants and discovering potential therapeutic drugs, but they have a lower degree of similarity to the neural processes involved in cognitive function compared with humans. Historically, aquatic models began with the use of goldfish, but more recently the models have focused on zebrafish because of the great flowering of molecular information about neural processes in this species. Invertebrate models including flatworms (*C. elegans*) and insects (drosophila) show promise, but these need much more development to become generally useful.

Animal models of cognitive impairment play crucial roles in the characterization of toxicants that cause cognitive dysfunction and the identification of potential new drugs for treating cognitive dysfunction, as well as providing critical insight into the neural bases of cognitive function and dysfunction. There are a variety of important issues specific to each model and in general across models that must be considered if these models are to be used productively.

Section I

Pharmacologic Models

2 Muscarinic Receptor Antagonists in Rats

Alvin V. Terry, Jr.
University of Georgia College of Pharmacy
and
Medical College of Georgia

CONTENTS

INTRODUCTION

The importance of cholinergic activity in the brain to learning and memory function was first recognized more than 30 years ago, when relatively low doses of certain muscarinic acetylcholine-receptor antagonists (e.g., the belladonna alkaloids atropine and scopolamine) were found to induce transient cognitive deficits in young human volunteers that resembled those observed in elderly (unmedicated) subjects.[1] This work and a number of subsequent clinical studies indicated that antimuscarinics disrupt attention,[2-4] the acquisition of new information, and the consolidation of memory.[1,4,5] Later studies found that scopolamine could alter certain features of the human electroencephalogram (e.g., delta, theta, alpha, and beta activity) in a fashion that mimics some of the changes observed in patients with Alzheimer's disease (AD) (reviewed by Ebert and Kirch[6]).

In support of these cited pharmacological investigations, a considerable number of studies of the brains of elderly people and of AD patients have shown damage or abnormalities in forebrain cholinergic projections that are important to memory structures (e.g., cortex, hippocampus), and these results correlate well with the level of cognitive decline.[7,8] Interestingly, scopolamine appears to negatively affect cognitive performance to a greater extent in elderly subjects than in younger subjects,[9-11] and it impairs subjects with AD more dramatically than nondemented elderly subjects.[12] Similarly, aged rodents display cognitive impairments in many learning and memory tasks,[13] manifest cholinergic deficits,[14-16] and are more sensitive to the disruptive effects of scopolamine than young rats.[16] Moreover, lesions in young animals that damage cholinergic input from the basal forebrain (e.g., nucleus basalis magnocellularis, medial septum/diagonal band) to the neocortex or hippocampus disrupt performance of a variety of memory-related tasks. These findings, particularly the early studies cited, led to the development of the so-called cholinergic hypothesis, which essentially states that a loss of cholinergic function in the central nervous system (CNS) contributes significantly to the cognitive decline associated with advanced age and AD (reviewed by Bartus[17]). Using this hypothesis as a guiding principle, scopolamine (and to a lesser extent other antimuscarinic agents) has been used extensively as an amnestic drug in animals to model the cognitive dysfunction associated with human dementia and Alzheimer's disease. While a few studies have been conducted using scopolamine as an amnestic agent in nonhuman primates (e.g., Aigner and Mishkin[18] and Terry et al.[19]), the majority of studies have been conducted in rodents, particularly rats. In addition, scopolamine-reversal experiments in rodents have been used extensively as an initial screening method to identify therapeutic candidates for cognitive disorders such as AD.

MEMORY-RELATED TASK IMPAIRMENT IN RATS BY SCOPOLAMINE AND OTHER ANTIMUSCARINIC AGENTS

Table 2.1 provides an overview (with a few representative references) of the wide variety of memory-related tasks that have been shown to be impaired by the nonselective antimuscarinic agent, scopolamine. Such tests encompass a wide range of behavioral procedures from classical conditioning (Pavlovian) tasks (e.g., inhibitory avoidance and fear conditioning), to spatial learning tasks (e.g., water maze and radial arm maze), to methods that assess prepulse inhibition of the auditory gating response. Antimuscarinics have also been shown to disrupt performance of more-complex operant (working-memory related) procedures such as delayed matching and delayed nonmatching tasks, time estimation procedures, and most recently, models of executive function (e.g., attentional set shifting) in rats.

Similar to scopolamine, the nonselective muscarinic antagonist atropine, as well as several drugs with selectivity at muscarinic-receptor subtypes, disrupts performance of several of the same behavioral procedures (see Table 2.1). For example, the M_1-selective antagonist pirenzepine impairs learning of rats in such hippocampal-dependent tasks as delayed nonmatching to position[20] and delayed matching to position[21] as well as working memory in tests such as the radial arm maze,[22] spatial memory in the water maze,[23,24] and T-maze representational memory.[25] The passive-avoidance test is impaired by either systemic[26] or central injections[27,28] of pirenzepine, administered before the acquisition session. In the previously cited Andrews study,[21] the effects of pirenzepine were interpreted as more specific for spatial short-term memory than scopolamine, since delay-dependent disruption of performance

TABLE 2.1
Behavioral Studies Indicating Amnestic Effects
of Antimuscarinic Agents in Rats

Pharmacologic Agent	Behavioral Test	References
Scopolamine	Water maze	68,72,78,79
	Radial arm maze	80–82
	T-maze	83–85
	Y-maze	50,86,87
	Delayed matching to position	21,58,88,89
	Delayed nonmatching to position	45,59
	Delayed conditional time discrimination task	90
	Other delayed conditional discrimination tasks	61,65,91
	Passive avoidance	56,68,79,92
	Fear conditioning	93,94
	Novel object recognition	95–97
	Prepulse inhibition	30,98
	Extradimensional set shifting	60
Atropine	Water maze	99–101
	Radial arm maze	22
	Delayed conditional discrimination	102
	Fear conditioning	103
	Passive avoidance	104,105
Pirenzepine	Water maze	23,24
	Radial arm maze	22
	T-maze	25
	Delayed matching to position	21
	Delayed nonmatching to position	20
	Passive avoidance	26–28
Trihexyphenidyl	Passive avoidance	29
	Prepulse inhibition	30
Benztropine	Prepulse inhibition	30
Biperidine	Passive avoidance	29
Dicyclomine	Passive avoidance	32
	Fear conditioning	32

was noted with the former but not the latter compound. In fact, scopolamine induced a delay-independent disruption of all task parameters, including motivation and motor performance. Since the M_2-receptor antagonist AFDX 116 had no effect on task performance in the same study, the authors suggested that M_2 receptors are not responsible for the disruptive effects of muscarinic antagonists on spatial short-term memory. Regarding other selective muscarinic antagonists, Roldán et al.[29] showed that systemic administration of the M_1 antagonists biperidine and trihexyphenidyl impaired the consolidation of the passive-avoidance response in a dose-dependent manner. Trihexyphenidyl and benztropine (another M_1 antagonist) produced significant dose-dependent decreases in prepulse inhibition in a fashion similar to that of scopolamine,[30] while the M_1-selective antagonist dicyclomine[31] impaired passive-avoidance learning as well as contextual fear conditioning in rats.[32]

THE NATURE OF THE EFFECTS OF ANTIMUSCARINICS ON MEMORY PERFORMANCE (POTENTIAL LIMITATIONS)

While few would argue that muscarinic antagonists have negative effects on memory-related task performance, the specifics of such behavioral effects have been debated for years (see reviews in the literature[33–37]). Many investigators have interpreted the negative behavioral actions as impairment of learning (or state-dependent learning), working memory, short-term memory, encoding/consolidation, or retrieval. Conversely, others propose that the behavioral effects of antimuscarinic agents are not specific to learning and memory, since muscarinic receptors are also involved in the regulation of attention and arousal.[4,6,36,38] The effects of atropine in a nonspatial (visual discrimination) swimming task led Whishaw and Petrie[39] to conclude that cholinergic systems are involved in the selection of the movements or strategies that are prerequisite for learning, as opposed to learning and memory *per se*. Such (so-called) nonmnemonic effects were also deduced from the delay-independent performance deficits observed in several delayed-response tasks[40–43] as well as by deficits in both motor function and motivation.[44] The negative effects of scopolamine on the accuracy of short retention intervals in a delayed nonmatching-to-position task led Han et al.[45] to suggest that antimuscarinic agents produce general deficits in reference or procedural memory. Finally, some investigators completely reject any mnemonic explanations for the actions of antimuscarinic agents and argue that they primarily alter stimulus sensitivity, sensory discrimination, vision, perseveration, or habituation, etc. (reviewed by Collerton[46]).

The fact that a wide variety of noncholinergic agents have been observed to antagonize the behavioral effects of scopolamine further supports some of the arguments presented above (i.e., that the scopolamine amnesia model is not selective for memory function or even for the cholinergic system). For example, the nootropic agent piracetam has been observed to antagonize scopolamine in a passive-avoidance task[47] and in a delayed match-to-position task.[48] Estrogen replacement in ovariectomized rats significantly reduced deficits in the performance of a delayed matching-to-position (T-maze), with the deficits being produced by intrahippocampal, but not systemic, scopolamine administration.[49] A TRH (thyrotropin releasing hormone) analog, NS-3(CG3703), ameliorated scopolamine-induced impairments in a delayed nonmatching-to-sample task using a T-maze as well as a radial arm maze procedure.[50] The ethanolic extract of Indian *Hypericum perforatum* (an herbal agent) reversed scopolamine deficits in passive avoidance,[51] while the seed oil of *Celastrus paniculatus* (another Indian herbal agent) prevented scopolamine-related deficits in water-maze performance in rats.[52] Furthermore, a number of serotonin-receptor ligands (e.g., 5-HT$_3$, 5HT$_{1A}$, 5HT$_4$, 5-HT$_6$)[53–56] reverse scopolamine deficits in several memory-related tasks, as do drugs from a number of additional classes (e.g., amphetamine, fluoxetine, and strychnine, as reviewed by Blokland[36]).

Obviously, these effects do not add credence to the idea of using antimuscarinics to model AD, which is clearly characterized by deficits in cholinergic activity, learning, working memory, and executive function. However, in contrast to the aforementioned delayed-response tasks, there are several rat studies in which scopolamine was found to have delay-dependent effects (e.g., spatial alternation, delayed matching, and delayed nonmatching-to-sample tasks) that were interpreted as deficits in working-memory performance.[57–59] Interestingly, a recent study indicated that scopolamine (but not methylscopolamine) impaired both affective (reversal learning) and attentional set-shifting components in the rat, thus implicating muscarinic receptors in the CNS control of executive function.[60] Furthermore, the effects of antimuscarinics on attention (often discussed as nonmnemonic) are not necessarily a limitation to using antimuscarinics to model certain aspects of dementia, since attentional deficits are a common feature of the condition. Finally, the criticisms regarding the so-called reversal effects of noncholinergic agents cannot be considered altogether convincing until rigorous pharmacological investigations of each of the aforementioned compounds are completed. It is clear that several representative agents from the classes listed (e.g., estrogen, 5-HT ligands, and herbal extracts) indeed have direct or indirect effects on the cholinergic system (see the literature[52, 61–63]).

OTHER LIMITATIONS AND CRITICISMS OF THE USE OF ANTIMUSCARINICS AS AMNESTIC AGENTS

The use of antimuscarinic agents to model human dementia and the conclusions drawn from such studies have been criticized for a number of additional reasons (see Fibiger[34]). The most formidable challenge appears to arise from the limitations of using a unitary theoretical construct (i.e., central muscarinic-receptor blockade) to model memory-related dysfunction, considering the fact that cholinergic neurons (or their projections) and, in particular, muscarinic receptors are found in virtually every region of the CNS. Accordingly, the disruption of memory-related task performance by antimuscarinic agents could arise from a multitude of pharmacological actions. Such effects could also contribute to the inconsistent results that have often been observed across several types of memory-related tasks, as discussed previously. In light of the diverse brain lesions observed in conditions such as AD and the inherent complexities involved in learning and memory processing, single-transmitter hypotheses of dementia have been described as "worse than naïve" by some.[33] Thus, while the "face validity" of scopolamine amnesia models can be argued from the perspective that pronounced cholinergic dysfunction in the brain is a common feature of dementia, "construct validity" may be less convincing, since scopolamine impairments are acute, systemic, and (thought to be) primarily postsynaptic, whereas the pathology of dementia is characterized as chronic, brain-specific, and — in the context of cholinergic projections — primarily presynaptic (reviewed by Steckler[64]). Furthermore, "predictive validity" also appears to be limited, given that numerous investigational compounds have been observed to have the ability to reverse or attenuate scopolamine-related deficits in memory-related tasks, whereas few have been successful in clinical trials (see Sarter et al.[35]).

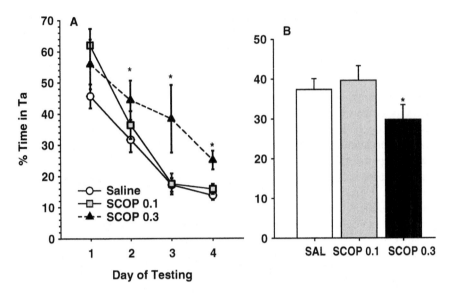

FIGURE 2.1 Dose-related effects of scopolamine on performance of rats in a water-maze test. Scopolamine hydrobromide (0.1 or 0.3 mg/kg) or saline was administered subcutaneously 30 min before testing. $N = 10$ to 12 rats per group. A: Results of hidden-platform test (days 1–4) show mean latencies to locate a hidden platform on each day of a water-maze task; bars represent the mean of four trials per day (in seconds ± SEM, the standard error of the mean). B: Results of probe trials (day 5) show the percentage of total time (± SEM) spent in the quadrant where the platform was located during the previous four days of testing. Note: asterisks (*) represent performance significantly different ($p < 0.05$) from saline performance. There was a significant dose-related impairment of performance ($p < 0.05$) in each test.

Somewhat less tenable criticisms are the so-called absence of dose-effect analyses in much of the published work and the failure of many investigators to use methylated (i.e., charged) forms of agents such as scopolamine and atropine to control for the effects of peripheral muscarinic blockade (reviewed by Fibiger[34]). While it is common for investigators to use a single dose of scopolamine in memory studies (in particular, the studies in which potential therapeutic agents are screened for their ability to prevent or reverse the amnestic effects of antimuscarinics), there are now a multitude of published studies across a number of behavioral paradigms in which dose-effect analyses have been conducted. As a result, it seems unnecessary for every investigator who performs a study with scopolamine to reestablish dose-effect curves, particularly in well-established learning and memory paradigms. Figure 2.1 through Figure 2.4 provide representative data from work in our laboratory, where dose-dependent effects of scopolamine (or clear dose-related trends) were observed in (three of the four) behavioral tasks ranging from spatial learning to auditory gating tasks to working-memory tasks. It should be noted that we did not detect delay-dependent impairments in performance of our delayed-response task (the Delayed Stimulus Discrimination task).[61,65]

The use of methylated (i.e., charged or quaternary amine) forms of antimuscarinic agents as controls is also well documented in the literature. Several studies

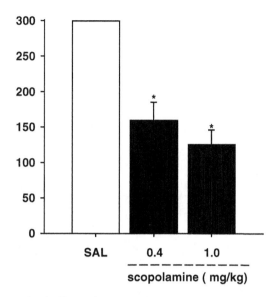

FIGURE 2.2 Dose-related effects of scopolamine on performance of rats in a step-through latency test. Scopolamine hydrobromide (0.4 or 1.0 mg/kg) or saline was administered subcutaneously 30 min before testing in a passive (inhibitory avoidance) task utilizing a 0.5-mA foot shock. $N = 15$ to 20 rats per group. Bars represent mean step-through latencies (in seconds ± SEM), i.e., retention trials, to enter the dark chamber 24 h after training trials. Note: asterisks (*) represent performance significantly different ($p < 0.05$) from saline performance. There was no significant difference between the two doses evaluated ($p > 0.05$ for dose effect).

using scopolamine methylbromide as a control reported that there was no effect of this analog on task performance, while other studies found that tertiary anticholinergic drugs ranged from 20 to 36 times more effective than quaternary analogs at impairing memory-related task performance. It should also be noted, however, that many of these studies involved spatial learning tasks and that significant effects of quaternary anticholinergics have been observed in a number of studies using food-motivated, delayed-response tasks (for a review, see Evans-Martin et al.[65]). Several plausible explanations for an effect of quaternary cholinergic antagonists on such behaviors have been presented. For example, scopolamine may reduce the motivation for food consumption by reducing salivary secretions, by altering the taste of the reward, or by altering gastrointestinal function. These effects have been suggested in studies of the scopolamine salts and conditioned taste aversion.[66] In humans, muscarinic antagonists induce other peripheral effects such as blurred vision (by impairing accommodation[67]), a factor that (if also present in rodents) could certainly impact the performance of tasks that require the discrimination of visual cues (e.g., water maze). We have, in fact, observed some impairment of the visible-platform task in the Morris water maze in previous experiments in our laboratory.[68]

Quaternary anticholinergics have been shown to inhibit the production of epinephrine by the adrenal medulla, which in turn decreases the release of glucose from the liver.[69,70] This may decrease the entry of glucose into the brain and thus the subsequent

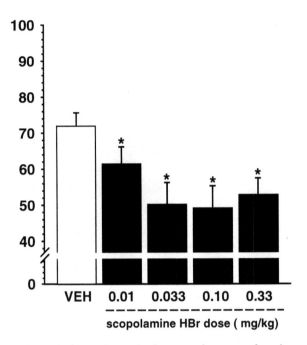

scopolamine HBr dose (mg/kg)

FIGURE 2.3 Dose-related effects of scopolamine on performance of rats in a test of prepulse inhibition of the auditory gating response. Scopolamine hydrobromide (0.01, 0.033, 0.10, or 0.33 mg/kg) or saline was administered subcutaneously 40 min before testing. $N = 10$ to 12 rats per group. The prepulse stimuli (75, 80, and 85 dB) inhibited the startle response to a 120-dB auditory stimulus in a fashion that was dependent on the decibel level (data not shown). The bars depicting the scopolamine effect are averaged across prepulse levels. Note: asterisks (*) represent performance significantly different ($p < 0.05$) from baseline (saline) performance. There was a significant dose-related impairment of the gating response ($p < 0.05$) in the test. Each value represents the mean ± S.E.M.

synthesis of acetylcholine or other important neurotransmitters. In support of this hypothesis, the effects of scopolamine on passive avoidance and spontaneous alternation have been shown to be attenuated or reversed by glucose or epinephrine.[71] Finally, there is credible evidence that behaviorally significant concentrations of quaternary anticholinergics penetrate the brain despite their polarized state *in vivo*.[21]

SCOPOLAMINE-REVERSAL STUDIES AND DRUG DISCOVERY

Notwithstanding the criticisms described in the preceding section, the scopolamine-reversal paradigm has retained popularity in drug discovery programs for the past 10 to 15 years for screening potential antidementia (or cognition-enhancing) compounds. It is likely that one of the reasons for the continued popularity of the model is that it can be set up relatively cheaply using rodents to perform tasks such as passive avoidance and the Morris maze, which makes the model amenable to high-throughput screening approaches. While the predictive validity of the model

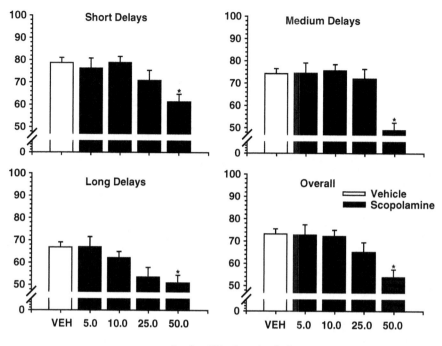

scopolamine HBr dose (µg/kg)

FIGURE 2.4 Dose-related effects of scopolamine on performance of rats in a delayed stimulus discrimination task (DSDT). Scopolamine hydrobromide (5, 10, 25, and 50 µg/kg) or saline was administered subcutaneously 30 min before testing. $N = 8$ rats per group. Rats were required to discriminate between a light and a tone and then, after a preprogrammed delay, to select a right or left lever to receive a food reward. Graphs 1 to 3: dose-effect curves for scopolamine hydrobromide on accuracy at short, medium, and long delays of the DSDT. Graph 4: dose-effect curves averaged across the delays. Each bar represents the mean ± SEM. Note: asterisks (*) represent performance significantly different ($p < 0.05$) from baseline (saline) performance. There was a significant dose effect ($p < 0.05$), but not a significant dose × delay interaction.

for drug development purposes has been challenged, it should be noted that all of the commonly prescribed antidementia compounds have been observed to attenuate scopolamine-induced deficits in attention, learning, and memory in rodents. For example, rivastigmine attenuated scopolamine-related deficits in a dose-dependent fashion in an operant delayed nonmatching-to-position (working memory) task in rats[59] and improved deficits in both reference and working-memory versions of a water-maze task.[72] Galantamine attenuated scopolamine-induced deficits in passive-avoidance tasks[73,74] as well as in T-maze and Morris water-maze tasks.[75] Donepezil (E2020) attenuated scopolamine-related impairments in a delayed match-to-position task, a five-choice serial-reaction time (sustained attention) task,[76] and an eight-arm radial arm maze task.[77] Finally, recent data implicating CNS muscarinic receptors in the control of executive function is likely to further increase the enthusiasm related to this model.[60]

CONCLUSIONS

For more than three decades, antimuscarinic agents — particularly scopolamine — have been used to investigate the role of the CNS cholinergic system in learning and memory processes in the mammalian brain and to screen potential treatments for dementia. For a variety of reasons (peripheral actions, nonmnemonic effects, lack of construct and predictive validity, reversal by noncholinergic agents), the scopolamine amnesia model has been criticized, although convincing arguments have also been presented to support the model. The popularity of the scopolamine amnesia model is likely to continue, especially in drug discovery programs for dementia-related therapeutics, in light of recent data implicating CNS muscarinic receptors in the control of executive function.[60]

REFERENCES

1. Drachman, D.A. and Leavitt, J., Human memory and the cholinergic system: a relationship to aging? *Arch. Neurol.*, 30, 113–121, 1974.
2. Wesnes, K. and Warburton, D.M., Effects of scopolamine on stimulus sensitivity and response bias in a visual vigilance task, *Neuropsychobiology*, 9, 154–157, 1983.
3. Wesnes, K. and Warburton, D.M., Effects of scopolamine and nicotine on human rapid information processing performance, *Psychopharmacology (Berl.)*, 82, 147–150, 1984.
4. Broks, P., Preston, G.C., Traub, M., Poppleton, P., Ward, C., and Stahl, S.M., Modelling dementia: effects of scopolamine on memory and attention, *Neuropsychologia*, 26, 685–700, 1988.
5. Petersen, R.C., Scopolamine induced learning failures in man, *Psychopharmacology (Berl.)*, 52, 283–289, 1977.
6. Ebert, U. and Kirch, W., Scopolamine model of dementia: electroencephalogram findings and cognitive performance, *Eur. J. Clin. Invest.*, 28, 944–949, 1998.
7. Perry, E.K., Blessed, G., Tomlinson, B.E., Perry, R.H., Crow, T.J., Cross, A.J., Dockray, G.J., Dimaline, R., and Arregui, A., Neurochemical activities in human temporal lobe related to aging and Alzheimer-type changes, *Neurobiol. Aging*, 2, 251–256, 1981.
8. Francis, P.T., Palmer, A.M., Sims, N.R., Bowen, D.M., Davison, A.N., Esiri, M.M., Neary, D., Snowden, J.S., and Wilcock, G.K., Neurochemical studies of early-onset Alzheimer's disease: possible influence on treatment, *N. Engl. J. Med.*, 313, 7–11, 1985.
9. Zemishlany, Z. and Thorne, A.E., Anticholinergic challenge and cognitive functions: a comparison between young and elderly normal subjects, *Isr. J. Psychiatry Relat. Sci.*, 28, 32–41, 1991.
10. Flicker, C., Ferris, S.H., and Serby, M., Hypersensitivity to scopolamine in the elderly, *Psychopharmacology (Berl.)*, 107, 437–441, 1992.
11. Terry, A.V., Jr. and Buccafusco, J.J., The cholinergic hypothesis of age and Alzheimer's disease-related cognitive deficits: recent challenges and their implications for novel drug development, *J. Pharmacol. Exp. Ther.*, 306, 821–827, 2003.
12. Sunderland, T., Tariot, P.N., Cohen, R.M., Weingartner, H., Mueller, E.A., III, and Murphy, D.L., Anticholinergic sensitivity in patients with dementia of the Alzheimer type and age-matched controls: a dose-response study, *Arch. Gen. Psychiatry*, 44, 418–426, 1987.

13. Ingram, D.K., Joseph, J.A., Spangler, E.L., Roberts, D., Hengemihle, J., and Fanelli, R.J., Chronic nimodipine treatment in aged rats: analysis of motor and cognitive effects and muscarinic-induced striatal dopamine release, *Neurobiol. Aging*, 15, 55–61, 1994.

14. Kubanis, P. and Zornetzer, S.F., Age-related behavioral and neurobiological changes: a review with an emphasis on memory, *Behav. Neural Biol.*, 31, 115–172, 1981.

15. Decker, M.W., The effects of aging on hippocampal and cortical projections of the forebrain cholinergic system, *Brain Res.*, 434, 423–438, 1987.

16. Gallagher, M. and Colombo, P.J., Ageing: the cholinergic hypothesis of cognitive decline, *Curr. Opin. Neurobiol.*, 5, 161–168, 1995.

17. Bartus, R.T., On neurodegenerative diseases, models, and treatment strategies: lessons learned and lessons forgotten a generation following the cholinergic hypothesis, *Exp. Neurol.*, 163, 495–529, 2000.

18. Aigner, T.G. and Mishkin, M., The effects of physostigmine and scopolamine on recognition memory in monkeys, *Behav. Neural Biol.*, 45, 81–87, 1986.

19. Terry, A.V., Jr., Buccafusco, J.J., and Jackson, W.J., Scopolamine reversal of nicotine enhanced delayed matching-to-sample performance in monkeys, *Pharmacol. Biochem. Behav.*, 45, 925–929, 1993.

20. Aura, J., Sirvio, J., and Riekkinen, P., Jr., Methoctramine moderately improves memory but pirenzepine disrupts performance in delayed non-matching to position test, *Eur. J. Pharmacol.*, 333, 129–134, 1997.

21. Andrews, J.S., Jansen, J.H., Linders, S., and Princen, A., Effects of disrupting the cholinergic system on short-term spatial memory in rats, *Psychopharmacology (Berl.)*, 115, 485–494, 1994.

22. Sala, M., Braida, D., Calcaterra, P., Leone, M.P., Comotti, F.A., Gianola, S., and Gori, E., Effect of centrally administered atropine and pirenzepine on radial arm maze performance in the rat, *Eur. J. Pharmacol.*, 194, 45–49, 1991.

23. Hagan, J.J., Jansen, J.H., and Broekkamp, C.L., Blockade of spatial learning by the M1 muscarinic antagonist pirenzepine, *Psychopharmacology (Berl.)*, 93, 470–476, 1987.

24. Hunter, A.J. and Roberts, F.F., The effect of pirenzepine on spatial learning in the Morris water maze, *Pharmacol. Biochem. Behav.*, 30, 519–523, 1988.

25. Messer, W.S., Jr., Bohnett, M., and Stibbe, J., Evidence for a preferential involvement of M1 muscarinic receptors in representational memory, *Neurosci. Lett.*, 116, 184–189, 1990.

26. Worms, P., Gueudet, C., Perio, A., and Soubrie, P., Systemic injection of pirenzepine induces a deficit in passive avoidance learning in rats, *Psychopharmacology (Berl.)*, 98, 286–288, 1989.

27. Ohnuki, T. and Nomura, Y., Effects of selective muscarinic antagonists, pirenzepine and AF-DX 116, on passive avoidance tasks in mice, *Biol. Pharm. Bull.*, 19, 814–818, 1996.

28. Caulfield, M.P., Higgins, G.A., and Straughan, D.W., Central administration of the muscarinic receptor subtype-selective antagonist pirenzepine selectively impairs passive avoidance learning in the mouse, *J. Pharm. Pharmacol.*, 35, 131–132, 1983.

29. Roldan, G., Bolanos-Badillo, E., Gonzalez-Sanchez, H., Quirarte, G.L., and Prado-Alcala, R.A., Selective M1 muscarinic receptor antagonists disrupt memory consolidation of inhibitory avoidance in rats, *Neurosci. Lett.*, 230, 93–96, 1997.

30. Jones, C.K. and Shannon, H.E., Muscarinic cholinergic modulation of prepulse inhibition of the acoustic startle reflex, *J. Pharmacol. Exp. Ther.*, 294, 1017–1023, 2000.

31. Giachetti, A., Giraldo, E., Ladinsky, H., and Montagna, E., Binding and functional profiles of the selective M1 muscarinic receptor antagonists trihexyphenidyl and dicyclomine, *Br. J. Pharmacol.*, 89, 83–90, 1986.

32. Fornari, R.V., Moreira, K.M., and Oliveira, M.G., Effects of the selective M1 muscarinic receptor antagonist dicyclomine on emotional memory, *Learn. Mem.*, 7, 287–292, 2000.

33. Izquierdo, I., Mechanism of action of scopolamine as an amnestic, *Trends Pharmacol. Sci.*, 10, 175–177, 1989.

34. Fibiger, H.C., Cholinergic mechanisms in learning, memory and dementia: a review of recent evidence, *Trends Neurosci.*, 14, 220–223, 1991.

35. Sarter, M., Hagan, J., and Dudchenko, P., Behavioral screening for cognition enhancers: from indiscriminate to valid testing: part I, *Psychopharmacology (Berl.)*, 107, 144–159, 1992.

36. Blokland, A., Acetylcholine: a neurotransmitter for learning and memory? *Brain Res. Rev.*, 21, 285–300, 1995.

37. McDonald, M.P. and Overmier, J.B., Present imperfect: a critical review of animal models of the mnemonic impairments in Alzheimer's disease, *Neurosci. Biobehav. Rev.*, 22, 99–120, 1998.

38. Parrott, A.C., The effects of transdermal scopolamine and four dose levels of oral scopolamine (0.15, 0.3, 0.6, and 1.2 mg) upon psychological performance, *Psychopharmacology (Berl.)*, 89, 347–354, 1986.

39. Whishaw, I.Q. and Petrie, B.F., Cholinergic blockade in the rat impairs strategy selection but not learning and retention of nonspatial visual discrimination problems in a swimming pool, *Behav. Neurosci.*, 102, 662–677, 1988.

40. Jones, D.N. and Higgins, G.A., Effect of scopolamine on visual attention in rats, *Psychopharmacology (Berl.)*, 120, 142–149, 1995.

41. Jakala, P., Sirvio, J., Jolkkonen, J., Riekkinen, P., Jr., Acsady, L., and Riekkinen, P., The effects of p-chlorophenylalanine-induced serotonin synthesis inhibition and muscarinic blockade on the performance of rats in a 5-choice serial reaction time task, *Behav. Brain Res.*, 51, 29–40, 1992.

42. Phillips, J.M., McAlonan, K., Robb, W.G., and Brown, V.J., Cholinergic neurotransmission influences covert orientation of visuospatial attention in the rat, *Psychopharmacology (Berl.)*, 150, 112–116, 2000.

43. Ruotsalainen, S., Miettinen, R., MacDonald, E., Koivisto, E., and Sirvio, J., Blockade of muscarinic, rather than nicotinic, receptors impairs attention, but does not interact with serotonin depletion, *Psychopharmacology (Berl.)*, 148, 111–123, 2000.

44. Stanhope, K.J., McLenachan, A.P., and Dourish, C.T., Dissociation between cognitive and motor/motivational deficits in the delayed matching to position test: effects of scopolamine, 8-OH-DPAT and EAA antagonists, *Psychopharmacology (Berl.)*, 122, 268–280, 1995.

45. Han, C.J., Pierre-Louis, J., Scheff, A., and Robinson, J.K., A performance-dependent adjustment of the retention interval in a delayed non-matching-to-position paradigm differentiates effects of amnestic drugs in rats, *Eur. J. Pharmacol.*, 403, 87–93, 2000.

46. Collerton, D., Cholinergic function and intellectual decline in Alzheimer's disease, *Neuroscience*, 19, 1–28, 1986.

47. Piercey, M.F., Vogelsang, G.D., Franklin, S.R., and Tang, A.H., Reversal of scopolamine-induced amnesia and alterations in energy metabolism by the nootropic piracetam: implications regarding identification of brain structures involved in consolidation of memory traces, *Brain. Res.*, 424, 1–9, 1987.

48. Christoffersen, G.R., von Linstow Roloff, E., and Nielsen, K.S., Effects of piracetam on the performance of rats in a delayed match-to-position task, *Prog. Neuropsychopharmacol. Biol. Psychiatry*, 22, 211–228, 1998.

49. Gibbs, R.B., Estrogen replacement enhances acquisition of a spatial memory task and reduces deficits associated with hippocampal muscarinic receptor inhibition, *Horm. Behav.*, 36, 222–233, 1999.

50. Ogasawara, T., Nakagawa, Y., Ukai, Y., Tamura, M., and Kimura, K., NS-3(CG3703), a TRH analog, ameliorates scopolamine-induced memory disruption in rats, *Pharmacol. Biochem. Behav.*, 51, 929–934, 1995.

51. Kumar, V., Singh, P.N., Muruganandam, A.V., and Bhattacharya, S.K., Effect of Indian *Hypericum perforatum* Linn. on animal models of cognitive dysfunction, *J. Ethnopharmacol.*, 72, 119–128, 2000.

52. Gattu, M., Boss, K.L., Terry, A.V., Jr., and Buccafusco, J.J., Reversal of scopolamine-induced deficits in navigational memory performance by the seed oil of *Celastrus paniculatus*, *Pharmacol. Biochem. Behav.*, 57, 793–799, 1997.

53. Barnes, J.M., Barnes, N.M., Costall, B., Deakin, J.F., Ironside, J.W., Kilpatrick, G.J., Naylor, R.J., Rudd, J.A., Simpson, M.D., Slater, P., and Tyers, M.B., Identification and distribution of 5-HT3 recognition sites within the human brainstem, *Neurosci. Lett.*, 111, 80–86, 1990.

54. Misane, I. and Ogren, S.O., Selective 5-HT1A antagonists WAY 100635 and NAD-299 attenuate the impairment of passive avoidance caused by scopolamine in the rat, *Neuropsychopharmacology*, 28, 253–264, 2003.

55. Lelong, V., Lhonneur, L., Dauphin, F., and Boulouard, M., BIMU 1 and RS 67333, two 5-HT4 receptor agonists, modulate spontaneous alternation deficits induced by scopolamine in the mouse, *Naunyn Schmiedebergs Arch. Pharmacol.*, 367, 621–628, 2003.

56. Foley, A.G., Murphy, K.J., Hirst, W.D., Gallagher, H.C., Hagan, J.J., Upton, N., Walsh, F.S., and Regan, C.M., The 5-HT(6) receptor antagonist SB-271046 reverses scopolamine-disrupted consolidation of a passive avoidance task and ameliorates spatial task deficits in aged rats, *Neuropsychopharmacology*, 29, 93–100, 2004.

57. Shannon, H.E., Bemis, K.G., Hendrix, J.C., and Ward, J.S., Interactions between scopolamine and muscarinic cholinergic agonists or cholinesterase inhibitors on spatial alternation performance in rats, *J. Pharmacol. Exp. Ther.*, 255, 1071–1077, 1990.

58. Buxton, A., Callan, O.A., Blatt, E.J., Wong, E.H., and Fontana, D.J., Cholinergic agents and delay-dependent performance in the rat, *Pharmacol. Biochem. Behav.*, 49, 1067–1073, 1994.

59. Ballard, T.M. and McAllister, K.H., The acetylcholinesterase inhibitor, ENA 713 (Exelon), attenuates the working memory impairment induced by scopolamine in an operant DNMTP task in rats, *Psychopharmacology (Berl.)*, 146, 10–18, 1999.

60. Chen, K.C., Baxter, M.G., and Rodefer, J.S., Central blockade of muscarinic cholinergic receptors disrupts affective and attentional set-shifting, *Eur. J. Neurosci.*, 20, 1081–1088, 2004.

61. Terry, A.V., Jr., Buccafusco, J.J., Jackson, W.J., Zagrodnik, S., Evans-Martin, F.F., and Decker, M.W., Effects of stimulation or blockade of central nicotinic-cholinergic receptors on performance of a novel version of the rat stimulus discrimination task, *Psychopharmacology (Berl.)*, 123, 172–181, 1996.

62. Terry, A.V., Jr., Williamson, R., Gattu, M., Beach, J.W., McCurdy, C.R., Sparks, J.A., and Pauly, J.R., Lobeline and structurally simplified analogs exhibit differential agonist activity and sensitivity to antagonist blockade when compared to nicotine, *Neuropharmacology*, 37, 93–102, 1998.

63. Kompoliti, K., Chu, Y., Polish, A., Roberts, J., McKay, H., Mufson, E.J., Leurgans, S., Morrison, J.H., and Kordower, J.H., Effects of estrogen replacement therapy on cholinergic basal forebrain neurons and cortical cholinergic innervation in young and aged ovariectomized rhesus monkeys, *J. Comp. Neurol.*, 472, 193–207, 2004.

64. Steckler, T., Animal models of cognitive disorders, in *Biological Psychiatry*, D'haenen, H.D., den Boer, J.A., and Willner, P., Eds., John Wiley & Sons, New York, 2002, pp. 215–233.

65. Evans-Martin, F.F., Terry, A.V., Jr., Jackson, W.J., and Buccafusco, J.J., Evaluation of two rodent delayed-response memory tasks: a method with retractable levers versus a method with closing doors, *Physiol. Behav.*, 70, 233–241, 2000.

66. Evenden, J.L., Lavis, L., and Iversen, S.D., Blockade of conditioned taste aversion by scopolamine and N-methyl scopolamine: associative conditioning, not amnesia, *Psychopharmacology (Berl.)*, 106, 179–188, 1992.

67. Parrott, A.C., Transdermal scopolamine: a review of its effects upon motion sickness, psychological performance, and physiological functioning, *Aviat. Space Environ. Med.*, 60, 1–9, 1989.

68. Terry, A.V., Jr., Gattu, M., Buccafusco, J.J., Sowell, J.W., and Kosh, J.W., Ranitidine analog, JWS USC 75IX, enhances memory related task performance in rats, *Drug Dev. Res.*, 47, 97–106, 1999.

69. Wenk, G.L., An hypothesis on the role of glucose in the mechanism of action of cognitive enhancers, *Psychopharmacology (Berl.)*, 99, 431–438, 1989.

70. Rush, D.K. and Streit, K., Memory modulation with peripherally acting cholinergic drugs, *Psychopharmacology (Berl.)*, 106, 375–382, 1992.

71. Stone, W.S., Walser, B., Gold, S.D., and Gold, P.E., Scopolamine- and morphine-induced impairments of spontaneous alternation performance in mice: reversal with glucose and with cholinergic and adrenergic agonists, *Behav. Neurosci.*, 105, 264–271, 1991.

72. Bejar, C., Wang, R.H., and Weinstock, M., Effect of rivastigmine on scopolamine-induced memory impairment in rats, *Eur. J. Pharmacol.*, 383, 231–240, 1999.

73. Chopin, P. and Briley, M., Effects of four non-cholinergic cognitive enhancers in comparison with tacrine and galanthamine on scopolamine-induced amnesia in rats, *Psychopharmacology (Berl.)*, 106, 26–30, 1992.

74. Bores, G.M., Huger, F.P., Petko, W., Mutlib, A.E., Camacho, F., Rush, D.K., Selk, D.E., Wolf, V., Kosley, R.W., Jr., Davis, L., and Vargas, H.M., Pharmacological evaluation of novel Alzheimer's disease therapeutics: acetylcholinesterase inhibitors related to galanthamine, *J. Pharmacol. Exp. Ther.*, 277, 728–738, 1996.

75. Fishkin, R.J., Ince, E.S., Carlezon, W.A., Jr., and Dunn, R.W., D-cycloserine attenuates scopolamine-induced learning and memory deficits in rats, *Behav. Neural Biol.*, 59, 150–157, 1993.

76. Kirkby, D.L., Jones, D.N., Barnes, J.C., and Higgins, G.A., Effects of anticholinesterase drugs tacrine and E2020, the 5-HT(3) antagonist ondansetron, and the H(3) antagonist thioperamide, in models of cognition and cholinergic function, *Behav. Pharmacol.*, 7, 513–525, 1996.

77. Ogura, H., Kosasa, T., Kuriya, Y., and Yamanishi, Y., Donepezil, a centrally acting acetylcholinesterase inhibitor, alleviates learning deficits in hypocholinergic models in rats, *Methods Find. Exp. Clin. Pharmacol.*, 22, 89–95, 2000.

78. Buckton, G., Zibrowski, E.M., and Vanderwolf, C.H., Effects of cyclazocine and scopolamine on swim-to-platform performance in rats, *Brain Res.*, 922, 229–233, 2001.

79. Albiston, A.L., Pederson, E.S., Burns, P., Purcell, B., Wright, J.W., Harding, J.W., Mendelsohn, F.A., Weisinger, R.S., and Chai, S.Y., Attenuation of scopolamine-induced learning deficits by LVV-hemorphin-7 in rats in the passive avoidance and water maze paradigms, *Behav. Brain Res.*, 154, 239–243, 2004.

80. Dennes, R.P. and Barnes, J.C., Attenuation of scopolamine-induced spatial memory deficits in the rat by cholinomimetic and non-cholinomimetic drugs using a novel task in the 12-arm radial maze, *Psychopharmacology (Berl.)*, 111, 435–441, 1993.

81. Braida, D., Paladini, E., Griffini, P., Lamperti, M., Maggi, A., and Sala, M., An inverted U-shaped curve for heptylphysostigmine on radial maze performance in rats: comparison with other cholinesterase inhibitors, *Eur. J. Pharmacol.*, 302, 13–20, 1996.

82. Ormerod, B.K. and Beninger, R.J., Water maze versus radial maze: differential performance of rats in a spatial delayed match-to-position task and response to scopolamine, *Behav. Brain Res.*, 128, 139–152, 2002.

83. Spangler, E.L., Rigby, P., and Ingram, D.K., Scopolamine impairs learning performance of rats in a 14-unit T-maze, *Pharmacol. Biochem. Behav.*, 25, 673–679, 1986.

84. Givens, B. and Olton, D.S., Bidirectional modulation of scopolamine-induced working memory impairments by muscarinic activation of the medial septal area, *Neurobiol. Learn. Mem.*, 63, 269–276, 1995.

85. M'Harzi, M., Willig, F., Gieules, C., Palou, A.M., Oberlander, C., and Barzaghi, F., Ameliorating effects of RU 47213, a novel oral and long-lasting cholinomimetic agent, on working memory impairments in rats, *Pharmacol. Biochem. Behav.*, 56, 663–668, 1997.

86. Riedel, G., Wetzel, W., and Reymann, K.G., Computer-assisted shock-reinforced Y-maze training: a method for studying spatial alternation behaviour, *Neuroreport*, 5, 2061–2064, 1994.

87. Biggan, S.L., Ingles, J.L., and Beninger, R.J., Scopolamine differentially affects memory of 8- and 16-month-old rats in the double Y-maze, *Neurobiol. Aging*, 17, 25–30, 1996.

88. Kirkby, D.L., Jones, D.N., and Higgins, G.A., Influence of prefeeding and scopolamine upon performance in a delayed matching-to-position task, *Behav. Brain Res.*, 67, 221–227, 1995.

89. Higgins, G.A., Enderlin, M., Fimbel, R., Haman, M., Grottick, A.J., Soriano, M., Richards, J.G., Kemp, J.A., and Gill, R., Donepezil reverses a mnemonic deficit produced by scopolamine but not by perforant path lesion or transient cerebral ischaemia, *Eur. J. Neurosci.*, 15, 1827–1840, 2002.

90. Berz, S., Battig, K., and Welzl, H., The effects of anticholinergic drugs on delayed time discrimination performance in rats, *Physiol. Behav.*, 51, 493–499, 1992.

91. Kirk, R.C., White, K.G., and McNaughton, N., Low dose scopolamine affects discriminability but not rate of forgetting in delayed conditional discrimination, *Psychopharmacology (Berl.)*, 96, 541–546, 1988.

92. Elrod, K. and Buccafusco, J.J., An evaluation of the mechanism of scopolamine-induced impairment in two passive avoidance protocols, *Pharmacol. Biochem. Behav.*, 29, 15–21, 1988.

93. Young, S.L., Bohenek, D.L., and Fanselow, M.S., Scopolamine impairs acquisition and facilitates consolidation of fear conditioning: differential effects for tone vs. context conditioning, *Neurobiol. Learn. Mem.*, 63, 174–180, 1995.

94. Wallenstein, G.V. and Vago, D.R., Intrahippocampal scopolamine impairs both acquisition and consolidation of contextual fear conditioning, *Neurobiol. Learn. Mem.*, 75, 245–252, 2001.

95. Ennaceur, A. and Meliani, K., Effects of physostigmine and scopolamine on rats' performances in object-recognition and radial-maze tests, *Psychopharmacology (Berl.)*, 109, 321–330, 1992.

96. Pitsikas, N., Rigamonti, A.E., Cella, S.G., and Muller, E.E., The 5-HT 1A receptor antagonist WAY 100635 improves rats' performance in different models of amnesia evaluated by the object recognition task, *Brain Res.*, 983, 215–222, 2003.

97. Woolley, M.L., Marsden, C.A., Sleight, A.J., and Fone, K.C., Reversal of a cholinergic-induced deficit in a rodent model of recognition memory by the selective 5-HT6 receptor antagonist, Ro 04-6790, *Psychopharmacology (Berl.)*, 170, 358–367, 2003.

98. Stanhope, K.J., Mirza, N.R., Bickerdike, M.J., Bright, J.L., Harrington, N.R., Hesselink, M.B., Kennett, G.A., Lightowler, S., Sheardown, M.J., Syed, R., Upton, R.L., Wadsworth, G., Weiss, S.M., and Wyatt, A., The muscarinic receptor agonist xanomeline has an antipsychotic-like profile in the rat, *J. Pharmacol. Exp. Ther.*, 299, 782–792, 2001.

99. Whishaw, I.Q., Cholinergic receptor blockade in the rat impairs locale but not taxon strategies for place navigation in a swimming pool, *Behav. Neurosci.*, 99, 979–1005, 1985.

100. Nilsson, O.G. and Gage, F.H., Anticholinergic sensitivity in the aging rat septohippocampal system as assessed in a spatial memory task, *Neurobiol. Aging*, 14, 487–497, 1993.

101. Fontana, D.J., Daniels, S.E., Wong, E.H., Clark, R.D., and Eglen, R.M., The effects of novel, selective 5-hydroxytryptamine (5-HT)4 receptor ligands in rat spatial navigation, *Neuropharmacology*, 36, 689–696, 1997.

102. Elsmore, T.F., Parkinson, J.K., Leu, J.R., and Witkin, J.M., Atropine effects on delayed discrimination performance of rats, *Pharmacol. Biochem. Behav.*, 32, 971–975, 1989.

103. Carnicella, S., Pain, L., and Oberling, P., Cholinergic effects on fear conditioning II: nicotinic and muscarinic modulations of atropine-induced disruption of the degraded contingency effect, *Psychopharmacology*, 5, 5, 2005.

104. Prado-Alcala, R.A., Signoret-Edward, L., Figueroa, M., Giordano, M., and Barrientos, M.A., Post-trial injection of atropine into the caudate nucleus interferes with long-term but not with short-term retention of passive avoidance, *Behav. Neural Biol.*, 42, 81–84, 1984.

105. Zarrindast, M.R., Bakhsha, A., Rostami, P., and Shafaghi, B., Effects of intrahippocampal injection of GABAergic drugs on memory retention of passive avoidance learning in rats, *J. Psychopharmacol.*, 16, 313–319, 2002.

3 Nicotinic Receptor Antagonists in Rats

Cindy S. Roegge and Edward D. Levin
Duke University Medical Center

CONTENTS

INTRODUCTION

Nicotinic acetylcholine-receptor systems are critical neural components of cognitive functions. Nicotine and nicotinic agonists have been shown to improve cognition in rats in numerous studies [13–15, 27, 31, 59]. Similarly, nicotinic-receptor antagonists — the subject of this chapter — can cause cognitive impairments in rats. The use of nicotinic-receptor antagonists in an animal model of cognitive impairment has clinical relevance because it models the functional effect of nicotinic-receptor loss. Studies have shown that patients with Alzheimer's disease suffer a dramatic reduction in hippocampal and cortical nicotine-receptor density that parallels the cognitive decline associated with this disease [56–58]. Significant nicotinic-receptor loss also occurs in Parkinson's disease [9], and postmortem studies in schizophrenics show a decrease in the number of $\alpha7$ nicotinic receptors in the brain [8, 19]. Both of these diseases can also include cognitive decline. Thus animal studies wherein nicotinic receptors are blocked with antagonists can be useful in developing animal models of how these diseases — Alzheimer's, Parkinson's, and schizophrenia — affect cognition.

The predominant nicotinic-receptor antagonist used in cognitive studies with rats is mecamylamine, a noncompetitive antagonist that is not selective for particular nicotinic-receptor-subtype recognition sites. Mecamylamine has been shown to impair cognitive function on a variety of tasks, and this chapter reviews the effects of mecamylamine on task performance. In addition, a small number of studies have examined other nonspecific nicotinic antagonists, including chlorisondamine and d-tubocurarine, and the receptor-subunit-specific $\alpha4\beta2$ (dihydro-β-erythroidine hydrobromide [DHβE]) and $\alpha7$ (methyllycaconitine [MLA]) antagonists, and these will also be discussed. In addition, many studies use the peripheral nicotinic blocker hexamethonium to show the specificity of mecamylamine effects to the central nervous system (CNS).

RADIAL ARM MAZE (RAM)

Several studies in the Levin laboratory have illustrated the cognitive impairments of mecamylamine on the RAM (Table 3.1). Most of these studies have utilized the eight-arm RAM, which has eight arms radiating from a central platform. The RAM is a test of working memory, taxing the rat to remember which arms it has entered during the session to complete the maze efficiently. The measure of cognition used in most of these RAM studies is entries to repeat (ETR), which is the total number of arms entered before an arm is repeated. The RAM is a task that becomes more difficult as the session progresses. The ETR measure is sensitive in detecting when, as the task becomes increasingly difficult, the first breakdown in performance accuracy occurs. Drugs that impair cognition, such as mecamylamine, reduce ETR. The dose threshold for cognitive impairment on the RAM appears to be above 5 mg/kg. Doses from 0 to 5 mg/kg of mecamylamine do not typically yield statistically significant impairments in cognition, while a dose of 10 mg/kg of mecamylamine has been shown repeatedly to significantly impair memory function (Table 3.1). Mecamylamine, given systemically, also significantly increases the response latency, which is measured as seconds/arm entry (Table 3.1). However, the cognitive effects of mecamylamine on RAM performance seem to be centrally acting, since hexamethonium, a peripherally acting nicotinic antagonist, also increases the response latency but does not impair ETR [22].

One study [34] has examined the specificity of mecamylamine on working vs. reference memory. Reference memory across testing sessions involves memory for a subset of arms (4) never baited on the 16-arm RAM. Working memory within a testing session involves memory for a subset of arms (12) baited at the beginning of each session but not thereafter. In this testing paradigm, mecamylamine impaired working memory, but did not impair reference memory [34]. Since the 16-arm RAM working-reference memory task is a more difficult task than the 8-arm RAM working-memory task, the dose threshold for mecamylamine-induced impairment is lower, at 1.25 mg/kg (Table 3.1). This is an important demonstration of effect, since there may be less-specific effects of mecamylamine at higher doses.

The effects of mecamylamine infused directly into the brain have also been studied on the RAM. Infusion of mecamylamine into the lateral ventricles, substantia nigra, ventral tegmental area, or ventral hippocampus all cause RAM impairments

TABLE 3.1
Effects of Mecamylamine on Radial Arm Maze Performance

Rat Strain	Sex	Dose	Route[a]	Outcome[b]	Reference
8-Arm RAM					
Sprague-Dawley	Females	2.5 mg/kg	IP 20 min prior	– ETR	[26, 36, 37]
				– No. of arms entered in first eight entries	[22]
				– Latency	[22, 36]
				↑ Latency	[26, 37]
		≈3mg/kg/d	Implanted minipump	↑ ETR first week	[29]
				↑ No. of arms entered in first eight entries	
				– Entries to finish	
				– Latency	
		5 mg/kg	IP 20 min prior	– ETR	[25]
				– No. of arms entered in first eight entries	[22]
			SC 20 min prior	↓ ETR	[28]
			IP or SC 20 min prior	↑ Latency	[22, 25, 28]
		10 mg/kg	IP or SC 20 min	↓ ETR	[23–25, 28, 32, 37, 38]
				↓ No. of arms entered in first eight entries	[22]
				↑ latency	
		200 μg/side	Bilateral lateral ventricle 5 min prior	↓ ETR	[4]
				↓ Modified % correct	
				– Latency	
				– ETR in naïve exploratory behavior on RAM	
		1, 3.3, 10 μg/side	Nucleus accumbens 5 min prior	– ETR	[21]
				– Latency	
			SN or VTA 5 min prior	↓ ETR (10 μg/side)	[30]
				– Latency	
			Ventral hippocampus 5 min prior	↓ ETR (all doses)	[21]
				↓ Latency	
Aged Rats					
Sprague-Dawley	Females	2.5, 5, 10 mg/kg	IP 20 min prior	– ETR	[32]

TABLE 3.1
Effects of Mecamylamine on Radial Arm Maze Performance (continued)

Rat Strain	Sex	Dose	Route[a]	Outcome[b]	Reference
16-Arm RAM with Working Memory and Reference Memory					
Sprague-Dawley	Females	0.625 mg/kg	SC 20 min prior	− WM errors	[34]
				− RM errors	
				− Latency	
		1.25 mg/kg		↑ WM errors	
				− RM errors	
				↑ Latency	

[a] IP = intraperitoneal injection; SC = subcutaneous injection; SN = substantia nigra infusion; VTA = ventral tegmental area infusion.
[b] ETR = entries to repeat; WM = working memory; RM = reference memory; RAM = radial arm maze; ↑ = enhanced performance; ↓ = impaired performance; − = no effect.

[4, 21, 30]. However, mecamylamine infusions up to 10 µg/side into the nucleus accumbens did not impair RAM performance [21]. Unlike systemic mecamylamine injections, infusions of mecamylamine directly into the CNS did not increase response latency on the RAM (Table 3.1).

The effects of d-tubocurarine chloride (dTC), DHβE, and MLA have also been studied on the RAM [17]. The antagonists were injected directly into the ventral hippocampus-entorhinal (VHE) area via bilateral cannulae. Doses of dTC could not be found that would impair RAM performance without causing seizures [17]. Both DHβE and MLA cause significant dose-related decreases in ETR and increases in latency. However, the specificity of the selective antagonists is questioned, given the high doses required to induce significant impairments on the RAM [17]. Thus, another study [35] tested the effects of DHβE and MLA on the more difficult 16-arm RAM working-reference memory task. In that study, bilateral VHE injections of DHβE significantly increased both working- and reference-memory errors, while MLA only significantly increased working-memory errors. Neither drug, at the lower doses used on the 16-arm RAM, increased latency [35].

THREE-PANEL RUNWAY TASK

The effects of mecamylamine have been tested on the three-panel runway task that has four choice points where the rat must choose among three doors [40]. Both working- and reference-memory versions of the task were utilized. For the working-memory task, the correct door choices were held constant throughout the six-trial session but changed from one session to the next, while the reference-memory test held the correct door choices constant throughout all testing sessions. Doses of 10 mg/kg mecamylamine effectively impaired working but not reference memory [40]. Similarly, injections of 10 to 18 µg/side directly into the dorsal hippocampus also significantly increased working-memory errors but not reference-memory errors [40]. As with the RAM studies, mecamylamine is showing selective effects on working and not reference memory on the three-panel runway task. Mecamylamine, administered either systemically or intrahippocampally, increased response latency during the working- but not the reference-memory task (Table 3.2).

T-MAZE ALTERNATION

Three studies have examined the effects of mecamylamine on T-maze alternation, in which rats are given food reward for alternating between the two arms of a T-shaped maze (Table 3.2). Mecamylamine doses up to 6.3 mg/kg, base wt. (or ≈5.17 mg/kg, salt wt.), were ineffective in impairing the T-maze alternation performance [7]. Chlorisondamine, a nicotinic antagonist that does not easily enter the brain, injected directly into the lateral ventricles also did not affect T-maze alternation [21]. Another study confirms that mecamylamine does not impair T-maze alternation without delay until a dose of 10 mg/kg is given [39]. However, when a 30-sec delay was imposed, the 10-mg/kg mecamylamine dose actually resulted in fewer errors than the saline controls. Further, chronic mecamylamine infused via implanted osmotic minipumps delivering an approximate dose of 3 mg/kg/d did not cause consistent impairments on a T-maze alternation task that imposed delays of up to 40 sec between trials [33]. Chronic mecamylamine caused a slight decrease in percent correct, in particular at the 0-sec delay on week 4 of testing. However, chronic mecamylamine exposure also caused a significant but transient improvement in T-maze performance at the 10-sec delay during the first week of testing. A similar transient improvement was observed previously with chronic mecamylamine on the RAM [29]. Some of the lack of cognitive deficits on the T-maze can be explained by the lower doses used. Doses greater than 5 mg/kg of mecamylamine are needed to impair T-maze alternation without delays [7, 33]. Surprisingly, mecamylamine may improve T-maze alternation performance when delays are added [33, 39].

T-MAZE DISCRIMINATION

Mecamylamine causes cognitive impairments on the T-maze spatial discrimination task (Table 3.2). The T-maze discrimination task trained rats to respond to one spatial location: either the left or right arm. Mecamylamine doses of 5 or 10 mg/kg increased the number of errors on this task with and without the 30-sec imposed delay [39].

DELAYED (NON)MATCHING TO SAMPLE

The effects of mecamylamine have been studied in both the delayed matching to sample (DMTS) [1, 5] and delayed nonmatching-to-sample (DNMTS) tasks [11, 51]. These studies were conducted in two-lever operant boxes (Table 3.2). In the DMTS task, one of the levers is presented, and the rat must, after a delay, press the previously presented lever (match) to receive reward. Mecamylamine dose-dependently decreased choice accuracy in the DMTS task, and the threshold for cognitive impairment appears to be between 1 and 1.8 mg/kg (Table 3.2). Mecamylamine does significantly increase response latency [1, 5], but the effects on cognition seem to be central, since the peripherally acting nicotinic blocker hexamethonium did not decrease percent correct [1].

TABLE 3.2
Effects of Mecamylamine on Other Appetitive Tasks

Rat Strain	Sex	Dose	Route[a]	Outcome[b]	Reference
Three-Panel Runway					
Wistar	Males	3.2 mg/kg	IP	− WM or RM task	[40]
		10 mg/kg		↑ Errors during WM task	
				↑ Latency during WM task	
				− Errors during RM task	
				− Latency during RM task	
		3.2 μg/side	IH	− WM or RM task	
		10, 18 μg/side		↑ Errors during WM task	
				↑ Latency during WM task	
				− Errors during RM task	
				− Latency during RM task	
T-Maze Alternation					
Long-Evans	Males	0.2, 0.63, 2 mg/kg[c]	SC 20 min prior	− No. of errors	[7]
				− Response latency	
		6.3 mg/kg[c]		− No. of errors	
				↑ Response latency	
Sprague-Dawley	Males	1, 5 mg/kg	IP 20 min prior	− No. of errors	[39]
		10 mg/kg		↑ No. of errors	
				↑ Response latency	
Sprague-Dawley	Females	~3 mg/kg/day	chronically implanted minipump	↑ Percent correct at 10-sec delay week 1	[33]
				↓ Percent correct at 0-sec delay week 4	
T-Maze Discrimination					
Sprague-Dawley	Males	1 mg/kg	IP 20 min prior	− No. of errors	[39]
		5, 10 mg/kg		↑ No. of errors	
				↑ Response latency	
Delayed Matching to Sample					
Long-Evans	Males	0.2, 1 mg/kg	SC 30 min prior	− Percent correct	[1]
				− Response latency	
		5 mg/kg		↓ Percent correct	
				↑ Response latency	
Long-Evans	Males	1, 1.8, 3 mg/kg	IP 15 min prior	↑ Nose-poke IRT	[5]
		1.8, 3 mg/kg		↓ Matching accuracy	
Delayed Nonmatching to Sample					
Han:Wistar	Males	1 mg/kg	IP 30 min prior	− Percent correct	[51]
		3 mg/kg		↓ Percent correct	
Hooded	Males	2, 5 mg/kg	IP 30 min prior	↓ No. of trials completed	[11]
		5 mg/kg		↓ Percent correct	

TABLE 3.2
Effects of Mecamylamine on Other Appetitive Tasks (continued)

Rat Strain	Sex	Dose	Route[a]	Outcome[b]	Reference
Visual Signal Detection Task					
Sprague-Dawley	Females	1, 2, 4 mg/kg	SC 10 min prior	↓ Percent hit	[41]
		4 mg/kg		↓ Percent correct rejections	
				↑ No. of no-response trials	
Long-Evans	Males	1.8, 3,	IP 15 min prior	↓ Percent hit	[6]
		5.6 mg/kg		↓ Percent correct rejections	
		3, 5.6 mg/kg		↑ No. of false alarms	
		5.6 mg/kg		↑ No. of no-response trials	
				↑ Reaction time	
Long-Evans		1, 3,	20 min prior	↓ Vigilance index	[55]
		10 mg/kg			
		3, 10 mg/kg		↑ No. of omission errors	
Five-Choice Serial Reaction Time Task					
Hooded	Males	0.5, 1.6,	SC 30 min prior	– Percent correct	[53]
		5 mg/kg		↑ Percent of omissions	
		(base wt.)		↑ Correct response latencies	
				↓ Anticipatory responses	
Han:Wistar	Males	1, 3 mg/kg	IP 30 min prior	↓ No. of trials completed	[52]
				↑ Percent of omissions	
				↓ Premature hole responses	
Hooded Listar	Males	0.3, 1, 3,	SC 30 min prior	– Choice accuracy	[20]
		5 mg/kg			
Aged Rats					
Hooded Listar	Males	0.3, 1 mg/kg	SC 30 min prior	– Choice accuracy	[20]
	(15 months)	3, 5 mg/kg		↓ Choice accuracy	
Long-Evans	Males	2 mg/kg	SC 1 h prior	– Percent correct	[54]
	(20–25	8 mg/kg		↓ Percent correct	
	months)				
Tests of Response Inhibition					
Sprague-Dawley	Males	1–5 μg	unilateral VMH	↑ Responding during signaled nonreinforcement period	[50]
Wistar	Males	0.25, 0.5,	SC 15 min prior	– Effect on DRL-30 sec	[2]
		1 mg/kg			
		2 mg/kg		↑ Premature responses	
		4 mg/kg		↓ Reinforcement rate	
Sprague-Dawley	Males	10 μg/μl	IH 7 min prior	↑ Premature responses	[49]
				↓ No. of reinforcements	

[a] IP = intraperitoneal injection; IH = intrahippocampal injection; SC = subcutaneous injection; VMH = ventromedial nucleus of the hypothalamus.

[b] WM = working memory; RM = reference memory; IRT = interresponse time; DRL = differential reinforcement of low rate; ↑ = enhanced performance; ↓ = impaired performance; – = no effect.

[c] Mecamylamine doses in this study refer to the base. A 6.3-mg/kg base-wt. dose is equivalent to a 5.17-mg/kg salt-wt. dose.

Another study tested the effects of DHBE on a win-stay version of the RAM, which is similar to the DMTS. Rats are allowed to enter only one arm on the RAM, and this arm is baited and changes with each day of testing. The rats are then given delays up to 45 min and placed back on the maze with all eight arms available, but only the previously presented arm is baited (match). Entries into the nonbaited arms are scored as errors. Intraventricular injections of DHβE (30, 100, 300 nmol) dose-dependently increased the number of errors, with a significant increase over saline injection at the highest dose [10].

The DNMTS task is similar to DMTS, except the rat is rewarded for pressing the lever not previously presented (nonmatch). The threshold for cognitive impairment with mecamylamine on the DNMTS task appears to be 3 mg/kg, as both 1- and 2-mg/kg doses did not cause significant impairments, while 3- and 5-mg/kg doses did (Table 3.2).

TESTS OF SUSTAINED ATTENTION

Mecamylamine impairs performance on an operant visual-signal-detection task [6, 41, 55]. Rats are trained in a two-lever operant chamber to press one lever when a light is presented (signal trial) and the other lever when no light is presented (blank trial). Variable presignal intervals ranging from 0.3 to 24.4 sec require sustained attention from the rat in order to detect the signal and respond correctly. Mecamylamine dose-dependently decreased choice accuracy by reducing both percent hit and percent correct rejection (Table 3.2). In addition, when rats were divided following training into low- and high-accuracy groups, mecamylamine reduced the percent hit in the low-accuracy group at a lower dose than in the high-accuracy group [41].

The five-choice serial reaction time task (5-CSRTT) is similar to the visual-signal-detection task described above, but it is conducted in an operant chamber with five nose-poke holes. A randomly chosen hole is briefly illuminated, and the rat is rewarded for nose-poking the signaled hole. Again, a variable intertrial interval requires sustained attention from the rat in order to detect the signal and respond correctly. The impairments of mecamylamine on the 5-CSRTT seem to be less of an effect on cognition and more of an effect on motor activity. Mecamylamine seems to cause the rats to be less responsive, with fewer completed trials and fewer anticipatory responses [52, 53]. However, mecamylamine caused a decrease in the percent of omission errors, and it also increased the correct response latencies, indicating that rats were trading off speed for accuracy [52, 53]. There was a significant interaction of mecamylamine with age on the 5-CSRTT (Table 3.2). Mecamylamine doses that were ineffective in young rats (3 months) impaired choice accuracy in middle-aged rats (15 months) [20]. The threshold for cognitive impairment in aged rats seems to be 3 mg/kg (Table 3.2).

TESTS OF RESPONSE INHIBITION

Mecamylamine appears to impair response inhibition in rats (Table 3.2). Intrahypothalamic injections of mecamylamine significantly increased responding during the signaled nonreinforced portion of a multiple schedule paradigm [50]. In addition, mecamylamine also impaired performance on the differential-reinforcement-of-low-rate (DRL-30 sec) task (Table 3.2). For the DRL-30-sec task, rats are trained to press a lever every 30 sec for food reward, and premature responding is punished by resetting the timer. Mecamylamine (2 mg/kg) significantly increased the number of very premature responses (less than 6 sec) on the DRL-30-sec task, while the 4-mg/kg dose significantly lowered the reinforcement rate [2]. Similarly, dorsal and ventral hippocampal injections of mecamylamine significantly increased premature responses and decreased the number of reinforcements received [49]. It is interesting that mecamylamine decreases the number of anticipatory responses on the 5-CRSTT but increases the number of premature responses on the DRL-30-sec task.

WATER MAZE

Mecamylamine has also been shown to impair cognition in the water-maze task, in which the rat is required to learn and remember the spatial location of a hidden submerged platform in order to escape from the water. Impaired performance is noted by increased escape distances or latencies. In addition, a spatial probe trial is often run once rats are trained in which the platform is removed, and the time spent swimming in each quadrant is recorded. Riekkinen and colleagues [42–48] refer to this as spatial bias or increased time swimming in the quadrant where the platform was previously located. A dose of 2.5 mg/kg of mecamylamine appears to be subthreshold having no effect on escape distance or spatial bias [44, 46], but doses of 3 to 10 mg/kg effectively impair water-maze performance (Table 3.3). The effect on cognition may be somewhat confounded, since mecamylamine can alter swim speed. Decker and Majchrzak [12] found that mecamylamine doses of 3 and 10 mg/kg slowed swimming somewhat to a visible platform, while Riekkinen et al. [46, 48] have found increased swim speeds with doses of 7.5 and 10 mg/kg. However, the peripherally acting hexamethonium does not cause any significant impairments in water-maze performance [12, 46, 47], strengthening the argument that the effect of mecamylamine on water-maze performance is central. Further, intracerebroventricular (ICV) injections of mecamylamine significantly impaired water-maze performance without altering swim speed [10, 12].

Mecamylamine appears to impair acquisition but not retrieval of spatial information in the water maze. Decker and Majchrzak [12] trained rats for 4 days to a new platform location, giving mecamylamine before each training session and before the spatial probe trial on the fifth day, which tests the retrieval of spatial information. Systemic mecamylamine did not impair escape latency or spatial bias, but the highest dose of ICV mecamylamine (100 µg) did marginally decrease the time spent in the previous platform quadrant (spatial bias) [12]. Systemic mecamylamine injections given only prior to the retention spatial probe trial had no effect [46].

Intraventricular (ICV) injections of DHβE impair water-maze performance [10]. Male Long-Evans rats were given ICV injections of DHβE (0, 30, 100, or 300 nmol) daily during the 3 to 4 days of training on the water maze. On the day following training, spatial probe trials were given with and without the DHβE injections. Only the highest dose increased escape latency during training and decreased spatial bias on the spatial probe trial. The decrease in spatial bias was observed with and without

TABLE 3.3
Effects of Mecamylamine on Water-Maze Performance

Rat Strain	Sex	Dose	Route[a]	Outcome[b]	Reference
Han:Wistar	Males	2.5 mg/kg	IP 40 min prior	− Escape distance	[44, 46]
				− Swim speed	[44, 46]
				− Spatial bias	[46]
Long-Evans	Males	3 mg/kg	IP 20 min prior	↑ Escape latency	[12]
				− Swim speed (swam slightly slower to visible platform)	
Kuo:Wistar	Males	7.5 mg/kg	IP 40 min prior	↑ Escape distance	[43]
				↑ Swim speed	
			IP 30 min prior	↑ Escape distance	[44]
				− Swim speed	
		10 mg/kg	IP 30 min prior	↑ Escape distance	[42, 46–48]
				↓ Spatial bias	[46]
				− Or ↑ swim speed	[46, 47]
		30 mg/kg		↑ Escape distance	[48]
Long-Evans	Males	10 mg/kg	IP 20 min prior	↑ Escape latency	[12]
				↓ Spatial bias	
				− Swim speed (swam slightly slower to a visible platform)	
Long-Evans	Males	10, 30, 100 μL	ICV injection	↑ Escape latency	[12]
				↓ Spatial bias	
				− Swimming to a visible platform	
Aged Rats					
Han-Wistar	Males (24–26 months)	10 mg/kg	IP 30 min prior	↑ Escape distance	[45]
				− Spatial bias	

[a] IP = intraperitoneal injection; ICV = intracerebroventricular injection.
[b] ↑ = enhanced performance; ↓ = impaired performance; − = no effect.

drug injection prior to testing. None of the doses of DHβE changed the swim speed or the escape latency to a visible platform [10].

AVOIDANCE LEARNING

Riekkinen and colleagues have also tested the effects of mecamylamine on a passive-avoidance task. In this task, the rat is placed in the bright side of a two-compartment box and is given a foot shock when it enters the dark compartment. Retention is tested 24 h later by placing the rat in the bright compartment, and learning is evident by longer retention latencies to enter the dark compartment. Riekkinen et al. found that 7.5-mg/kg injections of mecamylamine significantly reduced the retention latencies [43, 44], while apparently, a higher dose of 10 mg/kg did not [46–48]. However, a mecamylamine dose of 30 mg/kg did significantly reduce the retention latency [48]. Another study showed that a mecamylamine dose of 25 mg/kg effectively

TABLE 3.4
Effects of Mecamylamine on Avoidance Learning

Rat Strain	Sex	Dose	Route[a]	Outcome[b]	Reference
Long-Evans	Males	5 mg/kg	IP 30 min prior	– Drinking latency	[18]
Kuo:Wistar	Males	7.5 mg/kg	IP 30 or 40 min prior	↓ Retention latency	[43, 44]
		10 mg/kg	IP 30 min prior	– Retention latency	[46–48]
		30 mg/kg		↓ Retention latency	[48]
Wistar	Males	0.5, 5 mg/kg	SC 20 min prior	– Retention latency	[16]
		25 mg/kg		↓ Retention latency	
Sprague-Dawley	Males	10 µg/µl	IH 7 min prior	↑ Avoidance responses	[49]
				↓ Shock presentations	
Weanling Rats					
London black		5 µg/µL	VHE injection	↓ Retention latency PND11, 13, 16, 20	[3]
		50 µg/µL		↓ Retention latency PND11–16	
		100 µg/µL		↓ Retention latency PND11–20	

[a] IP = intraperitoneal injection; SC = subcutaneous injection; VHE = ventral hippocampus-entorhinal area injection.
[b] ↑ = enhanced performance; ↓ = impaired performance; – = no effect; PND = postnatal day.

decreased retention time in the passive-avoidance task, while the peripheral blocker hexamethonium did not [16].

Mecamylamine injected into the hippocampus improved performance in well-trained rats during the Sidman avoidance task (Table 3.4). Rats were first trained with a 7-sec light (CS) preceding a shock (US). The shock occurred every 28 sec, so the rat could avoid the shock by crossing to the other side of the shuttle box any time within the 21 sec preceding the light or during the light presentation. Rats were trained to criterion with the light cue, and then they were trained to criterion in the uncued Sidman procedure without the light cue. Following that, mecamylamine was administered. Mecamylamine significantly increased avoidance responses and decreased shock presentations compared with baseline performance [49]. It is unclear why mecamylamine improves performance on this avoidance task and not on the passive-avoidance task described above. Perhaps it is a function of the rats being well trained in the Sidman avoidance task.

Goldberg et al. [18] tested the effects of mecamylamine on fear learning in rats using conditioned suppression. In this task, food-restricted rats were first trained to lick for a milk solution to a criterion of 100 licks/min. Mecamylamine was then given before a session in which a foot shock was given and no milk solution was available. Then, in a following session without shock and with the milk solution available, the time to finish 100 licks was recorded and termed the drinking latency. Mecamylamine did not impair fear learning. In fact, it caused a subtle nonsignificant increase in drinking latency.

Lastly, one study examined the effects of nicotinic blockade during the development of the passive-avoidance task in young weanling rats. For this study, mecamylamine was injected directly into the ventral hippocampus-entorhinal (VHE) area, and it impaired the acquisition of a passive-avoidance task in preweanling rats and also reduced the resistance to extinction [3].

CONCLUSIONS

The effective mecamylamine dose for the majority of tasks appears to be above 5 mg/kg. The most sensitive tasks for cognitive impairments of mecamylamine were the visual-signal-detection task, the 16-arm RAM, DMTS, DNMTS, and the water maze. Doses as low as 3 mg/kg of mecamylamine caused impairments in the DNMTS task [51] and the water maze [12]. Mecamylamine doses of 1.25 and 1.8 mg/kg caused cognitive impairments on the 16-arm RAM [34] and DMTS task [5], respectively, and a dose as low as 1 mg/kg caused significant impairments on the visual-signal-detection task [6, 41, 55]. Using more-sensitive tasks can be useful, since higher doses of mecamylamine have less specific effects, making it harder to tease out the cognitive impairments from other side effects. Mecamylamine doses of 10 mg/kg have been reported to cause some sedation, slowed response, and ptosis (drooping of the upper eyelid) (e.g., [22, 55]).

Mecamylamine seems to have its greatest impairments during the acquisition of the task rather than during consolidation or retention of the task [12, 46]. It also seems to affect working memory and not reference memory [34, 40]. Mecamylamine had varying effects in aged rats, depending on the task given. It was ineffective in aged rats on the RAM [32] but exacerbated the effects on the 5-CRSTT [20].

REFERENCES

1. Andrews, J.S., Jansen, J.H., Linders, S., and Princen, A., Effects of disrupting the cholinergic system on short-term spatial memory in rats, *Psychopharmacology (Berl.)*, 115, 485–494, 1994.
2. Bizot, J.C., Effects of various drugs including organophosphorus compounds (OPC) and therapeutic compounds against OPC on DRL responding, *Pharmacol. Biochem. Behav.*, 59, 1069–1080, 1998.
3. Blozovski, D., Deficits in passive avoidance learning in young rats following mecamylamine injections in the hippocampo-entorhinal area, *Exp. Brain Res.*, 50, 442–448, 1983.
4. Brucato, F.H., Levin, E.D., Rose, J.E., and Swartzwelder, H.S., Intracerebroventricular nicotine and mecamylamine alter radial arm maze performance in rats, *Drug Dev. Res.*, 31, 18–23, 1994.
5. Bushnell, P.J., Levin, E.D., and Overstreet, D.H., Spatial working and reference memory in rats bred for autonomic sensitivity to cholinergic stimulation: acquisition, accuracy, speed, and effects of cholinergic drugs, *Neurobiol. Learn. Mem.*, 63, 116–132, 1995.
6. Bushnell, P.J., Oshiro, W.M., and Padnos, B.K., Detection of visual signals by rats: effects of chlordiazepoxide and cholinergic and adrenergic drugs on sustained attention, *Psychopharmacology (Berl.)*, 134, 230–241, 1997.

7. Clarke, P.B. and Fibiger, H.C., Reinforced alternation performance is impaired by muscarinic but not by nicotinic receptor blockade in rats, *Behav. Brain Res.*, 36, 203–207, 1990.

8. Court, J., Spurden, D., Lloyd, S., McKeith, I., Ballard, C., Cairns, N., Kerwin, R., Perry, R., and Perry, E., Neuronal nicotinic receptors in dementia with Lewy bodies and schizophrenia: alpha-bungarotoxin and nicotine binding in the thalamus, *J. Neurochem.*, 73, 1590–1597, 1999.

9. Court, J.A., Piggott, M.A., Lloyd, S., Cookson, N., Ballard, C.G., McKeith, I.G., Perry, R.H., and Perry, E.K., Nicotine binding in human striatum: elevation in schizophrenia and reductions in dementia with Lewy bodies, Parkinson's disease and Alzheimer's disease and in relation to neuroleptic medication, *Neuroscience*, 98, 79–87, 2000.

10. Curzon, P., Brioni, J.D., and Decker, M.W., Effect of intraventricular injections of dihydro-beta-erythroidine (DH beta E) on spatial memory in the rat, *Brain Res.*, 714, 185–191, 1996.

11. Deacon, R.M.J., Pharmacological studies of a rat spatial delayed nonmatch-to-sample task as an animal model of dementia, *Drug Dev. Res.*, 24, 67–79, 1991.

12. Decker, M.W. and Majchrzak, M.J., Effects of systemic and intracerebroventricular administration of mecamylamine, a nicotinic cholinergic antagonist, on spatial memory in rats, *Psychopharmacology (Berl.)*, 107, 530–534, 1992.

13. Decker, M.W., Majchrzak, M.J., and Arneric, S.P., Effects of lobeline, a nicotinic receptor agonist, on learning and memory, *Pharmacol. Biochem. Behav.*, 45, 571–576, 1993.

14. Decker, M.W., Curzon, P., Brioni, J.D., and Arneric, S.P., Effects of ABT-418, a novel cholinergic channel ligand, on place learning in septal-lesioned rats, *Eur. J. Pharmacol.*, 261, 217–222, 1994.

15. Decker, M.W., Brioni, J.D., Bannon, A.W., and Arneric, S.P., Diversity of neuronal nicotinic acetylcholine receptors: lessons from behavior and implications for CNS therapeutics, *Life Sci.*, 56, 545–570, 1995.

16. Elrod, K. and Buccafusco, J.J., Correlation of the amnestic effects of nicotinic antagonists with inhibition of regional brain acetylcholine synthesis in rats, *J. Pharmacol. Exp. Ther.*, 258, 403–409, 1991.

17. Felix, R. and Levin, E.D., Nicotinic antagonist administration into the ventral hippocampus and spatial working memory in rats, *Neuroscience*, 81, 1009–1017, 1997.

18. Goldberg, M.E., Sledge, K., Hefner, M., and Robichaud, R.C., Learning impairment after three classes of agents which modify cholinergic function, *Arch. Int. Pharmacodyn. Ther.*, 193, 226–235, 1971.

19. Guan, Z.Z., Zhang, X., Blennow, K., and Nordberg, A., Decreased protein level of nicotinic receptor alpha7 subunit in the frontal cortex from schizophrenic brain, *Neuroreport*, 10, 1779–1782, 1999.

20. Jones, D.N., Barnes, J.C., Kirkby, D.L., and Higgins, G.A., Age-associated impairments in a test of attention: evidence for involvement of cholinergic systems, *J. Neurosci.*, 15, 7282–7292, 1995.

21. Kim, J.S. and Levin, E.D., Nicotinic, muscarinic and dopaminergic actions in the ventral hippocampus and the nucleus accumbens: effects on spatial working memory in rats, *Brain Res.*, 725, 231–240, 1996.

22. Levin, E.D., Castonguay, M., and Ellison, G.D., Effects of the nicotinic receptor blocker mecamylamine on radial arm maze performance in rats, *Behav. Neural Biol.*, 48, 206–212, 1987.

23. Levin, E.D., McGurk, S.R., Rose, J.E., and Butcher, L.L., Reversal of a mecamylamine-induced cognitive deficit with the D2 agonist, LY 171555, *Pharmacol. Biochem. Behav.*, 33, 919–922, 1989.

24. Levin, E.D., McGurk, S.R., South, D., and Butcher, L.L., Effects of combined muscarinic and nicotinic blockade on choice accuracy in the radial arm maze, *Behav. Neural Biol.*, 51, 270–277, 1989.

25. Levin, E.D. and Rose, J.E., Anticholinergic sensitivity following chronic nicotine administration as measured by radial arm maze performance in rats, *Behav. Pharmacol.*, 1, 511–520, 1990.

26. Levin, E.D., Rose, J.E., McGurk, S.R., and Butcher, L.L., Characterization of the cognitive effects of combined muscarinic and nicotinic blockade, *Behav. Neural Biol.*, 53, 103–112, 1990.

27. Levin, E.D., Nicotinic systems and cognitive function, *Psychopharmacology (Berl.)*, 108, 417–431, 1992.

28. Levin, E.D., Briggs, S.J., Christopher, N.C., and Rose, J.E., Persistence of chronic nicotine-induced cognitive facilitation, *Behav. Neural Biol.*, 58, 152–158, 1992.

29. Levin, E.D., Briggs, S.J., Christopher, N.C., and Rose, J.E., Chronic nicotinic stimulation and blockade effects on working memory, *Behav. Pharmacol.*, 4, 179–182, 1993.

30. Levin, E.D., Briggs, S.J., Christopher, N.C., and Auman, J.T., Working memory performance and cholinergic effects in the ventral tegmental area and substantia nigra, *Brain Res.*, 657, 165–170, 1994.

31. Levin, E.D., *Drug Dev. Res.*, 38, 188–195, 1996.

32. Levin, E.D. and Torry, D., Acute and chronic nicotine effects on working memory in aged rats, *Psychopharmacology (Berl.)*, 123, 88–97, 1996.

33. Levin, E.D., Christopher, N.C., and Briggs, S.J., Chronic nicotinic agonist and antagonist effects on T-maze alternation, *Physiol. Behav.*, 61, 863–866, 1997.

34. Levin, E.D., Kaplan, S., and Boardman, A., Acute nicotine interactions with nicotinic and muscarinic antagonists: working and reference memory effects in the 16-arm radial maze, *Behav. Pharmacol.*, 8, 236–242, 1997.

35. Levin, E.D., Bradley, A., Addy, N., and Sigurani, N., Hippocampal alpha 7 and alpha 4 beta 2 nicotinic receptors and working memory, *Neuroscience*, 109, 757–765, 2002.

36. McGurk, S.R., Levin, E.D., and Butcher, L.L., Radial arm maze performance in rats is impaired by a combination of nicotinic-cholinergic and D2 dopaminergic antagonist drugs, *Psychopharmacology (Berl.)*, 99, 371–373, 1989.

37. McGurk, S.R., Levin, E.D., and Butcher, L.L., Nicotinic-dopaminergic relationships and radial arm maze performance in rats, *Behav. Neural Biol.*, 52, 78–86, 1989.

38. McGurk, S.R., Levin, E.D., and Butcher, L.L., Impairment of radial arm maze performance in rats following lesions involving the cholinergic medial pathway: reversal by arecoline and differential effects of muscarinic and nicotinic antagonists, *Neuroscience*, 44, 137–147, 1991.

39. Moran, P.M., Differential effects of scopolamine and mecamylamine on working and reference memory in the rat, *Pharmacol. Biochem. Behav.*, 45, 533–538, 1993.

40. Ohno, M., Yamamoto, T., and Watanabe, S., Blockade of hippocampal nicotinic receptors impairs working memory but not reference memory in rats, *Pharmacol. Biochem. Behav.*, 45, 89–93, 1993.

41. Rezvani, A.H., Bushnell, P.J., and Levin, E.D., Effects of nicotine and mecamylamine on choice accuracy in an operant visual signal detection task in female rats, *Psychopharmacology (Berl.)*, 164, 369–375, 2002.

42. Riekkinen, M., Riekkinen, P., Sirvio, J., and Riekkinen, P., Jr., Effects of combined methysergide and mecamylamine/scopolamine treatment on spatial navigation, *Brain Res.*, 585, 322–326, 1992.

43. Riekkinen, M., Sirvio, J., and Riekkinen, P., Jr., Pharmacological consequences of nicotinergic plus serotonergic manipulations, *Brain Res.*, 622, 139–146, 1993.

44. Riekkinen, M. and Riekkinen, P., Jr., Effects of THA and physostigmine on spatial navigation and avoidance performance in mecamylamine and PCPA-treated rats, *Exp. Neurol.*, 125, 111–118, 1994.

45. Riekkinen, M. and Riekkinen, P., Jr., Nicotine and D-cycloserine enhance acquisition of water maze spatial navigation in aged rats, *Neuroreport*, 8, 699–703, 1997.

46. Riekkinen, P., Jr., Sirvio, J., Aaltonen, M., and Riekkinen, P., Effects of concurrent manipulations of nicotinic and muscarinic receptors on spatial and passive avoidance learning, *Pharmacol. Biochem. Behav.*, 37, 405–410, 1990.

47. Riekkinen, P., Jr., Riekkinen, M., Sirvio, J., and Riekkinen, P., Effects of concurrent nicotinic antagonist and PCPA treatments on spatial and passive avoidance learning, *Brain Res.*, 575, 247–250, 1992.

48. Riekkinen, P., Jr., Riekkinen, M., and Sirvio, J., Cholinergic drugs regulate passive avoidance performance via the amygdala, *J. Pharmacol. Exp. Ther.*, 267, 1484–1492, 1993.

49. Ross, J.F. and Grossman, S.P., Intrahippocampal application of cholinergic agents and blockers: effects on rats in differential reinforcement of low rates and Sidman avoidance paradigms, *J. Comp. Physiol. Psychol.*, 86, 590–600, 1974.

50. Ross, J.F., McDermott, L.J., and Grossman, S.P., Disinhibitory effects of intrahippocampal of intrahypothalamic injections of anticholinergic compounds in the rat, *Pharmacol. Biochem. Behav.*, 3, 631–639, 1975.

51. Ruotsalainen, S., MacDonald, E., Miettinen, R., Puumala, T., Riekkinen, P., Sr., and Sirvio, J., Additive deficits in the choice accuracy of rats in the delayed non-matching to position task after cholinolytics and serotonergic lesions are non-mnemonic in nature, *Psychopharmacology (Berl.)*, 130, 303–312, 1997.

52. Ruotsalainen, S., Miettinen, R., MacDonald, E., Koivisto, E., and Sirvio, J., Blockade of muscarinic, rather than nicotinic, receptors impairs attention, but does not interact with serotonin depletion, *Psychopharmacology (Berl.)*, 148, 111–123, 2000.

53. Stolerman, I.P., Mirza, N.R., Hahn, B., and Shoaib, M., Nicotine in an animal model of attention, *Eur. J. Pharmacol.*, 393, 147–154, 2000.

54. Taylor, G.T., Bassi, C.J., and Weiss, J., Limits of learning enhancements with nicotine in old male rats, *Acta Neurobiol. Exp. (Wars.)*, 65, 125–136, 2005.

55. Turchi, J., Holley, L.A., and Sarter, M., Effects of nicotinic acetylcholine receptor ligands on behavioral vigilance in rats, *Psychopharmacology (Berl.)*, 118, 195–205, 1995.

56. Whitehouse, P.J., Martino, A.M., Antuono, P.G., Lowenstein, P.R., Coyle, J.T., Price, D.L., and Kellar, K.J., Nicotinic acetylcholine binding sites in Alzheimer's disease, *Brain Res.*, 371, 146–151, 1986.

57. Whitehouse, P.J., Martino, A.M., Wagster, M.V., Price, D.L., Mayeux, R., Atack, J.R., and Kellar, K.J., Reductions in [3H]nicotinic acetylcholine binding in Alzheimer's disease and Parkinson's disease: an autoradiographic study, *Neurology*, 38, 720–723, 1988.

58. Whitehouse, P.J. and Kalaria, R.N., Nicotinic receptors and neurodegenerative dementing diseases: basic research and clinical implications, *Alzheimer Dis. Assoc. Disord.*, 9 (Suppl. 2), 3–5, 1995.

59. Woodruff-Pak, D.S., Li, Y.T., and Kem, W.R., A nicotinic agonist (GTS-21), eyeblink classical conditioning, and nicotinic receptor binding in rabbit brain, *Brain Res.*, 645, 309–317, 1994.

4 Involvement of the NMDA System in Learning and Memory

Amir H. Rezvani
Duke University Medical Center

CONTENTS

INTRODUCTION

Since its discovery in the early 1950s, the N-methyl-D-aspartate (NMDA) receptor system in the brain has been implicated in many fundamental functions, including neuronal plasticity, neurotoxicity, learning, and memory (Riedel et al., 2003). The aim of this chapter is to summarize the current findings on the role of glutamate-receptor systems in learning and memory in different animal models and humans. The structure, localization, and pharmacology of these receptors will not be discussed in this chapter. For those who are interested in these particular topics, we refer them to an excellent recent review (Riedel et al., 2003).

ANIMAL STUDIES

Researchers have provided increasing evidence that the NMDA-receptor systems generally, and glutamate-mediated long-term potentiation (LTP) in particular, may play a crucial role in the processes of learning and memory formation. NMDA receptors in the brain have been implicated and been shown to play a crucial role in various types of learning. These receptors have been demonstrated to be involved in Pavlovian fear conditioning (Xu et al., 2001), eyeblink conditioning (Thompson and Disterhoft, 1997), spatial learning (Morris et al., 1986; Shimizu et al., 2000; Tsien et al., 1996), working and reference memory (Levin et al., 1998; May-Simera and Levin, 2003), place preference (Swain et al., 2004), passive-avoidance learning (Danysz et al., 1988), olfactory memory (Si et al., 2004; Maleszka et al., 2000), and reversal learning (Harder et al., 1998).

NMDA

It has been suggested that the activation of the NMDA receptor is required for long-term potentiation (LTP) in the hippocampus, amygdala, and medial septum (Izquierdo, 1994; Rockstroh et al., 1996; Scatton et al., 1991). This mechanism has been implicated in memory formation; the involvement of the glutamate-receptor system and LTP is strongly linked to new learning and memory in animal models (Lozano et al., 2001; Scheetz and Constantine-Paton, 1994; Tang et al., 1999, 2001; Wong et al., 2002). Both lesion studies and pharmacological manipulations in experimental animals suggest that the NMDA-receptor system may be important in the induction of memory formation, but not for the maintenance of memories (Constantine-Paton, 1994; Izquierdo, 1991; Izquierdo and Medina, 1993; Liang et al., 1993; Quartermain et al., 1994; Rickard et al., 1994). Indeed, it has been shown that NMDA-receptor blockade after learning a task had no effect on memory performance in humans, whereas blockade of receptors before learning resulted in memory impairment (Hadj Tahar et al., 2004; Oye et al., 1992; Rowland et al., 2005).

NMDA alone, systemically administered in rats, has been shown to potentiate cognitive functions (Hlinak and Krejci, 2002; Koek et al., 1990). A recent study (Hlinak and Krejci, 2003) investigated whether systemically administered NMDA can prevent amnesia induced by an NMDA antagonist dizocilpine (MK-801). Using a modified elevated plus maze paradigm, it was demonstrated that NMDA administered subcutaneously immediately after the acquisition session protected the mice against amnesia induced by MK-801 given shortly before the retention session.

TRANSGENIC AND MUTANT MICE

Studying transgenic and mutant mice has provided more evidence in support of the involvement of the NMDA in cognition. Mutant mice lacking the NMDA-receptor subunit NR2A have shown reduced hippocampal LTP and spatial learning (Sakimura et al., 1995). Also, transgenic mice lacking NMDA receptors in the CA1 region of the hippocampus show both defective LTP and severe deficits in both spatial and nonspatial learning (Shimizu et al., 2000; Tsien et al., 1996). On the other hand, genetic enhancement of NMDA-receptor function results in superior learning and memory. Recently, in an interesting study, Tang et al. (2001) confirmed the role of the NMDA-receptor system in LTP in the hippocampus and in learning and memory. Using the NR2B transgenic (Tg) lines of mice, in which the NMDA-receptor function is enhanced via the NR2B subunit transgene in neurons of the forebrain, they demonstrated both larger LTP in the hippocampus and superior learning and memory in naïve NR2B Tg mice (Tang et al., 1999; Tang et al., 2001; Wong et al., 2002).

In the novel-object recognition task, however, enriched NR2B Tg mice exhibited much longer recognition memory (up to 1 week) compared with that (up to 3 days) of naïve NR2B Tg mice. Together, these findings confirm and support the important role of the NMDA receptor in memory.

NMDA ANTAGONISTS

NMDA-receptor antagonists such as MK-801, ketamine, phencyclidine (PCP), 2-amino-5-phosphonopentanoate (AP5), and selective mGlu5-receptor antagonist 2-methyl-6-(phenylethynyl)-pyridine (MPEP) have been extensively used to study the role of NMDA in learning and memory. Both acute and subchronic administration of NMDA-receptor antagonists have been shown to impair performance on tasks that seem to depend upon hippocampal or amygdaloid functions (Izquierdo and Medina, 1993; Jentsch and Roth, 1999; Morris et al., 1986). These tasks include passive avoidance (Benvenga and Spaulding, 1988; Kesner and Dakis, 1993; Murray and Ridley, 1997; Venero and Sandi, 1997), acquisition of Morris water maze (Heale and Harley, 1990), and delayed alteration (Verma and Moghaddam, 1996).

The noncompetitive, highly specific N-methyl-D-aspartate NMDA-receptor antagonist dizocilpine (MK-801) (E. H. Wong et al., 1986) has been shown to induce dose-dependent impairment of learning and memory (Benvenga and Spaulding, 1988; Butelman, 1990; Carey et al., 1998; de Lima et al., 2005; Hlinak and Krejci, 1998, 2003; May-Simera and Levin, 2003; Murray and Ridley, 1997; Murray et al., 1995; Venero and Sandi, 1997). It has also been shown that MK-801 selectively disrupts reversal learning in rats using serial reversal tack (van der Meulen et al., 2003). Further, several studies have revealed that MK-801 administration impairs different aspects of learning and memory in the elevated plus maze in rodents (Hlinak and Krejci, 1998, 2000, 2002). Recently, the effects of NMDA-receptor blockade on formation of object-recognition memory were examined in rats. It was found that MK-801 impaired both short- and long- term retention of object-recognition memory when given either before or after training. These results suggest that NMDA-receptor activation is necessary for formation of object-recognition memory (de Lima et al., 2005). Amnesic effects of MK-801 in mice have also been reported. MK-801 injected intravenously in mice before a training trial in a passive-avoidance task produced an amnesic effect similar to that produced by the standard amnesic agent scopolamine, yet the potency of MK-801 was 40 times that of scopolamine. Effects of MK-801 on corticosterone facilitation of long-term memory have been examined in chicks. It has been shown that long-term memory formation for a weak passive-avoidance task in day-old chicks is facilitated by corticosterone administration (Venero and Sandi, 1997) and that intracerebral infusion of MK-801 prevents the facilitating effect of corticosterone when MK-801 is given before the training trial. These results support the view that corticosterone facilitates the formation of long-term memory in this particular learning model through the modulation of the NMDA-receptor system in the brain.

In addition to memory impairment, MK-801 has been shown to induce changes in motor activity (Ford et al., 1989). Therefore, it is important to know whether the MK-801 effects upon memory are secondary to its effect on motor disturbance. In an attempt to address this important issue, Carey et al. (1998) examined the effects of MK-801 upon retention of habituation to a novel environment and locomotor activity. It was demonstrated that a low dose of 0.1 mg/kg MK-801, which did not affect locomotor activity, severely interfered with retention of the novel environment. This observation suggests dissociation between the effects of MK-801 on memory and locomotor activity.

The eyeblink classical conditioning paradigm is an extensively used measure of associate learning and memory (Woodruff-Pak et al., 2000). The contribution of the NMDA-receptor system in the brain to classical eyeblink conditioning has been investigated pharmacologically in rabbits and mice. It has been shown that MK-801 slows the rate of acquisition during delay conditioning (Thompson and Disterhoft, 1997).

Using eyeblink classical conditioning in mice, Takatsuki et al. (2001) demonstrated the role of NMDA receptors in acquisition of the conditioned response (CR). Further, these researchers have shown that the contribution of these receptors to extinction is much smaller than their contribution to acquisition in mouse eyeblink conditioning. In these studies, it was shown that MK-801 impaired acquisition of the CR during mouse eyeblink conditioning in a task-dependent manner.

Learning impairments induced by glutamate blockade using MK-801 have also been reported in nonhuman primates. Acquisition and reversal learning of visual-discrimination tasks and acquisition of visuospatial discrimination tasks were assessed in marmosets using the Wisconsin General Test Apparatus. It was shown that MK-801 impaired acquisition of visuospatial (conditional) discrimination (Harder et al., 1998). Lesions of the fornix (Harder et al., 1996) or hippocampus (Ridley et al., 1988, 1995) have been shown to produce a specific and severe impairment on visuospatial tasks. Thus, it could be suggested that it is the effect of MK-801 on glutamatergic corticohippocampal projections that is responsible for the visuospatial impairment (Harder et al., 1998).

Selective impairment of learning and blockade of LTP by other NMDA-receptor antagonists has been reported. It has been shown that blockade of NMDA sites with the drug AP5 does not detectably affect synaptic transmission in the hippocampus, but prevents the induction of LTP. Interestingly, chronic intracerebroventricular infusion of D,L-AP5 (which blocks LTP *in vitro* and *in vivo*) selectively impaired the acquisition of place learning in the Morris water maze, a type of learning that is dependent on normal hippocampal functioning. These results further suggest that the NMDA-receptor system is involved in spatial learning and supports the hypothesis that LTP is involved in some forms of learning (Morris et al., 1986).

Recently, it was also demonstrated that MPEP at a relatively high dose, but not at low dose, impaired working memory and instrumental learning. MPEP administration also caused a transient increase in dopamine release in the prefrontal cortex and nucleus accumbens. MPEP exposure also augmented the effect of MK-801 on cortical dopamine release, locomotion, and stereotypy. Pretreatment with low (3 mg/kg) MPEP enhanced the detrimental effects of MK-801 on cognition (Homayoun et al., 2004). These results suggest that an mGlu5-receptor antagonist such as MPEP plays a major role in regulating NMDA-receptor-dependent cognitive functions.

To investigate the involvement of the NMDA receptors in different stages of memory consolidation, Tronel and Sara (2003) recently examined the effect of a competitive NMDA-receptor antagonist, 2-amino-5-phosphonovalerate (APV), on odor-reward associative learning in rats. It was shown that the blockade of NMDA receptors by APV injected intracerebroventricularly immediately after training induced a profound and enduring amnesia, but had no effect when the treatment was delayed 2 hours after training. More specifically, it was shown that the blockade of NMDA receptors in the prelimbic region of the frontal cortex, but not into the hippocampus, impaired memory formation of the odor-reward association in rats. These results confirm the role of NMDA receptors in the early stage of consolidation of a simple odor-reward associative memory and underlie the role of the frontal cortex in consolidation of long-term memory (Tronel and Sara, 2003).

There have also been reports of amnesia in passive-avoidance task induced by posttrial administration of APV into the hippocampus or the amygdala (Ferreira et al., 1992; Zanatta et al., 1996). However, in a recent study, Santini et al. (2001) demonstrated that the amnesia observed 24 hours after systemic administration of NMDA antagonist D(-)-3-(2-carboxypiperazine-4-yl)-propyl-1-phosphonic acid (CPP) was transient. The authors argue that the so-called rescued memory at 48 hours supports the existence of late waves of NMDA activity promoting memory consolidation.

Pretraining administration of NMDA-receptor antagonists has been shown to produce anterograde amnesia in Pavlovian fear conditioning (Kim et al., 1991; Xu and Davis, 1992), spatial learning (Hauben et al., 1999), and passive avoidance (Danysz et al., 1988). Intercranial administration of NMDA-receptor antagonist AP5, without impairing performance processes, produced anterograde amnesia when given before training in goldfish (Xu et al., 2001).

Phencyclidine (PCP), a noncompetitive NMDA-receptor antagonist, has been shown to produce both positive and negative symptoms of schizophrenia as well as cognitive defects in healthy humans (Javitt and Zukin, 1991; Tamminga, 1998). PCP has been used to further investigate the role of the NMDA-receptor system in learning and memory. The performance of rats and mice (Podhorna and Didriksen, 2005) in the Morris water maze and spatial continuous recognition memory task have all been shown to be impaired following acute PCP exposure. PCP-treated animals maintained the original learned rule, and they were only impaired in abolishing it and establishing a new rule. This suggests that acute PCP administration at doses of 1.0 and 1.5 mg/kg was able to significantly impair complex cognitive tasks without disrupting simple rule-learning parameters (Abdul-Monim et al., 2003). Recently, it was shown that repeated administration of PCP failed to produce enduring memory impairment in an eight-arm radial arm maze in rats or mice (Li et al., 2003).

Haloperidol failed to ameliorate the deficit in reversal task performance induced by PCP. In contrast, the new atypical antipsychotic ziprasidone produced a significant improvement in impairment of the reversal task performance induced by PCP (Abdul-Monim et al., 2003). Consistent with these findings, it has been demonstrated that in a novel objective recognition test, repeated administration of PCP significantly decreased exploratory preference in the retention test session but not in the training test session. PCP-induced deficits were significantly improved by subsequent sub-chronic administration of clozapine but not haloperidol (Hashimoto et al., 2005).

NMDA TRANSPORTER INHIBITORS

Blockade of glutamate uptake by transporter inhibitors has also been used to study the role of NMDA in memory formation. Recently, it was demonstrated that pre-training injections of a glutamate-transporter inhibitor L-trans-2,4-PDC (L-trans-2,4-pyrrolidine dicarboxylate) had no effect on acquisition and short-term (1 h) memory but impaired long-term (24 h) associative olfactory memory in a dose-dependent manner in the honeybee. This effect was found to be transient, and amnesic animals could be retrained 48 h after injections (Maleszka et al., 2000). Using the same species, Si et al. (2004) examined the behavioral effects of L-trans-2,4-PDC and antagonists memantine (low affinity) and MK-801 (high affinity) on learning and memory. Consistent with previous findings (Maleszka et al., 2000), L-trans-2,4-PDC exposure induced amnesia in the honeybee. Similar to L-trans-2,4-PDC, both pretraining and pretesting injections of MK-801 led to an impairment of long-term (24 h) memory, but had no effect on short-term (1 h) memory of an olfactory task. Interestingly, the L-trans-2,4-PDC-induced amnesia was "rescued" by memantine injected either before training or before testing, suggesting that memantine is able to restore memory recall rather than memory formation or storage (Si et al., 2004). Although the role of the glutamatergic system in the central nervous system of invertebrates is poorly understood and somewhat controversial, this result suggests a role for glutamatergic transmission in memory processing in this organ-ism. Studies by Si et al. (2004) are consistent with the idea that memantine and MK-801-sensitive receptors in the honeybee are involved in memory recall. It is worth mentioning that other neurotransmitters, in particular acetylcholine, have also been implicated in memory processes in the honeybee (Lozano et al., 2001; Shapira et al., 2001).

HUMAN STUDIES

The NMDA-receptor system in the brain has also been implicated in learning and in the process of new memory formation in humans. This has been demonstrated by several investigators using different NMDA antagonists. The NMDA-receptor antagonists ketamine and phencyclidine have been shown to evoke a range of symptoms and cognitive deficits that resemble those in schizophrenia (Honey et al., 2005; Krystal et al., 1994). Hypofunctionality of the glutamatergic system in the brain, and specifically hypofunction of the subpopulation of corticolimbic NMDA receptors, has been implicated in the pathophysiology of schizophrenia (Coyle, 1996; Olney and Farber, 1995; Tsai and Coyle, 2002). To test this hypothesis, ketamine has been used extensively for pharmacological manipulation of the NMDA receptor. Ketamine exposure leads to the blockade of NMDA receptors and consequent hypo-functionality of the glutamatergic system in the brain. The effects of the noncom-petitive NMDA antagonist ketamine on cognitive function in humans have been investigated by several workers (Honey et al., 2005; Krystal et al., 1994; Lisman et al., 1998; Malhotra et al., 1996; Newcomer and Krystal, 2001; Oye et al., 1992; Rockstroh et al., 1996; Schugens et al., 1997). Overall, these studies demonstrated that NMDA-receptor blockade in humans impairs learning and memory (Honey et al., 2005; Morgan et al., 2004; Rockstroh et al., 1996). Recently, the results of an fMRI (functional magnetic resonance imaging) study showed that acute ketamine exposure altered the brain response to executive demands in a verbal working-memory task. The results of this study suggest a task-specific effect of ketamine on working memory in healthy volunteers (Honey et al., 2004). In another fMRI study using a double-blind, placebo-controlled, randomized within-subjects design, it was demonstrated that ketamine exposure disrupted frontal and hippocampal contributions to encoding and retrieval of episodic memory (Honey et al., 2005). To explore the involvement of the glutamatergic system in memory processing in human subjects, Schugens et al. (1997) investigated the effects of a single dose of a low-affinity, noncompetitive blocker of the NMDA receptors (memantine [1-1minoadamantane derivative]) on learning and memory in young, healthy volunteers. Applying a double-blind placebo-controlled design, they found no significant effects of memantine on mood, attention, or immediate and delayed verbal and visuospatial memory. However, memantine delayed the acquisition of classical eyeblink conditioning and reduced the overall frequency of conditioned responses without affecting reflex or spontaneous eyeblinks.

Administration of SDZ EAA 494, a potent enantiomerically pure NMDA antagonist (Aebischer et al., 1989), has been shown to impair the memory process in humans. SDZ EAA 494 is a competitive, highly specific antagonist at the NMDA-type excitatory amino acid receptor. To investigate the effect of this NMDA antagonist on memory and attention in humans, SDZ EAA 494 was administered either acutely or as multiple doses over a course of 1 week. The assessment included simple and complex reaction time tests to assess attention, as well as verbal, nonverbal, and spatial memory tests with immediate and late recall. Verbal and nonverbal memory performance was significantly impaired after both acute and chronic administration of SDZ EAA 494. The reaction time and spatial-memory tests were not significantly affected.

Recently, in a double-blind study, it was demonstrated that amantadine, a low-affinity NMDA-receptor channel blocker, given orally to healthy young volunteers failed to block motor learning consolidation in subjects that had already learned the task (Hadj Tahar et al., 2004). It has also been shown that ketamine given to healthy human subjects impaired learning of spatial and verbal information, but not retrieval of information learned prior to ketamine administration (Rowland et al., 2005). In another double-blind placebo-controlled design with healthy human volunteers, Morgan et al. (2004) demonstrated that ketamine treatment produced a dose-dependent impairment to episodic and working memory. Ketamine also impaired recognition memory and procedural learning.

These results support the notions that (a) NMDA receptors are involved in new memory formation in humans and, as mentioned earlier in this chapter, (b) blockade of NMDA receptors after a task has been learned has no effect on memory in humans (Hadj Tahar et al., 2004; Oye et al., 1992; Rowland et al., 2005).

CONCLUSIONS

Both animal and human studies clearly indicate that the NMDA-receptor system in the brain, in addition to its roles in other brain functions, is conceivably involved in the processes of learning and memory formation. This notion has been supported by consistent results from studies investigating the effects of NMDA itself, NMDA antagonists, NMDA-transporter inhibitors, and NMDA-channel blockers in transgenic mice. As an animal model of schizophrenia, NMDA-antagonist-induced memory impairment is quite useful, since working-memory impairment has been suggested to be the core cognitive deficit in schizophrenia, leading to impairment in other domains of cognition (Goldman-Rakic and Selemon, 1997; Li et al., 2003). Therefore, an animal model of working-memory impairment produced by NMDA antagonists that parallels both the NMDA hypofunctionality in specific areas of the brain (Coyle, 1996) and the cognitive deficits found in schizophrenia, including response to pharmacological treatment, would be of considerable value in studying the pathophysiology and treatment of some aspects of schizophrenia.

REFERENCES

Abdul-Monim, Z., Reynolds, G.P., and Neill, J.C. (2003). The atypical antipsychotic ziprasidone, but not haloperidol, improves phencyclidine-induced cognitive deficits in a reversal learning task in the rat, *J. Psychopharmacol.*, 17, 57–65.

Aebischer, B., Frey, P., Haerter, H.P., Herrling, P.L., Mueller, W., Olverman, H.J. et al. (1989). Synthesis and NMDA antagonistic properties of the enantiomers of 4-(3-phosphonopropyl)piperazine-2-carboxylic acid (CPP) and of the unsaturated analogue (E)-4-(3-phosphonoprop-2-enyl)piperazine-2-carboxylic acid (CPP-ene), *Helvetica Chimica Acta*, 72, 1043–1047.

Benvenga, M.J. and Spaulding, T.C. (1988). Amnesic effect of the novel anticonvulsant MK-801, *Pharmacol., Biochem. Behav.*, 30, 205–207.

Butelman, E.R. (1990). The effect of NMDA antagonists in the radial arm maze task with an interposed delay, *Pharmacol., Biochem. Behav.*, 35, 533–536.

Carey, R.J., Dai, H., and Gui, J. (1998). Effects of dizocilpine (MK-801) on motor activity and memory, *Psychopharmacology*, 137, 241–246.

Constantine-Paton, M. (1994). Effects of NMDA receptor antagonists on the developing brain, *Psychopharmacol. Bull.*, 30, 561–565.

Coyle, J.T. (1996). The glutamatergic dysfunction hypothesis for schizophrenia, *Harv. Rev. Psychiatry*, 3, 241–253.

Danysz, W., Wroblewski, J.T., and Costa, E. (1988). Learning impairment in rats by N-methyl-D-aspartate receptor antagonists, *Neuropharmacology*, 27, 653–656.

de Lima, M.N., Laranja, D.C., Bromberg, E., Roesler, R., and Schroder, N. (2005). Pre- or post-training administration of the NMDA receptor blocker MK-801 impairs object recognition memory in rats, *Behavioural Brain Res.*, 156, 139–143.

Ferreira, M.B., Da Silva, R.C., Medina, J.H., and Izquierdo, I. (1992). Late posttraining memory processing by entorhinal cortex: involvement of NMDA and GABAergic receptors, *Pharmacol., Biochem. Behav.*, 41, 767–771.

Ford, L.M., Norman, A.B., and Sanberg, P.R. (1989). The topography of MK-801-induced locomotor patterns in rats, *Physiol. Behav.*, 46, 755–758.

Goldman-Rakic, P.S. and Selemon, L.D. (1997). Functional and anatomical aspects of prefrontal pathology in schizophrenia, *Schizophrenia Bull.*, 23, 437–458.

Hadj Tahar, A., Blanchet, P.J., and Doyon, J. (2004). Motor-learning impairment by amantadine in healthy volunteers, *Neuropsychopharmacology*, 29, 187–194.

Harder, J.A., Aboobaker, A.A., Hodgetts, T.C., and Ridley, R.M. (1998). Learning impairments induced by glutamate blockade using dizocilpine (MK-801) in monkeys, *Brit. J. Pharmacol.*, 125, 1013–1018.

Harder, J.A., Maclean, C.J., Alder, J.T., Francis, P.T., and Ridley, R.M. (1996). The 5-HT1A antagonist, WAY 100635, ameliorates the cognitive impairment induced by fornix transection in the marmoset, *Psychopharmacology*, 127, 245–254.

Hashimoto, K., Fujita, Y., Shimizu, E., and Iyo, M. (2005). Phencyclidine-induced cognitive deficits in mice are improved by subsequent subchronic administration of clozapine, but not haloperidol, *Eur. J. Pharmacol.*, 519, 114–117.

Hauben, U., D'Hooge, R., Soetens, E., and De Deyn, P.P. (1999). Effects of oral administration of the competitive N-methyl-D-aspartate antagonist, CGP 40116, on passive avoidance, spatial learning, and neuromotor abilities in mice, *Brain Res. Bull.*, 48, 333–341.

Heale, V. and Harley, C. (1990). MK-801 and AP5 impair acquisition, but not retention, of the Morris milk maze, *Pharmacol., Biochem. Behav.*, 36, 145–149.

Hlinak, Z. and Krejci, I. (1998). Concurrent administration of subeffective doses of scopolamine and MK-801 produces a short-term amnesia for the elevated plus-maze in mice, *Behavioural Brain Res.*, 91, 83–89.

Hlinak, Z. and Krejci, I. (2000). Oxiracetam prevents the MK-801 induced amnesia for the elevated plus-maze in mice, *Behavioural Brain Res.*, 117, 147–151.

Hlinak, Z. and Krejci, I. (2002). MK-801 induced amnesia for the elevated plus-maze in mice, *Behavioural Brain Res.*, 131, 221–225.

Hlinak, Z. and Krejci, I. (2003). N-methyl-D-aspartate prevented memory deficits induced by MK-801 in mice, *Physiological Res.*, 52, 809–812.

Homayoun, H., Stefani, M.R., Adams, B.W., Tamagan, G.D., and Moghaddam, B. (2004). Functional interaction between NMDA and mGlu5 receptors: effects on working memory, instrumental learning, motor behaviors, and dopamine release, *Neuropsychopharmacology*, 29, 1259–1269.

Honey, G.D., Honey, R.A., O'Loughlin, C., Sharar, S.R., Kumaran, D., Suckling, J. et al. (2005). Ketamine disrupts frontal and hippocampal contribution to encoding and retrieval of episodic memory: an fMRI study, *Cerebral Cortex*, 15, 749–759.

Honey, R.A., Honey, G.D., O'Loughlin, C., Sharar, S.R., Kumaran, D., Bullmore, E.T. et al. (2004). Acute ketamine administration alters the brain responses to executive demands in a verbal working memory task: an FMRI study, *Neuropsychopharmacology*, 29, 1203–1214.

Izquierdo, I. (1991). Role of NMDA receptors in memory, *Trends Pharmacological Sciences*, 12, 128–129.

Izquierdo, I. (1994). Pharmacological evidence for a role of long-term potentiation in memory, *FASEB J.*, 8, 1139–1145.

Izquierdo, I. and Medina, J.H. (1993). Role of the amygdala, hippocampus and entorhinal cortex in memory consolidation and expression, *Brazilian J. Medical Biological Res.*, 26, 573–589.

Javitt, D.C. and Zukin, S.R. (1991). Recent advances in the phencyclidine model of schizophrenia, *Am. J. Psychiatry*, 148, 1301–1308.

Jentsch, J.D. and Roth, R.H. (1999). The neuropsychopharmacology of phencyclidine: from NMDA receptor hypofunction to the dopamine hypothesis of schizophrenia, *Neuropsychopharmacology*, 20, 201–225.

Kesner, R.P. and Dakis, M. (1993). Phencyclidine disrupts acquisition and retention performance within a spatial continuous recognition memory task, *Pharmacol., Biochem. Behav.*, 44, 419–424.

Kim, J.J., DeCola, J.P., Landeira-Fernandez, J., and Fanselow, M.S. (1991). N-methyl-D-aspartate receptor antagonist APV blocks acquisition but not expression of fear conditioning, *Behavioral Neurosci.*, 105, 126–133.

Koek, W., Woods, J.H., and Colpaert, F.C. (1990). N-methyl-D-aspartate antagonism and phencyclidine-like activity: a drug discrimination analysis, *J. Pharmacol. Exp. Ther.*, 253, 1017–1025.

Krystal, J.H., Karper, L.P., Seibyl, J.P., Freeman, G.K., Delaney, R., Bremner, J.D. et al. (1994). Subanesthetic effects of the noncompetitive NMDA antagonist, ketamine, in humans: psychotomimetic, perceptual, cognitive, and neuroendocrine responses, *Arch. Gen. Psychiatry*, 51, 199–214.

Levin, E.D., Bettegowda, C., Weaver, T., and Christopher, N.C. (1998). Nicotine-dizocilpine interactions and working and reference memory performance of rats in the radial arm maze, *Pharmacol., Biochem. Behav.*, 61 (3), 335–340.

Li, Z., Kim, C. H., Ichikawa, J., and Meltzer, H.Y. (2003). Effect of repeated administration of phencyclidine on spatial performance in an eight-arm radial maze with delay in rats and mice. *Pharmacol., Biochem. Behav.*, 75 (2), 335–340.

Liang, K.C., Lin, M.H., and Tyan, Y.M. (1993). Involvement of amygdala N-methyl-D-aspartate receptors in long-term retention of an inhibitory avoidance response in rats, *Chin. J. Physiol.*, 36, 47–56.

Lisman, J.E., Fellous, J.M., and Wang, X.J. (1998). A role for NMDA-receptor channels in working memory, *Nat. Neurosci.*, 1, 273–275.

Lozano, V.C., Armengaud, C., and Gauthier, M. (2001). Memory impairment induced by cholinergic antagonists injected into the mushroom bodies of the honeybee, *J. Comp. Physiol.*, 187, 249–254.

Maleszka, R., Helliwell, P., and Kucharski, R. (2000). Pharmacological interference with glutamate re-uptake impairs long-term memory in the honeybee, *Apis mellifera*, *Behavioural Brain Res.*, 115, 49–53.

Malhotra, A.K., Pinals, D.A., Weingartner, H., Sirocco, K., Missar, C.D., Pickar, D. et al. (1996). NMDA receptor function and human cognition: the effects of ketamine in healthy volunteers, *Neuropsychopharmacology*, 14, 301–307.

May-Simera, H. and Levin, E.D. (2003). NMDA systems in the amygdala and piriform cortex and nicotinic effects on memory function, *Brain Res., Cognit. Brain Res.*, 17, 475–483.

Morgan, C.J., Mofeez, A., Brandner, B., Bromley, L., and Curran, H.V. (2004). Acute effects of ketamine on memory systems and psychotic symptoms in healthy volunteers, *Neuropsychopharmacology*, 29, 208–218.

Morris, R.G., Anderson, E., Lynch, G.S., and Baudry, M. (1986). Selective impairment of learning and blockade of long-term potentiation by an N-methyl-D-aspartate receptor antagonist, AP5, *Nature*, 319, 774–776.

Murray, T.K. and Ridley, R.M. (1997). The effect of dizocilpine (MK-801) on conditional discrimination learning in the rat, *Behavioural Pharmacol.*, 8, 383–388.

Murray, T.K., Ridley, R.M., Snape, M.F., and Cross, A.J. (1995). The effect of dizocilpine (MK-801) on spatial and visual discrimination tasks in the rat, *Behavioural Pharmacol.*, 6, 540–549.

Newcomer, J.W. and Krystal, J.H. (2001). NMDA receptor regulation of memory and behavior in humans, *Hippocampus*, 11, 529–542.

Olney, J.W. and Farber, N.B. (1995). NMDA antagonists as neurotherapeutic drugs, psychotogens, neurotoxins, and research tools for studying schizophrenia, *Neuropsychopharmacology*, 13, 335–345.

Oye, I., Paulsen, O., and Maurset, A. (1992). Effects of ketamine on sensory perception: evidence for a role of N-methyl-D-aspartate receptors, *J. Pharmacol. Experimental Ther.*, 260, 1209–1213.

Podhorna, J. and Didriksen, M. (2005). Performance of male C57BL/6J mice and Wistar rats in the water maze following various schedules of phencyclidine treatment, *Behavioural Pharmacol.*, 16, 25–34.

Quartermain, D., Mower, J., Rafferty, M.F., Herting, R.L., and Lanthorn, T.H. (1994). Acute but not chronic activation of the NMDA-coupled glycine receptor with D-cycloserine facilitates learning and retention, *Eur. J. Pharmacol.*, 257, 7–12.

Rickard, N.S., Poot, A.C., Gibbs, M.E., and Ng, K.T. (1994). Both non-NMDA and NMDA glutamate receptors are necessary for memory consolidation in the day-old chick, *Behavioral Neural Biol.*, 62, 33–40.

Ridley, R.M., Samson, N.A., Baker, H.F., and Johnson, J.A. (1988). Visuospatial learning impairment following lesion of the cholinergic projection to the hippocampus, *Brain Res.*, 456, 71–87.

Ridley, R.M., Timothy, C.J., Maclean, C.J., and Baker, H.F. (1995). Conditional learning and memory impairments following neurotoxic lesion of the CA1 field of the hippocampus, *Neuroscience*, 67, 263–275.

Riedel, G., Platt, B., and Micheau, J. (2003). Glutamate receptor function in learning and memory, *Behavioural Brain Res.*, 140, 1–47.

Rockstroh, S., Emre, M., Tarral, A., and Pokorny, R. (1996). Effects of the novel NMDA-receptor antagonist SDZ EAA 494 on memory and attention in humans, *Psychopharmacology*, 124, 261–266.

Rowland, L.M., Astur, R.S., Jung, R.E., Bustillo, J.R., Lauriello, J., and Yeo, R.A. (2005). Selective cognitive impairments associated with NMDA receptor blockade in humans, *Neuropsychopharmacology*, 30, 633–639.

Sakimura, K., Kutsuwada, T., Ito, I., Manabe, T., Takayama, C., Kushiya, E. et al. (1995). Reduced hippocampal LTP and spatial learning in mice lacking NMDA receptor epsilon 1 subunit, *Nature*, 373, 151–155.

Santini, E., Muller, R.U., and Quirk, G.J. (2001). Consolidation of extinction learning involves transfer from NMDA-independent to NMDA-dependent memory, *J. Neurosci.*, 21, 9009–9017.

Scatton, B., Carter, C., and Benavides, J. (1991). NMDA receptor antagonists: treatment of brain ischemia, *Drug News Persp.*, 4, 89–102.

Scheetz, A.J. and Constantine-Paton, M. (1994). Modulation of NMDA receptor function: implications for vertebrate neural development, *FASEB J.*, 8, 745–752.

Schugens, M.M., Egerter, R., Daum, I., Schepelmann, K., Klockgether, T., and Loschmann, P.A. (1997). The NMDA antagonist memantine impairs classical eyeblink conditioning in humans, *Neurosci. Lett.*, 224, 57–60.

Shapira, M., Thompson, C.K., Soreq, H., and Robinson, G.E. (2001). Changes in neuronal acetylcholinesterase gene expression and division of labor in honey bee colonies, *J. Mol. Neurosci.*, 7, 1–12.

Shimizu, E., Tang, Y.P., Rampon, C., and Tsien, J.Z. (2000). NMDA receptor-dependent synaptic reinforcement as a crucial process for memory consolidation, *Science*, 290, 1170–1174.

Si, A., Helliwell, P., and Maleszka, R. (2004). Effects of NMDA receptor antagonists on olfactory learning and memory in the honeybee (*Apis mellifera*), *Pharmacol., Biochem. Behav.*, 77, 191–197.

Swain, H.A., Sigstad, C., and Scalzo, F.M. (2004). Effects of dizocilpine (MK-801) on circling behavior, swimming activity, and place preference in zebrafish (*Danio rerio*), *Neurotoxicol. Teratology*, 26, 725–729.

Takatsuki, K., Kawahara, S., Takehara, K., Kishimoto, Y., and Kirino, Y. (2001). Effects of the noncompetitive NMDA receptor antagonist MK-801 on classical eyeblink conditioning in mice, *Neuropharmacology*, 41, 618–628.

Tamminga, C.A. (1998). Schizophrenia and glutamatergic transmission, *Crit. Rev. Neurobiol.*, 12, 21–36.

Tang, Y.P., Shimizu, E., Dube, G.R., Rampon, C., Kerchner, G.A., Zhuo, M. et al. (1999). Genetic enhancement of learning and memory in mice, *Nature*, 401, 63–69.

Tang, Y.P., Wang, H., Feng, R., Kyin, M., and Tsien, J.Z. (2001). Differential effects of enrichment on learning and memory function in NR2B transgenic mice, *Neuropharmacology*, 41, 779–790.

Thompson, L.T. and Disterhoft, J.F. (1997). N-methyl-D-aspartate receptors in associative eyeblink conditioning: both MK-801 and phencyclidine produce task- and dose-dependent impairments, *J. Pharmacol. Exp. Ther.*, 281, 928–940.

Tronel, S. and Sara, S.J. (2003). Blockade of NMDA receptors in prelimbic cortex induces an enduring amnesia for odor-reward associative learning, *J. Neurosci.*, 23, 5472–5476.

Tsai, G. and Coyle, J.T. (2002). Glutamatergic mechanisms in schizophrenia, *Ann. Rev. Pharmacol. Toxicol.*, 42, 165–179.

Tsien, J.Z., Huerta, P.T., and Tonegawa, S. (1996). The essential role of hippocampal CA1 NMDA receptor-dependent synaptic plasticity in spatial memory, *Cell*, 87, 1327–1338.

van der Meulen, J.A., Bilbija, L., Joosten, R.N., de Bruin, J.P., and Feenstra, M.G. (2003). The NMDA-receptor antagonist MK-801 selectively disrupts reversal learning in rats, *Neuroreport*, 14, 2225–2228.

Venero, C. and Sandi, C. (1997). Effects of NMDA and AMPA receptor antagonists on corticosterone facilitation of long-term memory in the chick, *Eur. J. Neurosci.*, 9, 1923–1928.

Verma, A. and Moghaddam, B. (1996). NMDA receptor antagonists impair prefrontal cortex function as assessed via spatial delayed alternation performance in rats: modulation by dopamine, *J. Neurosci.*, 16, 373–379.

Wong, E.H., Kemp, J.A., Priestley, T., Knight, A.R., Woodruff, G.N., and Iversen, L.L. (1986). The anticonvulsant MK-801 is a potent N-methyl-D-aspartate antagonist, *Proc. Natl. Acad. Sci. U.S.A.*, 83, 7104–7108.

Wong, R.W., Setou, M., Teng, J., Takei, Y., and Hirokawa, N. (2002). Overexpression of motor protein KIF17 enhances spatial and working memory in transgenic mice, *Proc. Natl. Acad. Sci. U.S.A.*, 99, 14500–14505.

Woodruff-Pak, D.S., Green, J.T., Coleman-Valencia, C., and Pak, J.T. (2000). A nicotinic cholinergic agonist (GTS-21) and eyeblink classical conditioning: acquisition, retention, and relearning in older rabbits, *Exp. Aging Res.*, 26, 323–336.

Xu, X. and Davis, R.E. (1992). N-methyl-D-aspartate receptor antagonist MK-801 impairs learning but not memory fixation or expression of classical fear conditioning in goldfish (*Carassius auratus*), *Behavioral Neurosci.*, 106, 307–314.

Xu, X., Russell, T., Bazner, J., and Hamilton, J. (2001). NMDA receptor antagonist AP5 and nitric oxide synthase inhibitor 7-NI affect different phases of learning and memory in goldfish, *Brain Res.*, 889, 274–277.

Zanatta, M.S., Schaeffer, E., Schmitz, P.K., Medina, J.H., Quevedo, J., Quillfeldt, J.A. et al. (1996). Sequential involvement of NMDA receptor-dependent processes in hippocampus, amygdala, entorhinal cortex and parietal cortex in memory processing, *Behavioural Pharmacol.*, 7, 341–345.

5 Animal Models and the Cognitive Effects of Ethanol

Merle G. Paule
National Center for Toxicological Research/FDA

CONTENTS

BACKGROUND

Ethanol (ethyl alcohol, alcohol) has been widely studied in many different formats and at many levels of complexity. Given the wealth of available information, this brief chapter cannot begin to recapitulate the literature on the many and varied facets of ethanol. For a solid introduction to the history of the use of ethanol as a research tool and details on its general absorption, distribution, metabolism, excretion, and toxicology, we refer the reader to the informative and interesting chapter in the recent edition of Goodman and Gilman [1].

Ethanol is unique among the psychoactive compounds available for human consumption: it is widely available, its use is legal and acceptable in many societies, and it takes relatively large (i.e., gram) quantities to exert pharmacological and other effects. Of all the psychoactive drugs available to humans, ethanol is implicated in the greatest number of accidents [2]. Decrements in human psychomotor performance appear to be due primarily to ethanol's effects on cognitive (central) mechanisms rather than motor (peripheral) components [2], and tasks that are more complex are likelier to be affected by small doses. The consensus is that ethanol causes deterioration in driving skills at or below blood alcohol concentrations of 0.05 g/100 ml and that such impairment increases significantly as these levels increase [2].

Perhaps importantly, the psychotropic effects of ethanol change along a dose continuum; thus, it can be classified as a stimulant, an anxiolytic, a sedative, a muscle relaxant, a euphoriant, and a general anesthetic that can cause coma and death if a large enough dose is given. While many of these effects would, for other drugs, be considered "therapeutic effects," the only recognized clinical use of systemically administered ethanol is the treatment of poisoning by methyl alcohol or ethylene glycol. Dehydrated ethanol might also be used via proximal injection to destroy nerves or ganglia to relieve long-lasting pain related to trigeminal neuralgia, inoperable cancer, etc. [1].

Ethanol is primarily a central nervous system (CNS) depressant; however, "behavioral disinhibition" as a result of the effects of ethanol on some brain areas/functions is often interpreted as "stimulation." Some researchers [3] have presented convincing evidence that, at low doses and soon after exposure, ethanol acts along a continuum of stimulation leading to depression. The observations supporting this thesis [4] show that, in rats trained to discriminate pentobarbital from amphetamine using a two-choice lever-pressing procedure, ethanol produced a greater number of amphetamine-appropriate responses than pentobarbital-appropriate responses when testing occurred within 15 min of exposure. Tests on humans [4] show that relatively low doses of ethanol (0.33 g/kg) can actually enhance performance in reaction time and visual search tasks: the results of these tests showed that accuracy was improved while response speed was not affected. Still others report [5] that the dose of ethanol is important, with memory facilitation occurring at low doses but impairment predominating at moderate to heavy doses. In addition, the effects of ethanol on memory may be different on the ascending arm, as opposed to the descending arm, of the blood alcohol curve.

Chronic exposure to ethanol can lead to tolerance, physical dependence, and alcoholism, and a number of animal models have been developed to study aspects of these important consequences of continued ethanol exposure [6]. When such exposure occurs during pregnancy, offspring can suffer teratological outcomes in the form of the fetal alcohol syndrome (FAS) or the fetal alcohol effects syndrome (FAES). Prenatal ethanol exposures in animal models have been shown to produce data in good congruence with respect to qualitative endpoints observed in humans [7, 8]. The focus of this chapter, however, is to present information on the acute — rather than chronic — effects of ethanol on the function of the CNS.

Because of the disruptive and debilitating effects of ethanol on a multitude of cognitive functions in animals, acute ethanol intoxication is used as a model of amnesia in which the effects of antiamnestic compounds are assessed [9, 10]. Generally, animals are treated with significant doses of ethanol (e.g., 2.0 g/kg) prior to training in a variety of tasks in which ethanol disrupts aspects of learning, and then potential antiamnestic agents are given concomitantly to determine whether the effects of ethanol can be attenuated [9]. In addition, the involvement of specific anatomical substrates in ethanol-induced amnesia have been explored [11], with the hippocampal formation being strongly implicated. Ethanol has been shown to affect specific neurotransmitter levels (i.e., reduced glutamate levels in the dorsal hippocampus), and such changes sometimes have strong relationships with alterations in important aspects of cognitive function such as spatial memory deficits [11].

BEHAVIORAL EFFECTS IN ANIMAL MODELS

RELEVANCE

The metabolism of ethanol by a variety of animals (e.g., squirrel monkeys) is generally similar to that of humans [12]. In addition, the effects of ethanol on similar behaviors — such as eye tracking — are similar in both monkeys and humans [13]. Thus, animal models can provide relevant information about specific effects of ethanol that might be expected to manifest in humans. This is particularly true when the tasks utilized are appropriate for use in both the animal models and humans [14, 15].

MECHANISM(S) OF ACTION OF ETHANOL

Subjective Effects: Drug Discrimination Studies

While the cellular/receptor mechanisms underlying ethanol's effects on neuronal function can be studied *in vitro* using a variety of neurotransmitter-receptor binding techniques, its psychoactive properties provide opportunity for extensive assessment using whole-animal models. Psychoactive compounds are thought to produce relatively specific interoceptive cues to which both human and animal subjects can attend. Both animals and people can be trained to make a specific response (e.g., press the rightmost of two response levers) in the presence of one drug (e.g., ethanol) and another response (press the leftmost lever) in the absence of that drug or in the presence of another drug. Once reliable responding has been established (the correct lever for the given drug condition is selected a high percentage of the time), then the ability of other compounds to substitute or generalize to the training drug(s) can be assessed. Such drug-discrimination studies have proved invaluable in providing insight into the neural substrates via which centrally acting drugs exert their effects. In addition, data obtained from drug-discrimination studies in animals have been found to be in good accord with those obtained in humans, in that findings are qualitatively similar across species and that potency relationships are quantitatively similar between most, but not all, other species and humans [16, 17]. Thus, data obtained from animal studies can be relevant to the human condition.

In rats trained to discriminate ethanol or pentobarbital from the "no drug" condition, both drugs substituted for each other, thus demonstrating the similarity, if not equivalence, of the internal or interoceptive cue of both agents and, thus, the mechanism of action. However, rats could also be trained to discriminate each drug from the other, demonstrating that the two drugs do differ in their interoceptive cues [18]. Other studies in rats showed that low doses of ethanol (0.10 g/kg) substituted more for amphetamine than pentobarbital, while higher doses of ethanol (0.20 g/kg) substituted for pentobarbital [3]. These results demonstrate that there are dose-dependent effects on different neurochemical systems and that biphasic effects can be observed. In squirrel monkeys trained to discriminate ethanol from saline, several ethanolic drinks (bourbon, gin, beer, vodka, and red wine) and both pentobarbital and barbital substituted for ethanol, whereas morphine did not [19]. Thus, ethanol shares properties with a variety of ethanolic drinks but not with the opiate, morphine. Interestingly, several other ethanolic drinks (cognac, scotch, and tequila) engendered a response different from that of ethanol, suggesting differences salient enough from the training dose of ethanol (1.6 g/kg) to be detected by some subjects. In novel drug-discrimination experiments in rats examining the early "excitatory" phase of ethanol's "subjective" effects (i.e., 6 min after administration) and the later or "sedative" phase (30 min after administration), it was demonstrated that the mu-opiate-receptor antagonist naloxone significantly attenuated discrimination of etha-nol's excitatory stimulus effects but not its sedative effects [20]. These data clearly indicate that the behavioral effects of ethanol are complicated and vary as a function of time after exposure.

Opioids

In studies to explore the potential role of endogenous opioid systems in the discrim-inative stimulus of ethanol, a series of opioid antagonists was administered to rats trained to discriminate ethanol from saline [21]. Of the two mu-opioid-receptor antagonists tested (naloxone and cyprodime), only naloxone — and only at a high dose — partially, but significantly, antagonized the ethanol cue; cyprodime was without effect. The delta-opioid-receptor antagonist naltrindole and the kappa-opi-oid-receptor antagonist norbinaltorphimine were both without effect in altering the ethanol cue; thus, it appears that, at most, the ethanol interoceptive stimulus is minimally dependent on its interaction with endogenous opioid systems.

Dopamine and Serotonin

In rats trained to discriminate ethanol from saline, a variety of dopaminergic and serotonergic agonists and antagonists were used to explore the involvement of these two systems in ethanol's effects [22]. None of the dopaminergic treatments utilized had an effect on the ethanol discrimination, but several 5-HT agonists (quipazine, 5-MeODMT [5-Methoxy-Nn-Dimethyltryptamine], buspirone, and 8-OH-DPAT [8-Hydroxy-2-di-n-propylamino-tetralin]) engendered intermediate ethanol-like responding, and the 5-HT1B-receptor agonist TFMPP (1-(3-trifluoromethylphenyl) piperazine) completely substituted for ethanol. In experiments designed to further explore the role of 5-HT receptors in the ethanol stimulus [23], the researchers used four 5-HT agonists with different selectivity for 5-HT1A, 5-HT1B, and 5-HT2C receptors to determine their ability to generalize to ethanol. The most selective 5-HT1B agonist tested (CGS 12066B [7-trifluoromethyl-4(4-methyl-1-piperazinyl) pyrnol [1,2-a]-quinoxaline dimaleate]) completely substituted for 1.0 g/kg ethanol, but not for higher training doses. The 5-HT1B/2C agonist mCPP (M-Chlorophenyl-piperazine) also substituted for the 1.0-g/kg training dose but not for higher training doses. The 5-HT1A/1B agonist RU-24969 substituted for all training doses of ethanol (although not to as great a degree for the highest training dose of 2.0 g/kg), and the 5-HT1A agonist 8-OH-DPAT did not substitute for any training dose of ethanol. The upshot of these studies is that agonists with 5-HT1B activity produce effects similar to those of relatively low (1.0 g/kg) and intermediate (1.5 g/kg) training doses of ethanol but not higher doses.

In studies to explore the role of 5-HT3 receptors in the mediation of the interoceptive stimulus of ethanol in rats, Stefanski et al. [24] administered the 5-HT3 antagonists tropisetron and ondansetron, but they were unable to antagonize the ethanol cue. In addition, the 5-HT3-receptor agonist 1-(m-chlorophenyl)-biguanide did not generalize to ethanol. These data argue against a strong role for the 5-HT3 receptor in subserving the stimulus properties of ethanol.

Gamma-Aminobutyric Acid (GABA)

In mice trained to discriminate ethanol from saline, the inhalants toluene, halothane, and TCE (1,1,1,-trichloroethylene) and the benzodiazepine oxazepam all substituted for ethanol [25]. Thus, a gaseous anesthetic, some abused solvents, and an anxiolytic share behavioral effects with ethanol, and all of these compounds have been shown to produce pentobarbital-like discriminative stimulus effects, suggesting that these agents have a common mechanism of action at GABA receptors [26]. In later studies [27], it was further demonstrated that the volatile anesthetics desflurane, enflurane, isoflurane, and ether all produce ethanol-like discriminative stimuli in mice, furthering the tenet that these agents are all ethanol-like in their effects. In rats trained to discriminate either pentobarbital, ethanol, diazepam, or lorazepam from the no-drug condition, substitution experiments indicated that two pregnanolone-derived neuroactive steroids generalized completely to pentobarbital, ethanol, and diazepam but not lorazepam. These data indicated that endogenous steroids exhibit properties that are consistent with sedative/anxiolytic activities and that these effects are likely mediated through a nonbenzodiazepine GABA-A site [28].

In rats trained to discriminate ethanol or gamma-hydroxybutyrate (GHB) from water, both compounds were found to substitute completely for the other, albeit only over a narrow dose range [29]. Thus, at least under circumscribed doses, the mechanisms subserving the discriminative stimuli for each compound appear to be shared extensively.

In studies in monkeys trained to discriminate pentobarbital from saline, ethanol was found to generalize to pentobarbital, demonstrating that the ethanol and barbiturate internal cues are similar in the primate. In addition, it was shown via isobolographic analyses that the effects of pentobarbital and ethanol were dose additive, not synergistic [30], suggesting that the same receptor was involved in the effects of both agents. In rats trained to discriminate ethanol, dizocilpine (MK-801, a noncompetitive inhibitor of the N-methyl-D-aspartate [NMDA] subtype of glutamate receptor), and water in a three-choice drug-discrimination paradigm, it was demonstrated that the GABA-A mimetics (modulators) allopregnanolone, diazepam, and pentobarbital all substituted completely for ethanol, whereas phencyclidine (PCP, a noncompetitive inhibitor of the NMDA-receptor complex) substituted completely for dizocilpine. Neither RU-24969 (a 5-HT1A/1B agonist) nor TFMPP (a 5-HT1B agonist) completely substituted for either ethanol or dizocilpine. RU 24969 partially ($\approx$60%) substituted for ethanol. Thus, these data showed that it is possible to tease out the involvement of the NMDA system in ethanol's discriminative stimuli even while GABA-A receptor mechanisms remained involved [31].

Glutamate

MK-801 was shown to have ethanol-like stimulus properties in rats [32, 33], confirming similar observations in pigeons and bolstering the premise that part of the ethanol interoceptive cue is mediated by blockade of the NMDA-receptor complex. In rats trained to an NMDA discrimination, ethanol failed to antagonize the NMDA cue and did not substitute fully for either the competitive NMDA antagonist NPC 12626 or the noncompetitive antagonist phencyclidine (PCP); a maximum of about 50% PCP-appropriate responding indicated partial substitution [34]. It was concluded that ethanol's effects on NMDA discrimination were distinct from competitive antagonists but similar to those of noncompetitive antagonists.

Further support for the involvement of the NMDA receptor in mediating the interoceptive ethanol cue comes from generalization studies in rats whereby MRZ 2/579, a novel uncompetitive NMDA-receptor antagonist [35], dose dependently generalized to the ethanol cue. Bienkowski et al. [36] showed that in rats trained to discriminate 1.0 g/kg ethanol from saline, dizocilpine and CGP 37849 (another competitive NMDA-receptor antagonist) substituted partially, and CGP 40116 (a competitive NMDA-receptor antagonist) and the active D-stereoisomer of CGP 37849 completely substituted for ethanol. These same authors [37] assessed the ability of N-methyl-D-aspartate and D-cyloserine (a partial agonist at the glutamate binding site) to antagonize the discriminative stimulus of ethanol. Neither compound antagonized the ethanol interoceptive cue in the rat, indicating that at least some agonists at the NMDA receptor do not block ethanol's discriminative stimulus.

Ethanol is known to be a potent inhibitor of the NMDA receptor in a variety of brain areas. Patch-clamp recordings from cells in culture have shown that ethanol does not appear to interact with NMDA at either the glutamate recognition site of the receptor or at any of the known modulatory sites such as the polyamine or glycine site [38]. Ethanol does not cause blockade of open ion channels and does not interact with magnesium ions at the site where Mg^{2+} causes open channel block. Molecular biological techniques have shown that the ability of ethanol to inhibit responses to NMDA is dependent on the subunit makeup of the NMDA receptor, with the NR1/NR2A and NR1/NR2B combinations being preferentially sensitive to inhibition by ethanol [38]. Chronic exposure to ethanol increases the number of NMDA receptors and facilitates the receptor function, which is thought to cause withdrawal-related seizures after cessation of exposure [38].

In further studies on the pharmacology of the ethanol-discriminative stimulus in rats, Bienkowski et al. [39] assessed the effects of compounds from another class of NMDA-receptor antagonists: glycine-, strychnine-insensitive receptor (glycine B site) antagonists. The generalization of the selective glycine B site antagonists, L-701,324 and MRZ 2/576, and memantine (a noncompetitive antagonist at the NMDA-receptor ion-channel site) were assessed in animals trained to discriminate ethanol from saline. Memantine and L-701,324 substituted for ethanol, whereas MRZ 2/576 produced only half-maximal ($\approx 50\%$) ethanol-appropriate responding. Glycine did not antagonize the ethanol stimulus. Thus, it appeared that glycine-, strychnine-insensitive site antagonists may induce some ethanol-like stimuli in rats. However additional studies indicated that neither the glycine antagonists L-701,324 or MRZ 2/502, the polyamine-site antagonist arcaine, nor the polyamine-site ligand spermidine substituted for ethanol [40]. These authors also demonstrated dose-dependent generalization of the NMDA-receptor ion-channel blockers dizocilpine, memantine, and phencyclidine (PCP) as well as the sigma1-receptor antagonists (+)–pentazocine and (+)–N-allyl-normetazocine (NANM) to ethanol. Thus it appears that some of the acute effects of ethanol are mediated via both NMDA receptors and sigma1-binding sites.

Using compounds thought to inhibit the release of glutamate (lamotrigine and riluzole), it was demonstrated that lamotrigine but not riluzole substituted for the ethanol discriminative stimulus in rats [41]. The authors suggested that the noted difference between these two compounds to generalize to ethanol was likely due to the ability of lamotrigine, but not riluzole, to inhibit voltage-gated calcium channels.

Calcium Channels

In rats trained to discriminate ethanol from water, the L-type calcium channel antagonist isradipine was found to block ethanol discrimination in a dose-dependent manner [42]. It was also demonstrated that isradipine completely and dose-dependently inhibited the ethanol cue in a two-choice maze procedure, whereas the dopaminergic antagonist haloperidol did not [42]. Thus, these data argue for a role of L-type calcium channels in the ethanol interoceptive cue.

Acetylcholine

It has been shown in rats that neither nicotine nor the nicotinic acetylcholinergic-receptor antagonist, mecamylamine, substituted for ethanol and that mecamylamine did not antagonize the ethanol stimulus [43]. Therefore it seems unlikely that the neurotransmitter acetylcholine plays a significant role in the ethanol-discriminative stimulus.

System Interactions

Given that a preponderance of the literature suggests an involvement of both GABA-A and NMDA receptors in the ethanol discriminative stimulus, subjects were trained to discriminate a combination of the GABA-A agonist, diazepam, and the NMDA antagonist, ketamine, from vehicle [44]. In animals so trained, diazepam and ketamine both generalized completely to the mixture, and ethanol was almost completely substituted. These findings were taken to indicate that simultaneous GABA-A agonism and NDMA antagonism produce a greater ethanol-specific discriminative stimulus than activation of either component individually.

Recently, it has been demonstrated that several neuroactive steroids are potent modulators of a number of membrane receptors, including GABA-A, NMDA,

5-HT3, and sigma1 receptors. To further explore the pharmacology of these compounds, drug-discrimination procedures in rats were used to assess the generalization of pregnanolone to a variety of other agents [45]. Neither the opiate agonist morphine nor the negative GABA-A modulator, dehydroepiandrosterone, substituted for the pregnanolone cue, whereas all of the tested GABA-A-positive modulators did; these included allopregnanolone, epipregnanolone, androsterone, pentobarbital, midazolam, and zolpidem. Direct GABA-A-site agonists, including muscimol, did not generalize to pregnanolone, but ethanol and the sigma1-receptor agonist SKF-10047 generalized completely. The NMDA-receptor antagonist MK-801 partially substituted, but the 5-HT3 antagonist tropisetron did not. The 5-HT3 agonist SR 57227A completely substituted for pregnanolone, but another 5-HT3 agonist (m-chlorophenylbiguanide) produced only partial substitution. These observations suggest that (a) positive GABA-A modulation, but not direct agonism, confers a discriminative stimulus effect similar to that of pregnanolone and (b) antagonism of NMDA receptors and activation of 5-HT3 and sigma1 receptors modulate stimulus effects similar to the pregnanolone cue. Here, the observation that ethanol generalized completely to pregnanolone provided further evidence that the interoceptive cue of ethanol likely involves both the GABA-A- and NMDA-receptor systems.

It was shown that for rats self-administering ethanol, the interoceptive cue was mediated by both the GABA-A- and NMDA-receptor systems, as evidenced by the ability of both pentobarbital (GABA-A-receptor modulator) and MK-801 or dizocilpine (a noncompetitive NMDA-receptor antagonist) to completely substitute for ethanol in rats trained to discriminate ethanol from saline [46].

Anatomy

The involvement of specific brain areas in the discriminative stimulus properties of ethanol has also been explored to some degree with direct application of the noncompetitive NMDA antagonist MK-801 into the nucleus accumbens core (AcbC) or the CA1 region of the hippocampus, resulting in complete dose-related substitution for the ethanol discriminative stimulus in rats [47]. Infusion of MK-801 into either the amygdala or the prelimbic cortex (PrLC) did not substitute for ethanol, and neither did injection of the competitive NMDA antagonist CPP into the AcbC. The direct GABA-A agonist muscimol resulted in full ethanol substitution in a dose-related manner when injected into either the AcbC or the amygdala but not the PrLC. The findings from these studies support the notion that ethanol's discriminative stimulus properties are mediated centrally by NMDA and GABA-A receptors in specific limbic brain regions and point to a strong role for the interplay between ionotropic GABA-A and NMDA receptors in the nucleus accumbens [47].

In rats, stimulation of GABA-A receptors in the nucleus accumbens (by local infusion of muscimol) results in full substitution for the ethanol discriminative stimulus. Blockade of GABA-A receptors in the amygdala (by local infusion of bicuculline) attenuates the ability of GABA-A agonism in the nucleus accumbens to substitute for the ethanol cue [48]. Thus, it would appear that the ethanol-like stimulus effects of GABA-A agonism in the nucleus accumbens are modulated by GABA-A-receptor activity in the amygdala.

Summary of the Discriminative Stimulus Properties of Ethanol

Some authors have postulated that the effects of lower doses of ethanol (≤ 0.5 g/kg in rats) are mediated via GABAergic systems; that at intermediate doses (0.75 to 2.0 g/kg) several other neurotransmitter systems are affected; and that at high doses nonspecific effects emerge, probably involving even more neurotransmitter systems [49]. A nice summary of the dose dependence of the ethanol interoceptive cue and its interaction with the GABA-A, 5-HT1, and NMDA systems can be found in Grant and Colombo [50], who suggest that at a lower ethanol dose of 1.0 g/kg, the discriminative stimulus is primarily subserved by GABA-A interaction, which decreases with dose such that at 2.0 g/kg, the stimulus is primarily subserved by the NMDA system. The 5-HT1 contribution appears to be present to about the same degree at several doses: 1.0, 1.5, and 2.0 g/kg [50]. Thus, the 5-HT1 mechanisms appear to occur at most doses, whereas GABA-A involvement appears greatest at lower doses and NMDA involvement appears greatest at higher doses. A review of ethanol's interactions with a host of neurotransmitter gated ion channels [51] describes potential molecular mechanisms that may be relevant to its effects on the neurotransmitter systems identified via drug-discrimination studies.

EFFECTS OF ETHANOL ON OTHER ASPECTS OF COGNITION: ATTENTION, LEARNING, AND MEMORY

Attention/Impulsivity

Assessments of attention are often made using some form of continuous performance test (CPT; see Rosvold et al. [52]). This type of test is thought to provide metrics of attentional capacity (or vigilance) and requires subjects to respond in a selective fashion to a series of stimuli (usually visual in the form of shapes, colors, or lights for animals, and numbers or letters for humans) presented rapidly and for very short durations (e.g., 500 msec). Generally, subjects must attend to specific stimuli and respond accordingly when one is detected. In human tests, subjects might be asked to push a response key every time they see an X appear on a monitor. More difficult versions might require a response to an X only if it occurs immediately after an O. For the rat equivalent, subjects might be required to respond to one of several locations every time they detect illumination of a stimulus light at that location. The usual measures for this type of task include omission errors (misses), commission errors (false alarms), and latencies. Omission errors are thought to indicate deficits in attention or vigilance; commission errors are interpreted by many to represent aspects of impulsivity, since they are thought to result from anticipatory or incomplete processing of stimuli [53]; and response latencies are thought to provide metrics of processing requirements, psychomotor speed, or the level of difficulty of the discrimination. The application of signal detection analyses to CPT data can also provide measures of target stimulus discriminability and other measures such as response bias and strategy [53].

While research on the effects of ethanol on CPT performance in animals is scarce, Rezvani and Levin employed a similar operant visual-signal-detection task to examine the acute effects of ethanol in rats [54]. In that study, ethanol impaired performance by decreasing percent correct rejections and percent hits, suggesting an effect to impair sustained attention. Data from human CPT studies suggest that if the task is of sufficient difficulty, even small doses ($\approx 0.035\%$ breath alcohol concentration) can produce measurable alterations in task performance. These effects include decreased target stimulus discrimination, increased commission errors, and changes in response strategies. Such effects have even been noted in the absence of significant effects on percent correct detections or response latencies [53]. The observed effect of ethanol to increase errors of commission could be interpreted to mean that ethanol increases aspects of impulsivity. Along those lines, performance under differential reinforcement of low response rate (DRL) schedules has also been interpreted by some to yield metrics of impulsivity [55]. For DRL tasks, subjects are required to withhold a response — say a lever press for food — for a specific duration (e.g., 10 sec). Responding too early resets the time clock so that the possibility of reinforcement is delayed by another 10 sec. Treatments that are thought to increase "impulsive" behavior would be predicted to cause an increase in the number of early responses in this task. Thus, based on the interpretation that ethanol increased impulsivity in the 1999 study by Dougherty et al. [53], one might predict that ethanol should increase premature responses in DRL tasks. However, Popke et al. [56] reported that ethanol, over a wide range of doses, did not increase early responses in rats performing a DRL 10- to 14-sec task, and thus the prediction did not hold. On the other hand, nicotine did significantly increase premature responding in rats performing the exact same DRL task [57], and ethanol significantly enhanced this effect. Whether DRL responding can provide insight into aspects of behavior that somehow inform mechanisms underlying human impulsivity remains to be determined, but the findings for nicotine and nicotine plus ethanol are intriguing.

Conditioning

In adult rats, acute ethanol administration at 2.0 g/kg disrupts conditioning of a visual but not an olfactory discrimination, suggesting that the olfactory system in the rat might be much more resistant to chemical effects in this species. In addition, the 5-HT3-receptor antagonist MDL 72222 was not able to prevent the disruptive effects of ethanol on the conditioning of the visual discrimination, although it did ameliorate ethanol-induced hyperlocomotion [58]. Thus, while the effects of ethanol to increase locomotor activity appeared to be mediated by serotonergic involvement at the 5-HT3 receptor, its effects on visual discrimination (much like its ability to produce discriminative interoceptive stimuli, as noted previously) do not appear to involve those receptors.

In a similar study in preweanling (gestational day [GD] 21) rats, it was demonstrated that ethanol (1.5 g/kg) disrupted visual (brightness) but not olfactory conditioning [59]. These findings in young animals also support the proposition that the olfactory system is more resistant to perturbation than the visual system in the rat. Because the olfactory system becomes functional at birth, while the visual system does not become functional until approximately 15 days of age [59], it is possible that the differences in sensitivity noted in these younger animals are simply a reflection of the maturity of the two systems, with the more immature visual system being more susceptible to disruption at this age. Alternatively, it could be that because the olfactory system is likely so much more important to the rat than the visual system, it may be that the olfactory system is much more protected from perturbation than is the visual system.

Learning and Memory

Learning — the acquisition of new information — is critically important to the survival of organisms. A popular learning assessment procedure, the repeated acquisition (RA) of response sequences or behavioral chains [60], has been used in both human subjects [61] and animals [14, 15, 62] to repeatedly study learning in the same subjects, often for long periods of time. In these procedures, subjects are required to learn a new sequence of responses during each test session. Typically, a panel containing several response manipulanda (levers, press plates, response keys, etc.) is presented to the subject, and the subject must acquire (learn) a specific sequence of responses to obtain reinforcers (typically food for animals; money, credits, or other secondary reinforcers for humans). The correct sequence of responses changes such that subjects must learn new response sequences during every test session. In many RA studies, performance components alternate with acquisition components within the same test session. In the performance component, the correct response sequence is invariant (does not change from session to session), and responding in this component serves as a control for the motoric and motivational requirements of the acquisition component, in which the correct response sequence changes during each session. Ethanol significantly impairs accuracy (increases errors) in the learning component at doses that do not affect responding in the performance component [61, 63], suggesting a relatively selective effect of ethanol to disrupt active learning or new acquisition while leaving motoric function, motivation, and long-term procedural memory intact.

Working/Short-Term Memory

Working memory can be considered to be a type of short-term memory that can be described as the moment-to-moment maintenance, monitoring, and processing of information [64]. In the laboratory setting, working memory generally refers to the capacity to retain information across trials within a test session [65, 66]. Thus, the ability of subjects to perform spatial-alternation tasks (i.e., if a left choice was correct on one trial, then the right choice will be correct on the next trial) is an indication of working-memory integrity. In studies in rats, it has been shown that ethanol disrupts accuracy of performance in such tasks at plasma levels that would be within the legal limits for people driving cars [67]. It is interesting in this context that these authors also found that caffeine potentiated many of the adverse effects of ethanol (decreased accuracy) while normalizing other aspects of performance (intertrial intervals, length of pausing).

Working/short-term memory function in animals is typically studied using delayed-recall tasks that require subjects to discriminate and "encode" specific stimuli and then use that information to make choice responses that occur later (usually over the seconds-to-minutes range). Such tasks often employ visual-discrimination tasks such as delayed matching-to-sample (DMTS) and delayed non-matching-to-sample (DNMTS) procedures or position-discrimination tasks such as delayed spatial alternation (DSA). In a typical DMTS task, subjects are shown an initial or "sample" stimulus, like a red dot. After some period of presentation, or when the subject indicates that the stimulus has been observed, the sample is removed and a "recall delay" intervenes prior to the presentation of two or more choice stimuli, one of which matches the sample stimulus for that trial. A response to the choice stimulus that "matches" the sample stimulus is correct. While visual forms of this task can be used in a variety of species with good visual acuity, spatial versions have been developed for rodent species (e.g., delayed matching or non-matching-to-position [68]).

Other types of tasks also used in the assessment of learning and memory include a variety of mazes (e.g., Morris and Cincinnati water mazes [69] and radial arm mazes [70], in which subjects are given repeated opportunities to learn how to navigate to a goal point or points).

The basic approach in all of these procedures is to present subjects with information (cues, stimuli) that is to be held in working memory and recalled a short time later. Efficient encoding of information is thought to be heavily dependent on the constructs of attention and vigilance: if subjects are not paying attention to a given stimulus, or if they are distracted during the encoding of that stimulus, then the likelihood that it will be efficiently encoded is diminished. Over the short term (seconds to minutes), generally within a given session, such retrieval is thought to be dependent on working memory, whereas recall over longer periods of several hours and beyond is thought to be dependent on consolidation of more-permanent or long-lasting memory. For the purposes of discussion here, we will be focusing more on the acute effects of ethanol on short-term rather than long-term-memory issues.

Correct choice responding at zero or very short recall delays (1 to 2 sec) is thought to be a good metric of processes associated with attention/encoding; if choice accuracy is good, then encoding obviously occurred with efficiency. As recall delays are increased, then processes associated with working memory or retrieval are thought to play increasingly important roles in performance. The slope of the percent-correct choice versus recall-delay curve is taken to represent a quantitative aspect of working memory or memory retrieval; the intercept of this line with the y-axis (zero recall delay) is thought to represent primarily attentional or encoding processes. Thus, it can be envisioned that a drug or chemical could change the intercept (affect encoding/attention) but not alter the slope of the line (have no effect on rate of memory decay). Alternatively, if the intercept is unaffected by treatment but the slope is altered, then it could be posited that attention/encoding processes remained unaffected but that working memory/retrieval processes were affected. If both the intercept and the slope are altered, then all processes might be affected.

To examine the effects of ethanol on working memory in rhesus monkeys, Mello [71] utilized a titrating delayed-matching-to-sample (TDMTS) task, in which the recall delay increased or decreased as a function of a subject's prior performance accuracy. While it was noted that performance accuracy decreased with increasing ethanol doses, working memory did not appear to be selectively impaired. That is, error frequency did not increase as a function of recall delay interval; errors generally clustered at the shorter delays, with animals performing accurately at longer delays, even with blood alcohol levels of >200 mg/100 ml. These observations support the notion that ethanol affected attentional/encoding processes to a greater degree than working memory/retrieval. In further support of this possibility, the findings of Melia and Ehlers [72] in 1989, obtained from squirrel monkeys performing a conditional discrimination task (correctness of choice dependent on the presence or absence of a conditional stimulus [sphere]), showed that the primary effects of ethanol were to decrease the discriminability of the stimuli-controlling behavior. This conclusion was based on a signal-detection analysis of their data and the demonstration that ethanol caused dose-related decreases in choice accuracy that were accompanied by decreases in the sensitivity of subjects to discriminate the stimuli in the absence of any response bias [72]. These findings are also in accord with those reported for humans, where ethanol caused dose-dependent declines in sensitivity of subjects to stimuli in a recognition task, with no accompanying change in response bias [73].

In other studies in monkeys (*Macaca mulatta*), it has been demonstrated that accuracy of recall was affected at a dose of ethanol (0.5 g/kg) that was lower than that needed (1.0 g/kg) to affect reaction times [74]. These same authors demonstrated that several aspects of eye movement (frequency of saccades; fixation periods [durations]; saccade excursion, velocity, and duration) are affected in a dose-dependent fashion, with several measures showing effects at relatively low doses (0.25 g/kg). These results were interpreted to mean that ethanol selectively affects cerebral substrates associated with visual attention [75]. However, subsequent studies in rats performing a delayed spatial-matching task showed that low doses of ethanol (0.25 and 0.50 g/kg) actually decreased the rate of forgetting [76]. Thus, processes associated with working memory or recall were clearly not adversely affected by ethanol at doses that affected attention or encoding.

In studies focusing more on aspects of long-term memory and consolidation processes, it was found that rats given high doses of ethanol (3 g/kg) immediately after training in either a shuttle (active) avoidance procedure or an inhibitory (passive) avoidance procedure performed no differently than controls when tested days later [77]. These results suggest that ethanol does not interfere with consolidation of short-term memory into long-term memory.

In a nonspatial visual matching-to-sample procedure in rats [78], it was determined that ethanol decreased both the ability of subjects to detect or respond to "sample" visual stimuli (decreased vigilance, attention, or encoding) and to retain information over time (caused a working-memory deficit). In addition, it was found that local infusion of ethanol directly into the medial septal area caused a selective decrease in choice accuracy as a function of recall delay without a concomitant decrement in sustained attention [78]. Thus, in this case, the medial septum appears to play an important role in working memory.

In an eight-arm radial-maze delayed test of working/short-term memory, rats were treated with ethanol prior to beginning their daily sessions. After entry into four of the eight arms, they were provided either a 15-sec or a 1-h delay prior to entry into the fifth arm. Ethanol was found to significantly impair recall at the 1-h delay but not the 15-sec delay [79], suggesting a disruption of short-term memory but not working memory.

COMPARATIVE SENSITIVITIES OF DIFFERENT COGNITIVE FUNCTIONS TO THE EFFECTS OF ETHANOL

In rodent studies in which subjects performed a battery of food-reinforced operant behavioral tasks designed to simultaneously model a variety of complex brain functions, it was shown that ethanol selectively impaired aspects of cognition at doses that did not affect the ability of subjects to respond. The battery of tasks [56] consisted of:

1. A temporal-response differentiation (TRD) task to assess time perception
2. A differential reinforcement of low response rate (DRL) task to assess time perception and response inhibition/impulsivity
3. An incremental repeated acquisition (IRA) task to monitor learning
4. A conditioned position responding (CPR) task to assess auditory, visual, and position discrimination
5. A progressive ratio (PR) task to assess appetitive motivation

For the TRD task, subjects had to hold a response lever down for a minimum of 10 sec but no more than 14 sec. For the DRL task, subjects had to withhold responding (lever press) for at least 10 sec but not more than 14 sec. For the IRA task, subjects had to learn a new sequence of lever presses during each session, where sequence lengths were incremented from easy (press the correct one out of three levers) to increasingly difficult (up to six correct lever presses required to complete the response chain) as subjects demonstrated mastery of the easier sequences. For the CPR task, subjects had to respond on a left lever after presentation of a low-frequency tone or a low-intensity visual stimulus and on a right lever after presentation of a high-frequency tone or a high-intensity visual stimulus. For the PR task, subjects had to increase the amount of work (number of lever presses) emitted for each subsequent reinforcer: the first food pellet "cost" one lever press, the second cost two, the third cost three, and so on.

The most alcohol-sensitive behaviors in the battery were the DRL (time perception and response inhibition/impulsivity), CPR (auditory and visual discrimination), and PR (motivation) tasks, all of which were significantly disrupted by the same dose of ethanol (1.5 g/kg). The TRD (time perception) task was significantly affected as the dose of ethanol was increased, but the IRA (learning) task was never significantly affected over the dose range tested (up to 3.0 g/kg). Thus, it appears that at doses of ethanol that affect motivation to respond for food, the ability to inhibit responding (i.e., to maintain "impulse" control) as well as sensory discrimination, time perception, and active learning capabilities remains unaffected, with learning function seemingly the most resilient to disruption by ethanol. Reports in human subjects have also indicated that time perception is not significantly affected by ethanol [80]. In an excellent summary of the ethanol sensitivity of a variety of tasks in a variety of species including humans, Newland [81] has shown that tremor- and response-duration measures in monkeys are some of the most sensitive, showing effects at blood concentrations as low as ≈ 0.3 mg/ml. In contrast, learning tasks, eye tracking, and driving simulations in humans require blood alcohol concentrations of ≈ 0.8 to 1.2 mg/ml.

AGE-RELATED DIFFERENCES IN SENSITIVITY TO ETHANOL

Given that *in vitro* administration of ethanol inhibits synaptic activity and plasticity more potently in hippocampal slices from immature rats than in older rats, studies were conducted in whole animals (rats) to determine whether the effects of ethanol on the acquisition of spatial memory (using a Morris water maze) were also age dependent [82]. It was found that pretreatment with ethanol significantly impaired spatial-memory acquisition in adolescents but not adults. Interestingly, ethanol did not impair the acquisition of nonspatial memory in either adolescents or adults, demonstrating the increased sensitivity of spatial memory in younger animals [82]. In later studies it was also demonstrated that rats exposed to bingelike alcohol episodes as adolescents exhibited enhanced susceptibility (compared with animals exposed as adults to alcohol binges) to the memory-impairing effects of alcohol when tested as adults [83]. In studies in which ethanol was given after training sessions in an odor-discrimination task, it was found again that, relative to adults, memory in adolescent rats is more strongly disrupted than that of adults. Here, ethanol was given after training to avoid confounding the effects of ethanol on memory with those on sensory and motivational influences, which might manifest when ethanol is given during training. These findings are of interest because adolescent rats are actually less sensitive than adults to a variety of noncognitive effects of ethanol, including hypothermia, muscle relaxation, hypnosis, and lethality [84]. These and other observations reviewed in Smith [85] make it clear that the effects of ethanol in young or periadolescent animals are not the same as those seen in adults, suggesting that younger subjects are more susceptible to the disruptive effects of ethanol than adults.

In yet other studies examining the effects of ethanol on spatial and nonspatial memory in adolescent and adult rats — in this case using an appetitive, dry (sandbox) maze — it was shown that nonspatial acquisition was unaffected by ethanol at either age, but that spatial acquisition was disrupted in adults but not adolescents. The authors of this study postulated that "the adolescent-associated development of stress-sensitive regions involved in spatial learning may have contributed to the differences observed between" their study [86] and the earlier study [82].

ENVIRONMENTAL AND OTHER INFLUENCES ON THE EFFECTS OF ETHANOL

It has been demonstrated under a variety of circumstances that the effects of ethanol are influenced by the environment and other factors. For example, under conditions where patterns and rates of lever pressing are strictly controlled by the scheduling of reinforcements around these responses, ethanol increases low rates of responding while decreasing high rates of responding in the same subjects [87, 88]. Others have demonstrated that the effects of ethanol can depend not only on the ongoing rate of a particular behavior, but also on the context in which the behavior occurs and the behavioral history of the subject [89, 90]. In addition, genetic differences can influence the effects of ethanol. For example, it is well known that ethanol can serve as a positive reinforcer in a variety of animal species and that it is self-administered by many [91]. Genetic differences, however, can be important with respect to the ability of ethanol to serve as a positive reinforcer, as evidenced by its ability to engender self-administration [90]. For example, studies in alcohol accepting (AA) and alcohol nonaccepting (ANA) mice have shown that ethanol readily serves as a reinforcer in AA mice but not in ANA mice. If genetic differences can influence the ability of ethanol to serve as a reinforcer, then it follows that genetic differences might also influence the effects of ethanol on other aspects of CNS function. These observations are critically important because, if they are not taken into consideration, interpretation of findings from seemingly easily interpretable behavioral studies can be easily confounded.

OVERVIEW

Ethanol clearly interacts with a variety of neurotransmitter systems, affecting different functional and anatomical systems to varying degrees. These effects depend on dose, time after administration, age, genetics, and a variety of other influences. Depending on the circumstances prevailing at the time of administration, ethanol can have seemingly opposite effects, e.g., stimulation versus sedation or performance enhancement versus disruption. The preponderance of the data support the proposition that ethanol acts to disrupt important aspects of cognitive function, primarily by degrading the discriminability of relevant stimuli. It is unclear at this time whether such effects result from or cause decreases in encoding or attentional properties. However, it does seem fairly clear that ethanol does not selectively disrupt working memory or learning at doses that adversely affect stimulus discriminability, attention, or encoding. The observation that important brain functions (such as learning and olfactory discrimination in rodents) are relatively insensitive to the adverse effects of ethanol suggests that the systems subserving these functions are greatly protected from chemical perturbation.

REFERENCES

1. Fleming, M., Mihic, S.J., and Harris, R.A., Ethanol, in *Goodman and Gilman's the Pharmacological Basis of Therapeutics*, 10th ed., Hardman, J.G., Limbird, L.E., and Gilman, A.G., Eds., McGraw-Hill, New York, 2001, p. 429.
2. Kerr, J.S. and Hindmarch, I., Alcohol, cognitive function and psychomotor performance, *Rev. Environ. Health*, 9, 117, 1991.
3. Schecter, M.D. and Lovano, D.M., Time-course of action of ethanol upon a stimulant-depressant continuum, *Arch. Intl. Pharmacodynamie Ther.*, 260, 189, 1982.
4. Maylor, E.A. and Rabbitt, P.M.A., Effects of alcohol on speed and accuracy in choice reaction time and visual search, *Acta Psychologica*, 65, 147, 1987.

5. Bruce, K.R. and Pihl, R.O., Forget "drinking to forget": enhanced consolidation of emotionally charged memory by alcohol, *Exp. Clin. Pyschopharm.*, 5, 242, 1997.
6. Altshuler, H.L., Behavioral methods for the assessment of alcohol tolerance and dependence, *Drug Alc. Depend.*, 4, 333, 1979.
7. Driscoll, C.D., Streissguth, A.P., and Riley, E.P., Prenatal alcohol exposure: comparability of effects in humans and animal models, *Neurotoxicol. Teratol.*, 12, 231, 1990.
8. Stanton, M.E. and Spear, L.P., Workshop on the qualitative and quantitative comparability of human and animal developmental neurotoxicity, Work Group I report: Comparability of measures of developmental neurotoxicity in humans and laboratory animals, *Neurotoxicol. Teratol.*, 12, 261, 1990.
9. Hiramatsu, M., Shiotani, T., Kameyama, T. et al., Effects of nefiracetam on amnesia animal models with neuronal dysfunctions, *Behav. Brain* Res., 82, 107, 1997.
10. Petkov, V.D., Belcheva, S., and Petkov, V.V., Behavioral effects of *Ginkgo biloba* L., *Panax ginseng* C.A. Mey., and Gincosan®, *Am. J. Chin. Med.*, 31, 841, 2003.
11. Shimizu, K., Matsubara, K., Uezono, T., Kimura, K., and Shiono, H., Reduced dorsal hippocampal glutamate release significantly correlates with the spatial memory deficits produced by benzodiazepines and ethanol, *Neuroscience,* 83, 701, 1998.
12. Kaplan, J.M., Hennessy, M.B., and Howd, R.A, Oral ethanol intake and levels of blood alcohol in the squirrel monkey, *Pharmacol. Biochem. Behav.*, 17, 111, 1982.
13. Ando, K., Johanson, C.E., and Schuster, C.R., The effects of ethanol on eye tracking in rhesus monkeys and humans, *Pharmacol. Biochem. Behav.*, 26, 103, 1987.
14. Paule, M.G., Validation of a behavioral test battery for monkeys, in *Methods of Behavioral Analysis in Neuroscience*, Buccafusco, J.J., Ed., CRC Press, Boca Raton, FL, 2001a, p. 281.
15. Paule, M.G., Using identical behavioral tasks in children, monkeys and rats to study the effects of drugs, *Curr. Therap. Res.*, 62, 820, 2001b.
16. Duka, T., Stephens, D.N., Russell, C. et al., Discriminative stimulus properties of low doses of ethanol in humans, *Psychopharmacology*, 136, 379, 1998.
17. Kamien, J.B., Bickel, W.K, Hughes, J.R. et al., Drug discrimination by humans compared to nonhumans: current status and future directions, *Psychopharmacology*, 111, 259, 1993.
18. Overton, D.A., Comparison of ethanol, pentobarbital, and phenobarbital using drug vs. drug discrimination training, *Psychopharmacology*, 53, 195, 1977.
19. York, J.L. and Bush, R., Studies on the discriminative stimulus properties of ethanol in squirrel monkeys, *Psychopharmacology*, 77, 212, 1982.
20. Shippenberg, T.S. and Altshuler, H.L., A drug discrimination analysis of ethanol-induced behavioral excitation and sedation: the role of endogenous opiate pathways, *Alcohol*, 2, 197, 1985.
21. Spanagel, R., The influence of opioid antagonists on the discriminative stimulus effects of ethanol, *Pharmacol. Biochem. Behav.*, 54, 645, 1996.
22. Signs, S.A. and Schechter, M.D., The role of dopamine and serotonin receptors in the mediation of the ethanol interoceptive cue, *Pharmacol. Biochem. Behav.*, 31, 55, 1988.
23. Grant, K.A., Colombo, G., and Gatto, G.J., Characterization of the ethanol-like discriminative stimulus effects of 5-HT receptor agonists as a function of ethanol training dose, *Psychopharmacology*, 133, 133,1997.
24. Stefanski, R., Bienkowski, P., and Kostowski, W., Studies on the role of 5-HT3 receptors in the mediation of the ethanol interoceptive cue, *Eur. J. Pharmacol.*, 309, 141, 1996.

25. Rees, D.C., Knisely, J.S., Breen, T.J. et al., Toluene, halothane, 1,1,1,-trichloroethylene and oxazepam produce ethanol-like discriminative stimulus effects in mice, *J. Pharmacol. Exp. Ther.*, 243, 931, 1987.

26. Balster, R.L. and Moser, V.C., Pentobarbital discrimination in the mouse, *Alc. Drug Res.*, 7, 233, 1987.

27. Bowen, S.E. and Balster, R.L., Desflurane, enflurane, isoflurane and ether produce ethanol-like discriminative stimulus effects in mice, *Pharmacol. Biochem. Behav.*, 57, 191, 1997.

28. Ator, N.A., Grant, K.A., Purdy, R.H. et al., Drug discrimination analysis of endogenous neuroactive steroids in rats, *Eur. J. Pharmacol.*, 241, 237, 1993.

29. Colombo, G., Agabio, R., Lobina, C. et al., Symmetrical generalization between the discriminative stimulus effects of gamma-hydroxy butyric acid and ethanol: occurrence within narrow dose ranges, *Physiol. Behav.*, 57, 105, 1995.

30. Massey, B.W. and Woolverton, W.L., Discriminative stimulus effects of combinations of pentobarbital and ethanol in rhesus monkeys, *Drug Alc. Depend.*, 35, 37, 1994.

31. Bowen, C.A. and Grant, K.A., Pharmacological analysis of the heterogeneous discriminative stimulus effects of ethanol in rats using a three-choice ethanol-dizocilpine-water discrimination, *Psychopharmacology*, 139, 86, 1998.

32. Schechter, M.D., Meehan, S.M., Gordon, T.L. et al., The NMDA receptor antagonist MK-801 produces ethanol-like discrimination in the rat, *Alcohol*, 10, 197, 1993.

33. Spanagel, R., Zieglgansberger, W., and Hundt, W., Acamprosate and alcohol, III: effects on alcohol discrimination in the rat, *Eur. J. Pharmacol.*, 305, 51, 1996.

34. Balster, R.L., Grech, D.M., and Bobelis, D.J., Drug discrimination analysis of ethanol as an N-methyl-D-aspartate receptor antagonist, *Eur. J. Pharmacol.*, 222, 39, 1992.

35. Holter, S.M., Danysz, W., and Spanagel, R., Novel uncompetitive N-methyl-D-aspartate (NMDA)-receptor antagonist MRZ 2/579 suppresses ethanol intake in long-term ethanol-experienced rats and generalizes to ethanol cue in drug discrimination procedure, *J. Pharmacol. Exp. Ther.*, 292, 545, 2000.

36. Bienkowski, P., Stefanski, R., and Kostowski, W., Competitive NMDA receptor antagonist, CGP 40116, substitutes for the discriminative stimulus effects of ethanol, *Eur. J. Pharmacol.*, 314, 277, 1996.

37. Bienkowski, P., Stefanski, R., and Kostowski, W., Discriminative stimulus effects of ethanol: lack of antagonism with N-methyl-D-aspartate and D-cycloserine, *Alcohol*, 14, 345, 1997.

38. Wirkner, K., Poelchen, W., Koles, L., Muhlberg, K., Scheiber, P., Allgaier, C., and Illes, P., Ethanol-induced inhibition of NMDA receptor channels, *Neurochem. Inter.*, 35, 153, 1999.

39. Bienkowski, P., Danysz, W., and Kostowski, W., Study on the role of glycine, strychnine-insensitive receptors (glycineB sites) in the discriminative stimulus effects of ethanol in the rat, *Alcohol*, 15, 87, 1998.

40. Hundt, W., Danysz, W., Holter, S.M. et al., Ethanol and N-methyl-D-aspartate receptor complex interactions: a detailed drug discrimination study in the rat, *Psychopharmacology*, 135, 44, 1998.

41. Hundt, W., Holter, S.M., and Spanagel, R., Discriminative stimulus effects of glutamate release inhibitors in rats trained to discriminate ethanol, *Pharmacol. Biochem. Behav.*, 59, 691, 1998.

42. Colombo, G., Agabio, R., Lobina, C. et al., Blockade of ethanol discrimination by isradipine, *Eur. J. Pharmacol.*, 265, 167, 1994.

43. Bienkowski, P, Piasecki, J., Koros, E. et al., Studies on the role of nicotinic acetylcholine receptors in the discriminative and aversive stimulus properties of ethanol in rats, *Eur. Neuropsychopharm.*, 8, 79, 1998.

44. Harrison, Y.E., Jenkins, J.A., Rocha, B.A. et al., Discriminative stimulus effects of diazepam, ketamine and their mixture: ethanol substitution patterns, *Behav. Pharmacol.*, 9, 31, 1998.

45. Engel, S.R., Purdy, R.H., and Grant, K.A., Characterization of discriminative stimulus effects of the neuroactive steroid pregnanolone, *J. Pharmacol. Exp. Ther.*, 297, 489, 2001.

46. Hodge, C.W., Cox, A.A., Bratt, A.M. et al., The discriminative stimulus properties of self-administered ethanol are mediated by GABA(A) and NMDA receptors in rats, *Psychopharmacology*, 154, 13, 2001.

47. Hodge, C.W. and Cox, A.A., The discriminative stimulus effects of ethanol are mediated by NMDA and GABA-A receptors in specific limbic brain regions, *Psychopharmacology*, 139, 95, 1998.

48. Besheer, J., Cox, A.A., and Hodge, C.W., Coregulation of ethanol discrimination by the nucleus accumbens and amygdala, *Alcoholism: Clin. Exp. Res.*, 27, 450, 2003.

49. Ryabinin, A.E., Role of hippocampus in alcohol-induced memory impairment: implications from behavioral and immediate early gene studies, *Psychopharmacology*, 139, 34, 1998.

50. Grant, K.A. and Colombo, G., Pharmacological analysis of the mixed discriminative stimulus effects of ethanol, *Alc. Alcohol.*, 2 (Suppl.), 445, 1993.

51. Lovinger, D.M., Alcohols and neurotransmitter gated ion channels: past, present and future, *N.-Schmiedeberg's Arch. Pharmacol.*, 356, 267, 1997.

52. Rosvold, H.E., Mirsky, A., Sarason, I., Bransome, E.D., Jr., and Beck, L.H., A continuous performance test of brain damage, *J. Consult. Psychol.* 20, 343, 1956.

53. Dougherty, D.M., Moeller, F.G., Steinberg, J.L., Marsh, D.M., Hines, S.E., and Bjork, J.M., Alcohol increases commission error rates for a continuous performance test, *Alcoholism: Clin. Exp. Res.*, 23, 1342, 1999.

54. Rezvani, A.H. and Levin, E.D., Nicotine-alcohol interactions and attentional performance on an operant visual signal detection task in female rats, *Pharmacol. Biochem. Behav.*, 76, 75, 2003.

55. McMillen, B.A., Means, L.W., and Matthews, J.N., Comparison of the alcohol-preferring rat to the Wistar rat in behavioral tests of impulsivity and anxiety, *Physiol. Behav.*, 63, 371, 1998.

56. Popke, E.J., Allen, S.R., and Paule, M.G., Effects of acute ethanol on indices of cognitive-behavioral performance in rats, *Alcohol*, 20, 187, 2000.

57. Popke, E.J., Fogle, C.M., and Paule, M.G., Ethanol enhances nicotine's effects on DRL performance in rats, *Pharmacol. Biochem. Behav.*, 66, 819, 2000.

58. Rajachandran, L., Spear, N.E., and Spear, L.P., Effects of the combined administration of the 5-HT3 antagonist MDL 72222 and ethanol on conditioning in the periadolescent and adult rat, *Pharmacol. Biochem. Behav.*, 46, 535, 1993.

59. Molina, J.C, Serwatka, J., Enters, E.K., Spear, L.P., and Spear, N., Acute intoxication disrupts brightness but not olfactory conditioning in preweanling rats, *Behav. Neurosci.*, 6, 846, 1987.

60. Cohn, J. and Paule, M.G., Repeated acquisition: the analysis of behavior in transition, *Neurosci. Biobehav. Rev.*, 19, 397, 1995.

61. Higgins, S.T., Bickel, W.K., O'Leary, D.K. et al., Acute effects of ethanol and diazepam on the acquisition and performance of response sequences in humans, *J. Pharmacol. Exp. Ther.*, 243, 1, 1987.

62. Popke, E.J., Allen, R.R., Pearson, E.C., Hammond, T.G., and Paule, M.G., Differential effects of two NMDA receptor antagonists on cognitive-behavioral development in non-human primates, I, *Neurotoxicol. Teratol.*, 23, 319–332, 2001.

63. Higgins, S.T., Rush, C.R., Hughes, J.R., Bickel, W.K., Lynn, M., and Capelies, M.A., Effects of cocaine and alcohol, alone and in combination, on human learning and performance, *J. Exp. Anal. Behav.*, 58, 87, 1992.

64. Baddeley, A.D. and Logie, R.H., Working memory: the multiple-component model, in *Models of Working Memory: Mechanisms of Active Maintenance and Executive Control*, Miyake, A. and Shah, P., Eds., Cambridge University Press, New York, 1999, p. 28.

65. Olton, D.S., Becker, J.T., and Handelmann, G.E., Hippocampal function: working memory or cognitive mapping, *Physiol. Psychol.*, 8, 239, 1980.

66. de Oliveira, R.W. and Nakamura-Palacios, E.M., Haloperidol increases the disruptive effect of alcohol on spatial working memory in rats: a dopaminergic modulation in the medial prefrontal cortex, *Psychopharmacology*, 170, 51, 2003.

67. Elsner, J., Alder, S., and Zbinden, G., Interaction between ethanol and caffeine in operant behavior of rats, *Psychopharmacology*, 96, 194, 1988.

68. Wiig, K.A and Burwell, R.D., Memory impairment on a delayed non-matching to position task after lesions of the perirhinal cortex in the rat, *Behav. Neurosci.*, 112, 4, 1998.

69. Williams, M.T., Moran, M.S., and Vorhees, C.V., Refining the critical period for methamphetamine-induced spatial deficits in the Morris water maze, *Psychopharmacology*, 168, 329, 2003.

70. Rezvani, A.H. and Levin, E.D., Nicotine-alcohol interactions and cognitive function in rats, *Pharmacol. Biochem. Behav.*, 72, 865, 2002.

71. Mello, N.K., Alcohol effects on delayed matching to sample performance by rhesus monkey, *Physiol. Behav.*, 7, 77, 1971.

72. Melia, K.F. and Ehlers, C.E., Signal detection analysis of ethanol effects on a complex conditional discrimination, *Pharmacol. Biochem. Behav.*, 33, 581, 1989.

73. Williams, J.L. and Rundell, O.H., Effect of alcohol on recall and recognition as functions of processing levels, *J. Stud. Alcohol*, 45, 10, 1984.

74. Fuster, J.M., Willey, T.J., Riley, D.M., and Ashford, J.W., Effects of ethanol on visual evoked responses in monkeys performing a memory task, *Electroenceph. Clin. Neurophys.*, 53, 621, 1982.

75. Fuster, J.M., Willey, T.J., and Riley, D.M., Effects of ethanol on eye movements in the monkey, *Alcohol*, 2, 611, 1985.

76. Melia, K.F., Koob, G.F., and Ehlers, C.L., Ethanol effects on delayed spatial matching as modeled by a negative exponential forgetting function, *Psychopharmacology*, 102, 391, 1990.

77. de Carvalho, L.P., Vendite, D.A., and Izquierdo, I., A near-lethal dose of ethanol, given intraperitoneally, does not affect memory consolidation of two different avoidance tasks, *Psychopharmacology*, 59, 71, 1978.

78. Givens, B. and McMahon, K., Effects of ethanol on nonspatial working memory and attention in rats, *Behav. Neurosci.*, 111, 275, 1997.

79. Oliveira, M.G.M., Kireeff, W., Hashizume, L.K., Bueno, O.F.A., and Masur, J., Ethanol decreases choice accuracy in a radial maze delayed test, *Brazilian J. Med. Biol. Res.*, 23, 547, 1990.

80. Tinklenberg, J.R., Kopell, B.S., Melges, F.T., and Hollister, L.E., Marihuana and alcohol: time production and memory functions, *Arch. Gen. Psychiatry*, 27, 812, 1972.

81. Newland, M.C., Operant behavior and the measurement of motor dysfunction, in *Neurobehavioral Toxicity: Analysis and Interpretation*, Weiss, B. and O'Donoghue, J.L., Eds., Raven Press, New York, 1994, p. 273.

82. Markwiese, B.J., Acheson, S.K., Levin, E.D. et al., Differential effects of ethanol on memory in adolescent and adult rats, *Alcoholism: Clin. Exp. Res.*, 22, 416, 1998.
83. White, A.M., Ghia, A.J., Levin, E.D. et al., Binge pattern ethanol exposure in adolescent and adult rats: differential impact on subsequent responsiveness to ethanol, *Alcoholism: Clin. Exp. Res.*, 24, 1251, 2000.
84. Land, C. and Spear, N.E., Ethanol impairs memory of a simple discrimination in adolescent rats at doses that leave adult memory unaffected, *Neurobiol. Lrn. Mem.*, 81, 75, 2004.
85. Smith, R.F., Animal models of periadolescent substance abuse, *Neurotoxicol. Teratol.*, 25, 291, 2003.
86. Rajendran, P. and Spear, L.P, The effects of ethanol on spatial and nonspatial memory in adolescent and adult rats studied using an appetitive paradigm, *Ann. N.Y. Acad. Sci.*, 1021, 441, 2004.
87. Glowa, J.R. and Barrett, J.E., Effects of alcohol on punished and unpunished responding of squirrel monkeys, *Pharmacol. Biochem. Behav.*, 4, 169, 1976.
88. Katz, J.L. and Barrett, J.E., Effects of ethanol on behavior under fixed-ratio, fixed-interval, and multiple fixed-ratio fixed-interval schedules in the pigeon, *Arch. Intl. Pharmacodynamie Therapie*, 234, 88, 1978.
89. Barrett, J.E. and Stanley, J.A., Effects of ethanol on multiple fixed-interval fixed-ratio schedule performances: dynamic interactions at different fixed-ratio values, *J. Exp. Anal. Behav.*, 34, 185, 1980.
90. McMillan, D.E. and Leander, J.D., Effects of drugs on schedule controlled behavior, in *Behavioral Pharmacology*, Glick, S.O. and Goldfarb, J., Eds., Mosby, St. Louis, 1976, p. 85.
91. Ritz, M.C., George, F.R., deFiebre, C.M., and Meisch, R.A., Genetic differences in the establishment of ethanol as a reinforcer, *Pharmacol. Biochem. Behav.*, 24, 1089, 1986.

Section II

Toxicologic Models

6 Animal Models of Cognitive Impairment Produced by Developmental Lead Exposure

Deborah C. Rice
Maine Center for Disease Control and Prevention

CONTENTS

Intermittent Schedules of Reinforcement..74
Learning 77
Memory...83
Attention 89
Sensory Dysfunction: Possible Contribution to "Cognitive" Effects 92
Conclusions...94
References..95

Lead is probably the most-studied environmental contaminant with respect to the effects of developmental exposure on cognition in children or animal models. It has been known since the 1940s that lead poisoning in children can result in permanent behavioral sequelae, including poor school performance, impulsive behavior, and short attention span [1], that were observations later replicated by other investigators [2–4]. Early in the 1970s, deficits in intelligence quotient (IQ), fine motor performance, and behavioral disorders such as distractibility and constant need for attention were observed in children who had never exhibited overt signs of toxicity [5, 6]. In 1979 Needleman et al. [7] reported decreased IQ and increased incidence of distractibility and inattention in middle-class children who had not been exposed to lead from paint.

Early studies of the effects of developmental exposure to lead in animals focused on determining deficits on a wide variety of tasks characterizing the constellation

73

of the effects of lead [8]. Exposures in various studies included postnatal, lifetime, *in utero*, or *in utero* plus postnatal. Researchers also sought to identify a dose or body burden that did not produce adverse effects. A series of experiments with monkeys in our laboratory documented adverse effects in a group of monkeys with peak blood lead concentrations averaging 15 µg/dl during infancy, with steady-state levels over most of the lifespan averaging 11 µg/dl. Animals with lower body burdens have apparently not been assessed. The current CDC (Centers for Disease Control) "level of concern" is 10 µg/dl for children, although it is clear that there are adverse effects on cognition at blood concentrations below 10 µg/dl [9, 10]. More recently, experimental researchers have focused on developing paradigms to explore the behavioral mechanisms responsible for the constellation of effects observed in previous studies; these studies generally used doses that are known to produce robust impairment.

INTERMITTENT SCHEDULES OF REINFORCEMENT

Intermittent schedules of reinforcement have been used in behavioral pharmacology and toxicology for over 50 years. Descriptions of these schedules are available from numerous sources (e.g., [11, 12]). Simple schedules such as fixed interval (FI), variable interval (VI), fixed ratio (FR), and differential reinforcement of low (response) rate (DRL) schedules are acquired reasonably rapidly by animals, and performance is similar across species, including humans. The schedule used most often in lead research was the FI, presumably because it offers a number of advantages. Although this schedule requires the subject to make only one response at the end of a specified (uncued) interval, FI performance is typically characterized by an initial pause followed by a gradually accelerating rate of response, terminating in reinforcement. The schedule does not differentially reinforce any particular response rate (other than no or very low rate of responding) and may therefore be sensitive to toxicant-induced differences in the rate of response. In addition, temporal discrimination can be examined by measuring the shape of the response pattern across the interval (e.g., quarter life or index of curvature). Lower doses of lead produced increased response rates on the FI schedule in rats and monkeys [13–21], whereas high doses resulted in lower response rates [22, 23]. In general, temporal discrimination *per se*, as measured by the pattern of responses across the interval, was not affected by lead (but see Rice [20] and Mele et al. [24]). When a time-out (TO) period (during which responses had no scheduled consequences) was included in the assessment of performance on the FI, lead exposure resulted in increased TO rates of response [18, 19].

Attention deficit hyperactivity disorder (ADHD) was associated with increased response rates on FI performance in 7- to 12-year-old boys, as well as a "bursting" pattern of response produced by a run of closely spaced responses separated by a short pause [25]. This pattern was also observed in 3-year-old monkeys exposed to lead from birth (Figure 6.1) [18]. Children with ADHD also responded more in the extinction (TO) portion of the schedule, as did lead-exposed monkeys. FI performance predicted poorer performance on a test of impulsivity in normal children [26, 27]: children with high response rates and shorter post-reinforcement pause times

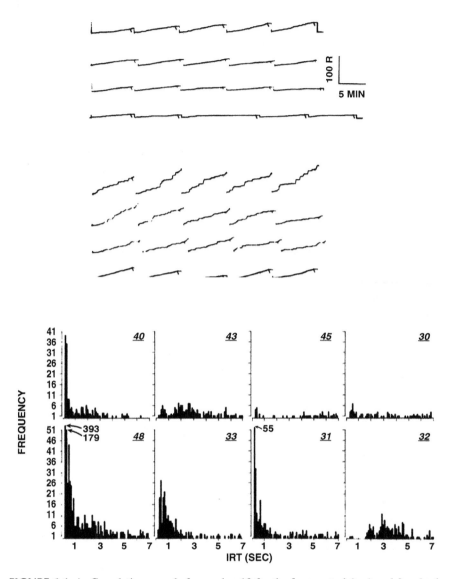

FIGURE 6.1 A: Cumulative records for session 10 for the four control (top) and four lead-treated (bottom) monkeys on an FI-TO schedule of reinforcement. Each lever press stepped the pen vertically; time is represented horizontally. The reinforced response in each FI was signaled by a downward deflection of the pen. The pen reset to baseline at the end of each TO period. The response rate in both the FI and TO periods is greater than controls for three of the four lead-treated monkeys. These monkeys also had a different pattern of response: they responded in bursts, causing the record to appear "steplike." B: Interresponse time (IRT) absolute frequency distribution histograms for control (top) and lead-treated (bottom) monkeys for session 7. Responses were divided into 100-msec bins, with all IRTs over 7 sec eliminated from the figure. The lead-treated monkeys in general had a much higher absolute frequency of short IRTs and a distribution skewed toward shorter IRTs. (Taken from Rice, D.C. et al., *Toxicol. Appl. Pharmacol.*, 51, 503, 1979. With permission.)

chose a smaller immediate reinforcer rather than a larger but delayed reinforcer. This was interpreted as evidence of impulsive behavior in the children with ADHD. Thus the FI schedule has numerous important advantages: equipment and computer programming required for the FI schedule (and other simple intermittent schedules) is minimal compared with more complicated testing procedures, acquisition of performance is relatively rapid, FI performance is similar across species, the FI schedule is sensitive to changes induced by lead (and other contaminants), and FI performance is predictive of performance on a test of impulsivity.

In contrast to the FI schedule, the DRL schedule requires a specified time between responses for reinforcement; responding before the specified time resets the contingency. Therefore the DRL schedule punishes failure of response inhibition. The DRL schedule also proved sensitive to developmental lead exposure. For example, DRL performance was assessed in groups of monkeys in which increased rates of response on the FI had been observed. The schedule required the monkey to space consecutive responses at least 30 sec apart to be reinforced. Monkeys with peak blood lead levels of 100 μg/dl and steady-state levels of 40 μg/dl exhibited a higher number of nonreinforced responses, a lower number of reinforced responses, and a shorter average time between responses over the course of the experiment than control monkeys [28]. Performance on this DRL schedule was also examined in a group of monkeys having steady-state blood lead levels of 11 or 13 μg/dl [29]. Lead-treated monkeys were able to perform the DRL task in a way that was indistinguishable from controls. However, they learned the task at a slower rate, as measured by the increment in reinforced responses and decrement in nonreinforced responses over the course of the early sessions. Increased rates of response [30] and increased frequencies of responses emitted close together (short interresponse times) [31] have been reported in rats performing on a DRL schedule with no previous exposure to intermittent schedules. Postnatal blood lead concentrations are also associated with failure to inhibit responding on a DRL schedule in 9-year-old children (Paul Stewart, personal communication).

Effects of lead were examined on schedules similar to the DRL that assessed temporal discrimination. Response duration performance was assessed on a task in which rats exposed to lead beginning at weaning were required to depress a lever for at least 3 sec to be reinforced [32]. Lead-treated rats depressed the lever for a shorter time than controls. In addition, introduction of a tone signaling the 3-sec interval was effective in improving performance of control but not treated rats. Infant monkeys exposed to lead *in utero*, with maternal blood lead concentrations of 61 or 72 μg/dl for two dose groups, were tested on a task in which responses were required to be emitted after 10 sec but before 15 sec had elapsed (a DRL with a limited hold contingency) [33]. Treated infants did not make premature responses (before 10 sec) but rather had an increased number of failures to respond before 15 sec. These results, observed at relatively high lead exposure, were similar to the decreased behavioral output observed on FI at high blood lead levels. However, there may also be a differential sensitivity of prenatal versus postnatal exposure.

The fixed ratio (FR) schedule requires the subject to emit a fixed number of responses to be reinforced and typically generates a high response rate. This schedule appears to be less sensitive to lead-induced changes than is the FI. Low doses of lead sometimes resulted in increased rates of response, often transiently, whereas higher doses decreased response rates. This was true for both rats [22, 34, 35] and monkeys [19, 36]. Unfortunately, performance on this schedule had apparently not been assessed in lead-exposed children.

In summary, the DRL task proved sensitive to lead exposure in both animals (rats and monkeys) and humans. FI performance was affected by lead in reproducible ways in animals in numerous studies. FI performance in children with ADHD was virtually identical to that in lead-treated animals, including a "bursting" response pattern. Performance on an FI schedule predicted performance on a relatively more complicated test of impulsivity in children. Therefore, it appears that these simple intermittent schedules predict important sequelae of lead exposure in children.

LEARNING

Developmental lead exposure was associated with decreased IQ in numerous studies [37]. Deficits in reading, math, spelling, language, and other academic skills were associated with increased childhood lead exposure [38–43]. Deficits in color naming were also associated with increased blood lead concentrations [44]. It is difficult to determine the degree to which poor performance in school is the result of learning deficits as opposed to attentional or other deficits in cognitive or sensory function. The experimental literature may help to elucidate the various behavioral mechanisms responsible for impaired cognitive functioning in children.

Deficits in acquisition of tasks (learning) have been demonstrated in experimental studies on a variety of tasks. Perhaps the simplest of these is visual discrimination. Rats exposed to a very high dose of lead (1000 mg/kg to the dam) during gestation and lactation were impaired on both a brightness and shape discrimination in a water-escape T-maze [45]. Lead-exposed rats had shorter swim times, and the authors suggested that the increased errors in the lead-treated group might result from a failure to attend to relevant discriminative cues, a hypothesis for which there would be substantial support in later studies.

Sheep exposed to lead *in utero* at maternal blood levels of 34 µg/dl (but not at 18 µg/dl) were assessed on a series of nonspatial visual discrimination tasks [46]. The first five discrimination problems were form discriminations and the sixth was a size discrimination. The lead-treated sheep were only impaired on the sixth problem, which was also the most difficult for the control group. It may be that the lead-exposed sheep were impaired on the last problem simply because it was difficult; alternatively, it may have been because the relevant stimulus dimension was changed from form to size. Similarly, rats exposed to lead prenatally were not impaired on a visual discrimination problem that was easy for control rats (vertical vs. horizontal stripes), whereas these lead-exposed rats were severely impaired on a difficult discrimination (bigger vs. smaller circle) [47]. As in the study in sheep, the discrimination that was the more difficult for controls also changed stimulus dimension from line orientation to size. These two studies provided a preview of two findings that would be consistently observed in later studies: difficult tasks are more sensitive to lead-induced impairment than easier ones, as are studies in which there is a change in the relevant stimulus-response class.

A strategy adopted early in the research on the developmental effects of lead was the introduction of two additional requirements to the visual discrimination task: the requirement for reversal performance on an already-learned discrimination task and the addition of irrelevant cues. In a discrimination reversal task, the formerly correct stimulus becomes the incorrect one, and vice versa. This task requires extinction of the previously learned response and the learning of a new (opposite) one. These requirements presumably make cognitive demands not required by the initial acquisition of the discrimination task. The introduction of irrelevant cues assesses reasoning and attentional processes, as well as providing the opportunity to change relevant stimulus dimension, further taxing cognitive abilities.

In the nonspatial version of the discrimination reversal task, the relevant stimulus dimension is form or color, for example, rather than the position of stimuli. Typically, the subject is required to perform a series of such reversals. This allows the degree of improvement in performance across reversals to be assessed, which is indicative of how quickly the subject learns that the rules of the game change in a predictable pattern. Nonspatial discrimination reversal performance was impaired by postnatal exposure in rhesus monkeys tested during infancy [48] and cynomolgus monkeys tested as juveniles [49]. Cynomolgus monkeys with blood lead levels of 15 or 25 μg/dl during infancy and steady-state levels of 11 or 13 μg/dl were impaired on a series of nonspatial discrimination reversal tasks with irrelevant cues as juveniles [50]. Lead-treated monkeys were not impaired on the acquisition of any of the three tasks; however, they were impaired over the set of reversals of a form discrimination, which was their introduction to a discrimination reversal task, and on a color discrimination with irrelevant cues, their introduction to irrelevant cues. Analysis of the kinds of errors made by treated monkeys revealed that they were attending to irrelevant cues in systematic ways, either responding on or avoiding a particular position or stimulus. This suggests that lead-treated monkeys were being distracted by these irrelevant cues to a greater degree than controls, which may have been responsible at least in part for their apparent learning deficit.

In a subsequent study on possible sensitive periods for deleterious effects produced by lead, monkeys were exposed to lead either continuously from birth, during infancy only, or beginning after infancy [51]. Lead levels were about 30 to 35 μg/dl when monkeys were exposed to lead and given access to infant formula, and 19 to 22 μg/dl when monkeys were dosed with lead after withdrawal of infant formula. These monkeys were tested as juveniles on the same nonspatial discrimination reversal tasks described above. Both the group dosed continuously from birth and the group dosed beginning after infancy were impaired over the course of the reversals in a way similar to that observed in the study discussed above, including increased distractibility by irrelevant cues. The higher exposure levels in this study were reflected in impairment on all three tasks, whereas in the previous study lead-treated monkeys were impaired on only the first two tasks. The group exposed only during infancy was unimpaired on these tasks. In addition, the group dosed continuously from birth was impaired in the acquisition of the task in which irrelevant cues were introduced; there were no other impairments in acquisition.

Performance on spatial discrimination reversal tasks, analogous to the nonspatial discrimination reversal tasks already described, also proved sensitive to disruption by developmental lead exposure. A subset of the monkeys in the Bushnell and Bowman study [48], in which effects on both spatial and nonspatial discrimination reversal had been found during infancy, exhibited impairment on a series of spatial discrimination reversal tasks with irrelevant color cues at 4 years of age, despite the fact that lead exposure had ceased at 1 year and blood lead levels at the time of testing were at control levels.

In the group of monkeys with stable blood lead levels of 11 or 13 $\mu g/dl$ discussed above [50], deficits were also observed on a series of three spatial discrimination reversal tasks, the first one with no irrelevant cues and the subsequent two with irrelevant cues of various types [52]. Treated monkeys were impaired relative to controls over the series of reversals in the presence, but not in the absence, of irrelevant stimuli. Moreover, the lower dose group was impaired only during the first task after the introduction of irrelevant cues but not on the second task with irrelevant cues, when irrelevant stimuli were familiar. As in the nonspatial discrimination reversal task, there was evidence that lead-exposed monkeys were attending to the irrelevant stimuli in systematic ways, suggesting that this behavior was responsible for, or at least contributing to, the impairment in performance. This is also suggested by the fact that lead-treated monkeys were impaired in the presence of but not in the absence of irrelevant stimuli. In the group of monkeys in which the relevance of the developmental period of exposure was being assessed, described above [51], spatial discrimination reversal performance was also assessed [53]. Treated monkeys were the most impaired over the series of reversals on the first task after the introduction of irrelevant cues, although performance was impaired on all three tasks. Contrary to the result of the nonspatial discrimination reversal task in which the group dosed only during infancy was unimpaired, all three dose groups were impaired to an equal degree. These data suggest that spatial and nonspatial tasks may be affected differentially, depending on the development period of lead exposure.

Deficits on visual discrimination problems have also been observed in the absence of the requirement for reversal performance under some circumstances, such as high blood lead levels or increased task difficulty. Infant Rhesus monkeys exposed *in utero* to lead, with maternal blood lead levels of 61 to 72 $\mu g/dl$ for two dose groups, were impaired on a three-choice consecutive form discrimination task [33]. In this task, one of three possible form stimuli was presented at each trial, and the monkey was required to respond to one but not the other two. Lilienthal et al. [54] studied the effects of developmental lead exposure on learning-set formation, in which a series of visual discrimination problems was learned sequentially. Rhesus monkeys were exposed to lead *in utero* and continuing during infancy at doses sufficient to produce blood lead concentrations up to 50 $\mu g/dl$ in the lower dose group and 110 $\mu g/dl$ in the high dose group. When tested as juveniles, both groups of lead-exposed monkeys displayed impaired improvement in performance across trials on any given problem, as well as impaired ability to learn successive problems more quickly as the experiment progressed. Such a deficit represents impairment in the ability to take advantage of previous exposure to a particular set of rules. This deficit is reminiscent of the failure of lead-treated monkeys to improve as quickly as controls over a series of discrimination reversals.

Concurrent discrimination performance was assessed in the group of monkeys described above in which the contribution of the developmental period of exposure to the behavioral toxicity of lead was explored by exposing them to lead continuously from birth, during infancy only, or beginning after infancy [55]. Monkeys were required to learn a set of six problems concurrently; after criterion was reached on all six pairs, a second set of six was introduced. All three treated groups learned more slowly than controls, although monkeys dosed during infancy only were less impaired than the other two groups. Treated monkeys were most impaired on the first task, upon introduction of a new set of contingencies. In addition, all three treated groups exhibited perseverative behavior, responding incorrectly more often than controls at the same position that had been responded on in the previous trial.

Rats exposed to lead via the dam's milk until weaning at 21 days of age, with blood lead levels following 20 days of exposure of 11 (control), 29, or 65 µg/dl, were impaired on the acquisition of a light-dark simultaneous visual discrimination (i.e., the discriminative stimuli were presented at the same time) at 120 days of age [56]. Treated groups were not impaired on a successive visual discrimination task beginning at 270 days of age, nor on what was termed a go/no-go discrimination task beginning at 330 days of age. In the successive discrimination, one of two possible stimuli were presented in a trial (light on or off), and the rat was required to turn left or right in a maze, depending on the stimulus. The go/no-go discrimination task was actually an FR 20/extinction (TO) task, in which responses were reinforced in the presence of a light but not in its absence. Following acquisition, a reversal was implemented in which responses were reinforced only in the absence of light. There was a trend toward retarded acquisition of the reversal that did not reach statistical significance. Because performance on the three tasks was tested at such different ages, it is impossible to know whether the tasks were differentially susceptible to disruption by lead or whether the effects of lactational exposure to lead were at least partially reversible.

Olfactory discrimination reversal was examined in rats exposed to lead during gestation and lactation, or during lactation only (two doses) [57]. Blood lead levels at weaning were very high: 130 to 160 µg/dl. Rats performed three reversals following initial acquisition. There were no differences in the total number of errors to criterion for the initial acquisition or the three reversals despite the high blood lead levels. However, analysis of error pattern during different phases of the acquisition of the reversals revealed differences between treated and control groups. Groups were not different during the initial phase of the reversal, in which responses were made predominantly to the previously rewarded stimulus. This was termed the "perseverative phase" by the authors. Subsequent performance was divided into a "chance" phase, a "postchance" phase (between 63 and 88% correct), and the "final" phase. The "postchance" phase was significantly longer for the treated groups. "Response bias" was defined as 12 or more successive responses on the same lever. The group exposed gestationally and lactationally exhibited response bias by this strict criterion, as well as more lenient criteria of strings of five or eight responses. (Note that in discussion of results from our laboratory, "perseveration" for position is defined as all additional incorrect responses after the first, a quite different definition, and less strict than the definition of lever bias by Garavan et al. [57].) The authors argue that the increased errors in the postchance phase are the result of lever bias and "impaired ability to associate cues and/or actions with effective consequences."

Similar results were found over a series of five olfactory discrimination reversals in rats exposed from conception onward with blood lead concentrations of 28 or 51 µg/dl [58]. Lead-treated groups were also impaired on an "extradimensional shift" task immediately following the olfactory reversal task, in which olfactory cues were present but the relevant stimulus domain was spatial (left or right odor-delivery port). These results are reminiscent of results in monkeys on discrimination reversal tasks in which the relevant stimulus dimension was changed. However, the results of this study are less straightforward, since treated groups were still impaired on the last reversal of the olfactory task. In the monkey studies, groups did not differ over a number of reversals before the shift.

In a study of the siblings described in the previous paragraph [59], rats exposed chronically to lead beginning prenatally were tested on a three-choice visual discrimination task as adults. Blood lead concentrations were 26 and 51 µg/dl both at birth and during adulthood in the two treated groups. Performance was analyzed during the "chance phase" and "postchance phase," the latter defined as the point at which percent correct was greater than 46% in a session. Treated groups required more trials to criterion for both the chance and postchance phases. The authors interpreted the results as indicative of an associative deficit, although deficits in attentional processes may be involved. It is interesting that rats were impaired over the entire course of the reversal in the visual discrimination but not the olfactory discrimination tasks. In contrast to humans, rats rely more heavily on the olfactory system than the visual system for information about the environment. (Rats have keen olfactory capabilities and rather poor spatial vision.) Therefore, the olfactory discrimination task may be more "natural" for them than visual discrimination tasks.

An intermittent schedule of reinforcement was used to examine the ability of squirrel monkeys exposed *in utero* to lead to change response strategy in response to changes in reinforcement density [60]. Monkeys whose mothers had blood lead levels of 21 to 79 µg/dl were tested at 5 to 6 years of age on a concurrent random interval–random interval schedule, in which two random interval (RI) schedules operated separately on two levers. (On an RI schedule, reinforcements are available at unpredictable times, but with some average time such as 15 sec.) Reinforcement densities were varied across the experiment in such a way that the left or right lever was programmed to produce a greater reinforcement density. Under steady-state conditions, monkeys exposed *in utero* to over 40 µg/dl lead in maternal blood were insensitive to the relative "payoff" on the two levers, and exhibited lever bias (responding on a favorite lever irrespective of schedule contingencies). When the relative reinforcement densities on the levers changed, control monkeys gradually switched their responding pattern to the appropriate ratio (e.g., 70% right, 30% left). In contrast, performance of these lead-exposed monkeys changed slowly, not at all, or in the wrong direction (Figure 6.2). Monkeys whose mothers had lower blood lead concentrations learned to apportion their responses appropriately, but they learned at a slower rate than controls. The results were interpreted as "insensitivity to changing reinforcement contingencies" (p. 6) and "insensitivity to changes in the

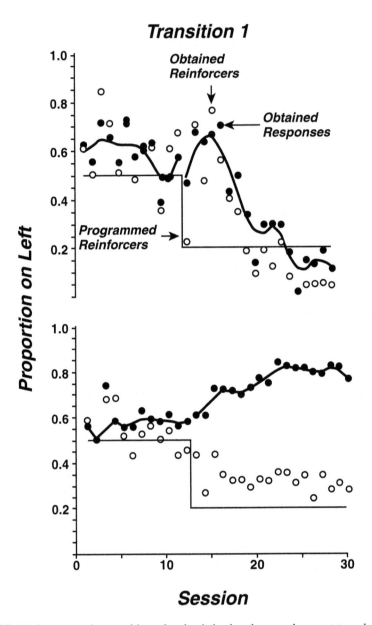

FIGURE 6.2 Representative transitions showing behavior change subsequent to a change in the reinforcement densities on the two levers for a control monkey (top) and lead-exposed monkey (bottom) on a concurrent RI-RI schedule. The ordinate is relative response rates on the left lever. The thin line shows programmed relative reinforcement rates. Open circles show obtained relative reinforcement rates on the left lever (left-lever reinforcers divided by all reinforcers). Filled circles show obtained relative response rates on the left lever, and the thick line is a smoothed (using Lowess smoothing) version of these data. Transition behavior was smooth for the unexposed monkey but pathological for the lead-exposed monkey. (Taken from Newland, M.C. et al., *Toxicol. Appl. Pharmacol.*, 126, 6, 1994. With permission.)

consequences of behavior" (p. 11). These results are consistent with results on other tasks, described above, in which lead-treated animals persisted (perseverated) in nonadaptive response patterns, seemingly unresponsive to changing environmental contingencies or the consequences of their own behavior.

The effect of lead exposure was assessed in a visual discrimination reversal task in 6- to 15-year-old children [61]. Pairs of twins discordant for blood lead concentrations were tested on a size discrimination task and one reversal. Average blood lead concentrations were 30 to 50 $\mu g/dl$ for the lower-lead twins and 43 to 80 $\mu g/dl$ for the higher-lead twins. The higher-lead twins had a lower percentage of correct responses and made more errors reaching a criterion of 100 correct responses. The testing time required was only 20 minutes. These results suggest that effects in animals are congruent with those in children on this task, and that this task might be a useful addition to testing paradigms in children.

The Cambridge Neuropsychological Testing Automated Battery (CANTAB) was used to assess cognitive function in 5.5-year-old children in relation to average lifetime blood lead concentrations [62]. This battery is a computer-based set of cognitive tests, including tests of attention, spatial and nonspatial memory, and executive function. Blood lead concentrations were associated with poorer performance on "intradimensional" and "extradimensional" shift. In this task, the original discrimination required attention to colored (filled) shapes. Following acquisition, a reversal for shape was instituted (intradimensional shift). Irrelevant stimuli (white lines) were then introduced. The stimulus class was then changed from filled shapes to white lines (extradimensional shift). This task is virtually identical to the nonspatial discrimination reversal task with irrelevant cues assessed with monkeys, and the congruence of effects in children and animal models is reassuring.

The evidence for learning impairment in animals exposed to lead is extensive, and the conditions under which it occurs are relatively well characterized. Lead exposure may produce impairment on acquisition on difficult discriminations or at higher lead levels. The requirement to reverse a previously learned discrimination, a change in the stimulus dimension or response class, or the introduction of novel stimuli (irrelevant cues) all may result in impaired performance, even at low blood lead concentrations. Lead-exposed children were also impaired when required to shift stimulus dimension [62]. However, most studies in children have used endpoints that were a terminal result of learning (school performance) or a compilation of a number of processes, including various types of learning (IQ). Therefore the experimental literature is more informative than the epidemiological literature in the elucidation of the behavioral mechanisms underlying the observed learning impairment produced by lead.

MEMORY

In contrast to the substantial evidence for deficits in learning produced by lead exposure in animals, interpretation of the studies designed to assess memory is more difficult. There is no question that lead produces impairment on such tasks, but whether the deficits are the result of impairment in memory is less clear. Rats exposed *in utero* or pre- plus postnatally to lead, with blood lead concentrations of 34 $\mu g/dl$, were impaired on the retention of a size discrimination task 42 days after initial task acquisition in the absence of deficits on initial acquisition [63]. Performance in a radial arm maze (RAM) was also assessed in these rats. The RAM apparatus consisted of a central compartment with eight alleys radiating from it like spokes of a wheel. Food reinforcers were placed in each arm of the maze. Treated groups took longer to eat all the pellets during the initial acquisition and a retention task four weeks later. However, treated groups were not different from controls on the number of arms visited before the first error (i.e., entering an alley already entered), nor did lead exposure affect the number of arms visited on the first eight choices. It therefore appears that the lead-treated rats exhibited no deficit in spatial memory, although they were impaired on retention of the size discrimination task. Lactational exposure to lead at levels that produced drastic effects on weight gain and overt signs of toxicity also had minimal effect on radial arm maze performance [30].

The effects of developmental lead exposure were assessed on performance in the Morris water maze [64, 65]. The Morris water maze requires the subject to learn the location of a submerged platform to escape submersion in a pool of water. In the first experiment, rats exposed to lead throughout gestation and lactation showed increased time to find the hidden platform at weaning, but not at 56 or 91 days of age. In the second experiment, rats were exposed during gestation and/or lactation at three times the dose as the first experiment, resulting in blood lead concentrations at weaning of 60 μg/dl. Performance on the Morris water maze was assessed beginning at 100 days of age. Deficits in escape latency and increased swim path length were observed in the group exposed prenatally only, but not groups exposed pre- plus postnatally or postnatally only, despite the high blood lead concentrations. The results of these studies taken together suggest that spatial memory may be relatively unaffected by lead exposure in rodents. The Morris water maze may be an insensitive test, since normal rats acquire the performance in four to five trials. However, the radial arm maze is a more difficult task that has proved sensitive to contaminants such as polychlorinated biphenyls (PCBs) [66, 67].

A task used in monkeys that is conceptually similar to the radial arm maze is the Hamilton Search Task. In this task, a row of boxes is baited with food and then closed. The monkey lifts the lids to obtain the food. The most efficient performance requires that each box be opened only once, necessitating that the monkey remember which boxes have already been opened. Monkeys exposed postnatally to doses of lead sufficient to produce blood lead levels of approximately 45 or 90 μg/dl or *in utero* at blood lead concentrations of 50 μg/dl were impaired in their ability to perform this task at 4 to 5 years of age [68]. These results were replicated in another group of monkeys exposed postnatally to higher lead levels and tested at 5 to 6 years of age [69]. However, error pattern was not analyzed, so it is impossible to know whether the results are due to a memory impairment *per se* or a nonadaptive response strategy such as perseveration or position bias.

A task that has proved particularly sensitive to disruption by lead exposure in monkeys is the delayed spatial alternation task. In this task, the subject is required to alternate responses between two positions; there are no cues signaling which position is correct on any trial. Delays may be introduced between opportunities to respond in order to assess spatial memory. Rhesus monkeys exposed to lead from birth to 1 year of age, with peak blood levels as high as 300 µg/dl and levels of 90 µg/dl for the remainder of the first year of life, were markedly impaired on this task as adults [68]. Delays between 0 and 40 sec were assessed within each session; a greater deficit was observed at shorter rather than longer delay values. This indicates that the poorer performance of the lead-exposed monkeys was not the result of a memory impairment, but rather some type of associative deficit.

In our laboratory, increasingly longer delays were introduced over successive sessions in adult monkeys from two studies, those with steady-state blood lead levels of 11 or 13 µg/dl [70] and the groups in which potential sensitive periods were assessed (dosed during infancy only, beginning after infancy, or continuously from birth [71]). In contrast to the study discussed above, the task included a "correction" procedure, such that if the monkey responded incorrectly on a button, a correct response on the opposite button was required before the alternation schedule resumed. Sessions consisted of 100 correct trials; thus each incorrect response extended the session. All treated groups in both studies were impaired on the acquisition of this task because of indiscriminate responding on both buttons. Treated monkeys were impaired at the beginning of the experiment (short delays), unimpaired at intermediate delay values, and increasingly more impaired at the 5- and 15-sec delays. In the study assessing sensitive periods, all three lead-exposed groups were impaired to an approximately equal degree, as was the case on the spatial version of the discrimination reversal task, thus providing further evidence of a lack of sensitive period for lead-induced impairment on spatial tasks. In addition, treated monkeys in this latter study responded more during the delay periods than did controls, indicating failure to inhibit inappropriate responding. However, analysis of error pattern revealed that this was not responsible for the increased number of errors in the treated group. Treated monkeys in both studies also displayed marked perseveration for position, responding on the same position repeatedly, in some instances for hours at a time (Figure 6.3). Because of the marked perseverative behavior displayed by some treated monkeys, it was actually not possible to assess memory capabilities at the 5- and 15-sec delay value. Memory impairment certainly does not account for the poor performance of the treated groups in these experiments.

Spatial delayed alternation performance was examined in rats with chronic postweaning exposure to lead, with blood lead levels of 19 and 39 µg/dl in the two treated groups [72]. Testing began at 52 weeks of age. Lead-treated groups were not impaired in the acquisition of the alternation task. Following acquisition, delays of 0, 10, 20, or 40 sec were presented within each session. Treated groups exhibited an impairment of constant magnitude across all delays, suggesting that the performance deficit was not the result of memory impairment. Exploration of error pattern revealed that the higher dose group exhibited position bias. Analysis of the error pattern with respect to the effect of the actual delay time consequent to the rat responding during the delay (which reset the time) and the influence of whether the previous response had been correct or incorrect revealed no lead-related differences. The observed position bias is consistent with effects in monkeys observed on a number of tasks, including spatial delayed alternation.

Improved performance on delayed alternation was observed in young and old rats but not in rats exposed as adults [73]. The training procedure in this study

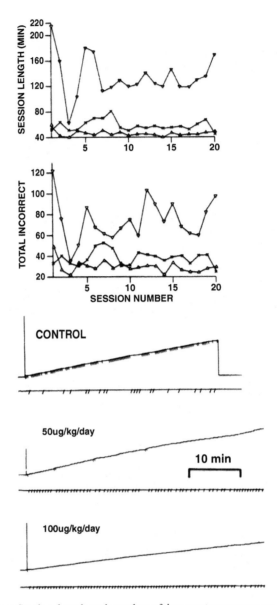

FIGURE 6.3 Top: Session length and number of incorrect responses on a delayed spatial alternation task in monkeys for all sessions at the 15-sec (longest) delay. Each point represents the geometric mean for the dose group: <control; % low dose; = high dose. Bottom: Sample cumulative records for the 15-sec delay for each of the treatment groups. Each response stepped the top pen upward. Each correct response is indicated by a downward deflection of the top pen; each incorrect response by a downward deflection of the lower pen. The control record is of average length, indicated by the top pen returning to baseline. The other two records represent only a small part of a single session, since a session consisted of 100 correct trials. Treated monkeys exhibited marked perseveration on a preferred position. (Taken from Rice, D.C. and Karpinski, K.F., *Neurotoxicol. Teratol.*, 10, 207, 1988. With permission.)

consisted of many sessions of a cued alternation procedure; i.e., the rat had only to respond on the lever associated with a cue light as it alternated between positions from trial to trial. The authors interpreted the improved performance of the lead-treated groups as perseveration of the alternation behavior as a result of the extensive training procedure.

A delayed matching to sample task was used to assess both spatial and nonspatial memory in a group of monkeys with preweaning blood lead values of 50 μg/dl and postweaning values of 30 μg/dl [74]. In the nonspatial version of the task, one of three colors appeared on a sample button on which the monkey responded a specified number of times, which turned off the stimulus and initiated a delay. After a delay period, one of the three colors appeared on each of the three test buttons, and the monkey was required to respond on the button corresponding to the sample color. Colors were balanced for position and correct choice across trials. For the spatial version, one of the three test buttons was lit green; response on that button a specified number of times turned it off and instituted a delay. Following the delay, all three buttons were lit green, and the monkey was required to respond on the sample position. The nonspatial task was assessed twice, with the spatial task in between. For the first assessment of the nonspatial task and spatial task, a series of sessions with increasing constant delays from 1 to 32 sec was instituted. Following the constant-delay sessions, a series of variable-delay sessions was instituted on each task, consisting of a short delay and increasingly long delays to determine the value at which performance reached chance. Lead-exposed monkeys were impaired on both the spatial and nonspatial versions of this task. They were not impaired in their ability to learn the matching tasks or at 0-sec delay. On the constant-delay sessions, treated monkeys had poorer overall performance than controls on the spatial but not the nonspatial task. On the variable-delay schedules, lead-exposed monkeys reached chance performance on the nonspatial task at shorter delays than controls the first time it was tested. When performance on the nonspatial task was assessed the second time, there was no overlap in the delay value at which chance performance was reached (Figure 6.4). The delay at which chance performance was reached in the spatial variable-delay sessions was marginally significant. Treated monkeys were not different from controls on the short delay in the variable-delay sessions.

Investigation of the error pattern revealed that for the nonspatial matching task, lead-exposed monkeys responded incorrectly on the position that had been responded on correctly on the previous trial. This type of behavior may be considered to represent perseverative behavior and is reminiscent of the perseverative errors in other groups on delayed alternation. On the other hand, it may be considered to be the result of increased distractibility by irrelevant cues by lead-treated monkeys, similar to the increased attention to irrelevant cues displayed in the discrimination reversal tasks. (These interpretations are not mutually exclusive.) This behavior is at least partly responsible for the poorer performance at long delays observed in lead-treated monkeys on the nonspatial matching-to-sample task, although other mechanisms may also play a part. The lack of interference from previous trials on the spatial version of the task, however, may indicate a pure deficit in spatial short-term memory. The fact that lead-exposed monkeys were not impaired at 0-sec delay or on the short delays on the variable-delay tasks, but reached chance performance

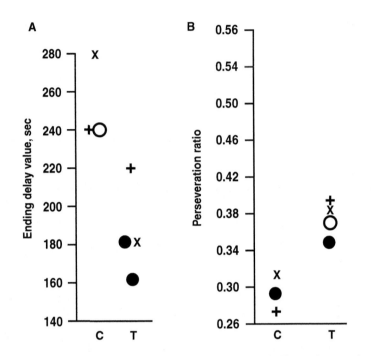

FIGURE 6.4 Delay value at which control (C) and lead-treated (T) monkeys performed at chance levels on a nonspatial delayed matching-to-sample task (A) and the ratio of incorrect responses made on the button that had been responded to correctly in the previous trial (B). Each symbol represents an individual monkey. These results suggest that perseveration for position interfered with performance in lead-exposed monkeys. (Taken from Rice, D.C., *Environ. Health Perspect.*, 104, 337, 1996, and from Rice, D.C., *Toxicol. Appl. Pharmacol.*, 75, 337, 1984. With permission.)

at shorter delay values, provides evidence overall for a deficit in memory in these monkeys. The deficit across the constant-delay sessions of the spatial task, on the other hand, may be a consequence of the change in relevant stimulus dimension from nonspatial to spatial.

Assessment of spatial learning and memory in the rat revealed an interesting pattern of errors responsible for the overall poorer performance of lead-treated subjects [75]. Rats were exposed to lead in drinking water beginning at weaning, with blood lead concentrations of 25 and 73 µg/dl, and tested beginning at 55 days of age on a task with two components. The repeated acquisition component of the schedule required the rat to learn a new sequence of lever presses every day. In the performance component, the rat was required to perform the same sequence of lever presses during every session. Significant impairment of performance was observed on the repeated acquisition component but not on the performance component in lead-exposed rats compared with controls. Analyses of error patterns revealed that the decrease in the percent of correctly completed sequences in lead-treated rats on the repeated acquisition component was the result of specific types of perseverative behavior.

Apparently little attention has been paid to the effects of lead on memory in epidemiological studies. Exploration of the effects of lead on working memory in children can be included as part of a battery of cognitive tests, such as an assessment of IQ. However, scores on various subtests of standard IQ tests (e.g., Digit Span) were often not reported in epidemiological studies of the effects of lead. The California Verbal Learning Test (CVLT) and Story Recall were assessed in the Boston prospective cohort at 10 years of age [76]. There was little evidence of effect on either test. Blood lead levels of the children at 24 months of age were marginally associated with poorer performance on Digit Span in the Wechsler Intelligence Scale for Children – Revised (WISC-R). Digit span score on the WISC-III was associated with concurrent blood lead concentrations in 7.5-year-old children [77], whereas lead was not associated with deficits on story memory or verbal learning. However, performance on Digit Span reflects attentional and higher-order sensory processes at least as much as it does memory.

Deficits as a function of lead exposure in children were observed on two tests of spatial memory on the CANTAB battery [62]. One test was logically equivalent to the Hamilton Search Task, described above, in which impairment was observed in lead-exposed monkeys [68, 69]. In the other task, the child was required to remember the sequence in which stimuli (boxes) changed color on the computer screen. The sequence varied from trial to trial. This task is logically equivalent to the repeated acquisition task on which deficits were observed in lead-exposed rats [75].

It is difficult to draw conclusions concerning the congruence of effects of lead on memory in animals and humans. Memory has been little explored in children, and most tests employed relied on language and assessed encoding and higher-order auditory processing rather than memory *per se*. In animals, perseverative responding to position and interference from responses on previous trials were observed in a number of studies in various tasks, and were responsible for, or at least contributed to, the poorer performance by lead-exposed animals. In addition, lead-treated animals were impaired to an equal degree across all delay values in some studies, suggesting a deficit other than memory impairment. Perhaps the strongest evidence for impaired memory is the results of the delayed matching to sample task in monkeys, in which treated monkeys were not impaired at short delays but reached chance performance at shorter delays than controls. However, the overall evidence for memory deficits produced by lead in experimental studies is not compelling.

ATTENTION

Nonadaptive behavior in children has repeatedly been interpreted as deficits in attention in human studies. Increased lead body burden was associated with increased inattentiveness, distractibility, impulsivity, and lack of persistence on teachers' and parents' rating scales [7, 41, 42, 77, 78]. Impairment on the "learning" studies in animals described above may result at least in part from deficits in attentional processes. The increased systematic response to irrelevant cues on discrimination tasks may reflect attentional deficits. Inability to inhibit responding, such as on the DRL task, may also represent a type of attention deficit. However, responding during delays on trial tasks was observed in some circumstances but not others. For example, in the delayed alternation task in which some lead-treated monkeys perseverated on the same response button for hours at a time, no increase in delay responses was observed at lower blood lead concentrations [70], and delay responses did not contribute to the increased error rate at higher blood lead concentrations [71]. Perseveration has also been interpreted as indicative of attentional impairment [75, 79].

The deficits in ability to change response strategy in response to new schedule contingencies, observed in the monkey studies, may also be considered to represent an attention deficit. For example, a task that has been used in lead-exposed children is the Wisconsin Card Sort Test (WCST). This task is logically equivalent to the discrimination reversal task with irrelevant cues used in monkeys, in that it requires the ability to extract general rules and change response strategy. In this task, correct responses depend on generalizing whether the relevant domain is color, number, or shape. The investigator can change the relevant stimulus class at any time; the subject must infer the rule by whether a series of responses is correct or incorrect. Errors on the WCST are considered to test ability to shift attention, which is considered one of the factors in at least one theoretical framework of attention [80, 81]. Increased total and perseverative errors at 10 years were related to 57-month blood lead concentrations in the Boston prospective study [76], and perseverative errors were related to dentine lead levels in the first or second grade in 19- and 20-year-olds [82]. An increased number of errors was also observed on the WCST in a study of 246 7.5-year-olds; perseverative errors were not different after control of confounders [77]. These results are similar to those observed in monkeys, in which the lead-exposed groups made more errors than controls in initial reversals within a set of reversals, and responded more than controls to previously relevant stimuli when the relevant stimulus domain was changed [50, 51].

The effects of lead on sustained attention in children were explored in a number of studies using reaction time tasks, either on a simple reaction time task, in which the subject is asked to respond as quickly as possible to a single stimulus, or on a vigilance task, in which the subject is required to respond to a target stimulus and refrain from responding to others. Stimuli are presented at variable intervals, and the task is sufficiently long to require sustained attention. Increased reaction times were observed in groups of children with blood lead concentrations of 17, 24, or 35 µg/dl compared with those with average lead levels of 7.4 µg/dl [42, 83], as well as at higher blood lead concentrations [84], with the greatest impairment occurring at a 12-sec compared with a 3-sec delay, and at the end of the task compared with the beginning. Increased reaction time compared with controls was also observed on a simple reaction time task in a group of 20-year-olds with a history of lead exposure as children [85]. In contrast, no effect on simple reaction time task was observed in the group of monkeys exposed to lead from birth, with preweaning blood lead values of 50 µg/dl and steady-state blood lead levels of 30 µg/dl, when they were adults [86]. Reaction times of lead-treated monkeys did not differ from those of controls over a number of delay values, although treated monkeys exhibited an increased incidence of holding the bar longer than the maximum 15 sec allowed.

They were also able to react as quickly as controls when required to respond as quickly as possible.

Performance on a vigilance task was assessed in two cohorts of German children [87–90]. Two different signal presentation rates (approximately 1 to 2 sec) were used. Correct responses, false hits (errors of commission), and failure to respond to correct stimuli (errors of omission) were analyzed. In the first study [88, 90], there was an indication of an effect in the absence of an effect on the German WISC-R as a function of tooth lead in 9-year-old children. In a subsequent study in 6- to 9-year-old children, a robust effect was observed [87]. Effects were greater at the higher signal rate. As in the previous study, there was a greater effect on errors of commission than on errors of omission. These results have been replicated using the same device in a population of Greek children with higher blood lead levels [84]. In a later study by the German group [91], 384 6-year-old children were assessed on components of the Neurobehavioral Evaluation System 2 (NES2), including simple reaction time, pattern memory, and a vigilance task, in addition to IQ on the WISC. Increased errors of omission and commission were related to increased blood lead levels (geometric mean = 4.25 μg/dl), even after controlling for IQ. No effect was observed on simple reaction time or the memory task. Sustained attention on a vigilance task was assessed using either visual or auditory stimuli in the study of 7.5-year-old children [77]. An increased number of incorrect responses was observed as a function of increased blood lead levels in the visual but not auditory version of the task, with no increase in errors of commission. (Errors of omission were apparently not measured.) In a study in first-graders in Denmark, performance was marginally associated with dentine lead levels [92]; errors of commission showed a greater correlation than errors of omission. In contrast, Bellinger et al. [82] found no effect in 19- or 20-year-olds on reaction time or on errors of commission or omission related to dentine levels in childhood.

Vigilance performance was studied in rats exposed to lead beginning at weaning, with blood lead concentrations of 16 or 28 μg/dl [93]. The low-dose group made more errors of commission during sessions with long or variable interstimulus intervals, whereas the high-dose group showed an increase in errors of commission only during sessions with long interstimulus intervals. These inconsistent effects were interpreted as evidence of no effect of lead on sustained attention.

Effects of lead exposure were reported on a vigilance task in rats exposed either during lactation only (two groups) or during gestation and lactation [94]. Blood lead concentrations in all three groups were very high on day 24 (131 to 158 μg/dl for the three dose groups), and may have been even higher by weaning at 30 days of age. These lead levels were sufficiently high to produce impaired body weight gain. Groups did not differ for the percent correct responses or on responses in the absence of a stimulus. Treated groups exhibited increased omission errors at stimulus delays greater than 0 sec, and on trials following an incorrect response. However, the magnitude of effect was modest, despite the very high blood lead levels.

An FR schedule was used as one component of a study to explore underlying behavioral mechanisms responsible for the observation that lead produces inappropriate responding on a wide array of tasks [95]. Exposure began at weaning, with blood lead concentrations of 11 and 29 μg/dl in the two dosed groups. Rats were trained, after 40 days of exposure, to respond on an FR 50 schedule of reinforcement (i.e., 50 responses were required for reinforcement). Following each FR, the rat could receive "free" reinforcements for waiting an increasingly long period of time (2, 4, 6, etc., sec). Lead-exposed rats exhibited faster response rates on the FR, and they waited a shorter period of time in the "free" reinforcement portion of the schedule before responding, which initiated another FR. The lead-treated rats actually received more reinforcements. The results were interpreted as indicative of "an inability to manage delays in reinforcement." It is also possible, however, that the pattern of responses of the lead-exposed rats was maintained by the higher reinforcement density produced (which the authors acknowledge). The results of this study are not necessarily inconsistent with other studies documenting failure to inhibit responding by lead-exposed subjects. In fact, the results of this study are more difficult to interpret because the response strategy adopted by the lead-exposed rats was adaptive rather than nonadaptive or neutral, as in most other studies.

The effects of lead on sustained attention as measured by simple reaction time or vigilance tasks seem inconsistent between humans and animals. Increased response time on simple reaction time tasks was observed in children in four studies, but not in a fifth study at very low blood lead concentrations and not in monkeys at relatively high blood lead levels. However, the monkeys had a substantial history of performance on a number of behavioral tasks, such that the simple reaction time task may have made minimal demands on attention in such experienced monkeys. On vigilance tasks, effects were observed in six studies in children, including one with very low lead concentrations, but not in a seventh. Two studies in rats were largely negative, including one at high blood lead concentrations. The reason for negative effects observed in rats is somewhat unclear. However, the number of sessions required to train the task was substantial in both studies. Each study included contingencies that resulted in a low rate of stimulus presentation. It may be that the extensive familiarity with the task or the low rate of stimulus presentation rendered the task insensitive to disruption by lead.

SENSORY DYSFUNCTION: POSSIBLE CONTRIBUTION TO "COGNITIVE" EFFECTS

It is clear that lead exposure in animals and humans may result in impairment in visual and auditory function. An inability to perceive or process sensory information would obviously interfere with performance on any particular test. More importantly, however, deficits in higher-order sensory processing would make it difficult for the child to learn and respond appropriately to the environment. In fact, it is difficult to define the point at which deficits in sensory processing may be defined as "cognitive" deficits. Whereas there have been studies of first-order sensory function in both animals and children, higher-order sensory processing has been studied in very few instances in either children or animals.

Elevated auditory thresholds in the range of speech frequencies were observed as a function of increasing blood lead concentrations in children in the Second National Health and Nutritional Survey (NHANES) II [96]. A study in Poland documented higher auditory thresholds across a range of frequencies as a function of blood lead levels, including blood lead concentrations below 10 μg/dl [97]. Monkeys exposed to lead from birth, with blood lead concentrations of 30 μg/dl at the time of testing, exhibited impaired detection of pure tones at 13 years of age, with the pattern of impairment being somewhat idiosyncratic [98]. A number of electrophysiological studies in animals demonstrated impairment at various parts of the auditory pathway in monkeys and other animals related to lead exposure [99–103], although electrophysiological changes have not been universally observed [104].

Pure-tone thresholds provide only basic, first-level information concerning auditory function. For example, an individual may have normal pure-tone detection and still have difficulty distinguishing speech. Speech generally comprises small but rapid changes in frequency and amplitude. Frequency and amplitude difference thresholds (i.e., the threshold for detection of a change in frequency or amplitude) have been assessed in various animal species, including monkeys and guinea pigs, and may be more sensitive to disruption by toxic exposure than pure tone detection thresholds [105]. A test that is used clinically to evaluate auditory (language) processing is the determination of the infant's ability to distinguish "ba" and "da" and, at a later age, "bi" and "di." Monkeys can also discriminate human speech sounds. In a study with rhesus monkeys, there was evidence that developmental lead exposure impaired the ability of young monkeys to discriminate speech sounds ("da" and "pa") based on an electrophysiological procedure [106]. Lead-exposed children are impaired on the Seashore Rhythm Test [7, 77], which requires the subject to discriminate whether pairs of tone sequences are the same or different. This is a simplified discrimination compared with analysis of speech sounds. Lead-exposed children also had a decreased ability to identify words when frequencies were filtered out (i.e., when information was missing [107]). Increased lead body burdens were associated with impaired language processing on difficult but not easy tasks [108], with impairment of the development of word recognition [109], and with impaired auditory comprehension [110]. The degree to which the deficits in language development are the result of deficits in higher-order auditory processing is unknown.

Lead also produces deficits in visual function. Altmann et al. [111] reported changes in visual evoked potentials in a cohort of over 3800 4-year-old children with average blood lead concentrations of 42 μg/dl, but no differences in visual acuity or spatial contrast sensitivity functions. In a study in infant monkeys with very high lead levels (300 to 500 μg/dl), one infant appeared to develop temporary blindness [112]. Monkeys exposed to lead during the first year of life with blood lead concentrations during exposure of 85 μg/dl, but not those with 60 μg/dl, had impaired scotopic (very low luminance) spatial contrast sensitivity at 3 years of age [113]. Both spatial and temporal (motion) contrast sensitivity were examined in monkeys with lifetime exposure to lead, with steady-state blood lead concentrations of 25 to 35 μg/dl [114]. Lead-exposed monkeys exhibited deficits in temporal vision at low and middle frequencies under low-luminance conditions, with no other impairments. Lead-induced changes in electrophysiologic responses have also been observed in monkeys exposed developmentally to lead, particularly under low-luminance conditions [100, 115, 116]. Consistent with these findings, rod function was found to be preferentially affected by lead in rats [103].

Only the most basic level of visual function has been assessed in either children or animals with respect to potential effects of lead exposure. Half of the primate brain is devoted to visual processing, yet neither experimental nor epidemiological researchers have explored the effects of lead exposure in higher-order visual processing.

Lead impairs primary sensory function in both animal models and children. The only study higher-order function in animals (discrimination of speech sounds) identified impairment produced by lead. There are also hints from studies in children of sensory system impairment. All of the "cognitive" endpoints assessed in children require higher-order processing in the visual or auditory systems, including tests of memory, attention, executive function, learning, or any other categorization of performance. Yet there has been virtually no exploration of higher-order sensory processing in either children or animal models; therefore, the degree to which deficits identified as cognitive are actually sensory is unknown. This failure is of considerable practical relevance, since intervention strategies may be different depending on the actual nature of the deficit.

CONCLUSIONS

Lead-induced impairment in learning has been observed in numerous studies in animals. Performance on simple discrimination problems is impaired at high blood lead concentrations. Introduction of a requirement to shift response strategies revealed lead-induced impairment, often in the absence of impairment on initial task acquisition. These results suggest that impairment in attentional processes (attentional shift) underlies the learning deficits, at least in some cases. Deficits in attentional shift were also observed in lead-exposed children. In contrast, the evidence for deficits in sustained attention in animals is weak, despite the fact that deficits on similar tasks were observed in lead-exposed children. Most studies of effects of lead on memory in animals were either confounded by perseveration of various sorts or provided evidence for an associative rather than a memory deficit. In fact, a number of investigators proposed that one of the effects of lead is a failure to associate behavior (response) and the consequences of that behavior. Memory has received little attention in epidemiological studies, and the tests that to some degree assessed working memory also assessed attention and higher-order sensory processing. The effects of lead on intermittent schedules of reinforcement in animals predicted performance in lead-exposed children or were consistent with effects in children with attentional deficits. Unfortunately, the effects of lead on higher-order sensory processing have received scant attention in epidemiological or experimental studies. It has been known for a couple of decades that lead impairs primary auditory and visual function. Nevertheless, the extent to which sensory impairment from lead exposure contributes to what is broadly labeled as "cognitive" deficits in children remains unknown.

REFERENCES

1. Byers, R.K. and Lord, E.E., Late effects of lead poisoning on mental development, *Am. J. Dis. Child*, 66, 471, 1943.
2. Thurston, D.L., Middelkamp, J.N., and Mason, E., The late effects of lead poisoning, *J. Pediatr.*, 47, 413, 1955.
3. Jenkins, C.D. and Mellins, B.B., Lead poisoning in children, *AMA Arch. Neurol. Psychiatry*, 77, 70, 1957.
4. Perlstein, M.A. and Attala, R., Neurological sequelae of plumbism in children, *Clin. Pediatr.*, 5, 282, 1966.
5. Lin-Fu, J.S., Undue absorption of lead among children — a new look at an old problem, *N. Engl. J. Med.*, 186, 702, 1972.
6. de la Burdé, B. and Choate, M.S., Early asymptomatic lead exposure and development at school age, *J. Pediatr.*, 87, 638, 1972.
7. Needleman, H.L., Gunnoe, C., Leviton, A., Reed, R., Peresie, H., Maher, C., and Barrett, P., Deficits in psychologic and classroom performance of children with elevated dentine lead levels, *N. Engl. J. Med.*, 300, 689, 1979.
8. Rice, D.C., Behavioral effects of lead: commonalities between experimental and epidemiological data, *Environ. Health Perspect.*, 104, 337, 1996.
9. Canfield, R.L., Henderson, M.A., Jr., Cory-Slechta, D.A., Cox, C., Jusko, T.A., and Lanphear, B.P., Intellectual impairment in children with blood lead concentrations below 10 µg per deciliter, *N. Engl. J. Med.*, 348, 1517, 2003.
10. Lanphear, B.P., Hornung, R., Khoury, J., Yolton, K., Baghurst, P., Bellinger, D.C., Canfield, R.L., Dietrich, K.N., Bornschein, R., Greene, T., Rothenberg, S.J., Needleman, H.L., Schnaas, L., Wasserman, G., Graziano, J., and Roberts, R., Low-level environmental lead exposure and children's intellectual function: an international pooled analysis, *Environ. Health Perspect.*, 113, 894, 2005.
11. Rice, D.C., Quantification of operant behavior, *Toxicol. Lett.*, 43, 361, 1988.
12. Cory-Slechta, D.A., Implications of changes in schedule-controlled behavior of rodents correlated with prolonged lead exposure, in *Neurobehavioral Toxicity: Analysis and Interpretation*, Weiss, B. and O'Donoghue, J., Eds., Raven Press, New York, 1994, p. 195.
13. Cory-Slechta, D.A., Weiss, B., and Cox, C., Delayed behavioral toxicity of lead with increasing exposure concentration, *Toxicol. Appl. Pharmacol.*, 71, 342, 1983.
14. Cory-Slechta, D.A., Weiss, B., and Cox, C., Performance and exposure indices of rats exposed to low concentrations of lead, *Toxicol. Appl. Pharmacol.*, 78, 291, 1985.
15. Cory-Slechta, D.A., Brockel, B.J., and O'Mara, D.J., Lead exposure and dorsomedial striatum mediation of fixed interval schedule-controlled behavior, *Neurotoxicology*, 23, 313, 2002.
16. Cory-Slechta, D.A. and Thompson, T., Behavioral toxicity of chronic postweaning lead exposure in the rat, *Toxicol. Appl. Pharmacol.*, 47, 151, 1979.
17. Cory-Slechta, D.A. and Pokora, M.J., Behavioral manifestations of prolonged lead exposure initiated at different stages of the life cycle: I. Schedule-controlled responding, *Neurotoxicology*, 12, 745, 1991.
18. Rice, D.C., Gilbert, S.G., and Willes, R.F., Neonatal low-level lead exposure in monkeys (*Macaca fascicularis*): locomotor activity, schedule-controlled behavior, and the effects of amphetamine, *Toxicol. Appl. Pharmacol.*, 51, 503, 1979.
19. Rice, D.C., Lead exposure during different developmental periods produces different effects in FI performance in monkeys tested as juveniles and adults, *Neurotoxicology*, 13, 757, 1992.

20. Rice, D.C., Schedule-controlled behavior in infant and juvenile monkeys exposed to lead from birth, *Neurotoxicology*, 9, 75, 1988.
21. Nation, J.R., Frye, G.D., Von Stultz, J., and Bratton, G.R., Effects of combined lead and cadmium exposure: changes in schedule-controlled responding and in dopamine, serotonin and their metabolites, *Behav. Neurosci.*, 103, 1108, 1989.
22. Angell, N.F. and Weiss, B., Operant behavior of rats exposed to lead before or after weaning, *Toxicol. Appl. Pharmacol.*, 63, 62, 1982.
23. Zenick, H. et al., Deficits in fixed-interval performance following prenatal and postnatal lead exposure, *Dev. Psychobiol.*, 12, 509, 1979.
24. Mele, P.C., Bushnell, P.J., and Bowman, R.E., Prolonged behavioral effects of early postnatal lead exposure in rhesus monkeys: fixed-interval responding and interactions with scopolamine and pentobarbital, *Neurobehav. Toxicol. Teratol.*, 6, 129, 1984.
25. Sagvolden, T., Aase, H., Zeiner, P., and Berger, D., Altered reinforcement mechanisms in attention-deficit/hyperactivity disorder, *Behav. Brain Res.*, 94, 61, 1998.
26. Darcheville, J.C., Riviere, V., and Wearden, J.H., Fixed-interval performance and self-control in children., *J. Exp. Anal. Behav.*, 57, 187, 1992.
27. Darcheville, J.C., Riviere, V., and Wearden, J.H., Fixed-interval performance and self-control in infants, *J. Exp. Anal. Behav.*, 60, 239, 1993.
28. Rice, D.C., Behavioral effects of lead in monkeys tested during infancy and adulthood, *Neurotoxicol. Teratol.*, 14, 235, 1992.
29. Rice, D.C. and Gilbert, S.G., Low lead exposure from birth produces behavioral toxicity (DRL) in monkeys, *Toxicol. Appl. Pharmacol.*, 80, 421, 1985.
30. Alfano, D.P. and Petit, T.L., Behavioral effects of postnatal lead exposure: possible relationship to hippocampal dysfunction, *Behav. Neurol. Biol.*, 32, 319, 1981.
31. Dietz, D.D., McMillan, D.E., Grant, L.D., and Kimmel, C.A., Effects of lead on temporally spaced responding in rats, *Drug Chem. Toxicity*, 1, 401, 1978.
32. Cory-Slechta, D.A. et al., Chronic postweaning lead exposure and response duration performance, *Toxicol. Appl. Pharmacol.*, 60, 78, 1981.
33. Hopper, D.L., Kernan, W.J., and Lloyd, W.E., The behavioral effects of prenatal and early postnatal lead exposure in the primate *Macaca fascicularis*, *Toxicol. Indust. Health*, 2, 1, 1986.
34. Cory-Slechta, D.A., Weiss, B., and Cox, C., Performance and exposure indices of rats exposed to low concentrations of lead, *Toxicol. Appl. Pharmacol.*, 71, 342, 1983.
35. Cory-Slechta, D.A., Prolonged lead exposure and fixed ratio performance, *Neurobehav. Toxicol. Teratol.*, 8, 237, 1986.
36. Rice, D.C., Schedule-controlled behavior in monkeys, in *Behavioral Pharmacology: The Current Status*, Seiden, L.S. and Balster, R.L., Eds., Alan R. Liss, New York, 1985, chap. 8.
37. Centers for Disease Control and Prevention, *A Review of Evidence of Health Effects of Blood Lead Levels <10 μg/dl in Children*, Centers for Disease Control and Prevention, National Center for Environmental Health, Atlanta, 2004.
38. Fulton, M., Raab, G., Thomson, G., Laxen, D., Hunter, R., and Hepburn, W., Influence of blood lead on the ability and attainment of children in Edinburgh, *Lancet*, 8544, 1221, 1987.
39. Fergusson, D.M., Fergusson, J.E., Horwood, L.J., and Kinzett, N.G., A longitudinal study of dentine lead levels, intelligence, school performance and behavior, part I: dentine lead levels and exposure to environmental risk factors, *J. Child Psychol. Psychiatry*, 29, 781, 1988.
40. Fergusson, D.M., Fergusson, J.E., Horwood, L.J., and Kinzett, N.G., A longitudinal study of dentine lead levels, intelligence, school performance and behavior, part II: dentine lead and cognitive ability, *J. Child Psychol. Psychiatry*, 29, 783, 1988.

41. Fergusson, D.M., Fergusson, J.E., Horwood, L.J., and Kinzett, N.G., A longitudinal study of dentine lead levels, intelligence, school performance and behavior, part III: dentine lead levels and attention/activity, *J. Child Psychol. Psychiatry*, 29, 811, 1988.

42. Yule, W., Lansdown, R., Millar, I., and Urbanowicz, M., The relationship between blood lead concentration, intelligence, and attainment in a school population: a pilot study, *Dev. Med. Child Neurol.*, 23, 567, 1981.

43. Leviton, A., Bellinger, D., Allred, E.N., Rabinowitz, M., Needleman, H., and Schoenbaum, S., Pre- and postnatal low-level lead exposure and children's dysfunction in school, *Environ. Res.*, 60, 30, 1993.

44. Canfield, R.L., Kreher, D.A., Cornwell, C., and Henderson, C.R., Jr., Low-level lead exposure, executive functioning, and learning in early childhood, *Neuropsychol. Dev. Cognit. C. Child Neuropsychol.*, 9, 35, 2003.

45. Zenick, H., Rodriquez, W., Ward, J., and Elkington, B., Influence of prenatal and postnatal lead exposure on discrimination learning in rats, *Pharmacol. Biochem. Behav.*, 8, 347, 1978.

46. Carson, T.L., Van Gelder, G.A., Karas, G.C., and Buck, W.B., Slowed learning in lambs prenatally exposed to lead, *Arch. Environ. Health*, 29, 154, 1974.

47. Winneke, G., Brockhaus, A., and Baltissen, R., Neurobehavioral and systemic effects of long-term blood lead elevation in rats, I: discrimination learning and open-field behavior, *Arch. Toxicol.*, 37, 247, 1977.

48. Bushnell, P.J. and Bowman, R.E., Reversal learning deficits in young monkeys exposed to lead, *Pharmacol. Biochem. Behav.*, 10, 733, 1979.

49. Rice, D.C. and Willes, R.F., Neonatal low-level lead exposure in monkeys (*Macaca fascicularis*): effect on two-choice nonspatial form discrimination, *J. Environ. Pathol. Toxicol.*, 2, 1195, 1979.

50. Rice, D.C., Chronic low-lead exposure from birth produces deficits in discrimination reversal in monkeys, *Toxicol. Appl. Pharmacol.*, 77, 201, 1985.

51. Rice, D.C. and Gilbert, S.G., Sensitive periods for lead-induced behavioral impairment (nonspatial discrimination reversal) in monkeys, *Toxicol. Appl. Pharmacol.*, 102, 101, 1990.

52. Gilbert, S.G. and Rice, D.C., Low-level lifetime lead exposure produces behavioral toxicity (spatial discrimination reversal) in adult monkeys, *Toxicol. Appl. Pharmacol.*, 91, 484, 1987.

53. Rice, D.C., Lead-induced behavioral impairment on a spatial discrimination reversal task in monkeys exposed during different periods of development, *Toxicol. Appl. Pharmacol.*, 106, 327, 1990.

54. Lilienthal, H., Winneke, G., Brockhaus, A., and Malik, B., Pre- and postnatal lead-exposure in monkeys: effects on activity and learning set formation, *Neurobehav. Toxicol. Teratol.*, 8, 265, 1986.

55. Rice, D.C., Effect of lead during different developmental periods in the monkey on concurrent discrimination performance, *Neurotoxicology*, 13, 583, 1992.

56. Hastings, L., Cooper, G.P., Bornschein, R.L., and Michaelson, I.A., Behavioral deficits in adult rats following neonatal lead exposure, *Neurobehav. Toxicol.*, 1, 227, 1979.

57. Garavan, H., Morgan, R.E., Levitsky, D.A., Hermer-Vazquez, L., and Strupp, B.J., Enduring effects of early lead exposure: evidence for a specific deficit in associative ability, *Neurotoxicol. Teratol.*, 22, 151, 2000.

58. Hilson, J.A. and Strupp, B.J., Analyses of response patterns clarify lead effects in olfactory reversal and extradimensional shift tasks: assessment of inhibitory control, associative ability, and memory, *Behav. Neurosci.*, 111, 532, 1997.

59. Morgan, R.E., Levitsky, D.A., and Strupp, B.J., Effects of chronic lead exposure on learning and reaction time in a visual discrimination task, *Neurotoxicol. Teratol.*, 22, 337, 2000.

60. Newland, M.C. et al., Prolonged behavioral effects of *in utero* exposure to lead or methyl mercury: reduced sensitivity to changes in reinforcement contingencies during behavioral transitions and in steady state, *Toxicol. Appl. Pharmacol.*, 126, 6, 1994.

61. Evans, H.L., Daniel, S.A., and Marmor, M., Reversal learning tasks may provide rapid determination of cognitive deficits in lead-exposed children, *Neurotoxicol. Teratol.*, 16, 471, 1994.

62. Canfield, R.L., Gendle, M.H., and Cory-Slechta, D.A., Impaired neuropsychological functioning in lead-exposed children, *Dev. Neuropsychol.*, 26, 513, 2004.

63. Muñoz, C., Garbe, K., Lilienthal, H., and Winneke, G., Persistence of retention deficit in rats after neonatal lead exposure, *Neurotoxicology*, 7, 569, 1986.

64. Jett, D.A., Kuhlmann, A.C., Farmer, S.J., and Gilarte, T.R., Age-dependent effects of developmental lead exposure on performance in the Morris water maze, *Pharmacol. Biochem. Behav.*, 57, 271, 1997.

65. Kuhlmann, A.C., McGlothan, J.L, and Guilarte, T.R., Developmental lead exposure causes spatial learning deficits in adult rats, *Neurosci. Lett.*, 233, 101, 1997.

66. Schantz, S.L., Moshtaghian, J., and Ness, D.K., Spatial learning deficits in adult rats exposed to ortho-substituted PCB congeners during gestation and lactation, *Fundam. Appl. Toxicol.*, 26, 117, 1995.

67. Schantz, S.L., Seo, B.-W., Wong, P.W., and Pessah, I.N., Long-term effects of developmental exposure to 2,2′,3,5′,6-pentachlorobiphenyl (PCB 95) on locomotor activity, spatial learning and memory and brain ryanodine binding, *Neurotoxicology*, 18, 457, 1997.

68. Levin, E.D. and Bowman, R.E., Long-term lead effects on the Hamilton Search Task and delayed alternation in adult monkeys, *Neurobehav. Toxicol. Teratol.*, 8, 219, 1986.

69. Levin, E. and Bowman, R., The effect of pre- or postnatal lead exposure on Hamilton Search Task in monkeys, *Neurobehav. Toxicol. Teratol.*, 5, 391, 1983.

70. Rice, D.C. and Karpinski, K.F., Lifetime low-level lead exposure produces deficits in delayed alternation in adult monkeys, *Neurotoxicol. Teratol.*, 10, 207, 1988.

71. Rice, D.C. and Gilbert, S.G., Lack of sensitive period for lead-induced behavioral impairment on a spatial delayed alternation task in monkeys, *Toxicol. Appl. Pharmacol.*, 103, 364, 1990.

72. Alber, S.A. and Strupp, B.J., An in-depth analysis of lead effects in a delayed spatial alternation task: assessment of mnemonic effects, side bias, and proactive interference, *Neurotoxicol. Teratol.*, 18, 3, 1996.

73. Cory-Slechta, D.A., Pokora, M.J., and Widzowski, D.V., Behavioral manifestations of prolonged lead exposure initiated at different stages of the life cycle, II: delayed spatial alternation, *Neurotoxicology*, 12, 761, 1991.

74. Rice, D.C., Behavioral deficit (delayed matching to sample) in monkeys exposed from birth to low levels of lead, *Toxicol. Appl. Pharmacol.*, 75, 337, 1984.

75. Cohn, J., Cox, C., and Cory-Slechta, D.A., The effects of lead exposure on learning in a multiple repeated acquisition and performance schedule, *Neurotoxicology*, 14, 329, 1993.

76. Stiles, K.M. and Bellinger, D.C., Neuropsychological correlates of low-level lead exposure in school-age children: a prospective study, *Neurotoxicol. Teratol.*, 15, 27, 1993.

77. Chiodo, L.M., Jacobson, S.W., and Jacobson, J.L., Neurodevelopmental effects of postnatal lead exposure at very low levels, *Neurotoxicol. Teratol.*, 26, 359, 2004.

78. Tuthill, R.W., Hair lead levels related to children's classroom attention-deficit behavior, *Arch. Environ. Health*, 51, 214, 1996.

79. Rice, D.C., Lead-induced changes in learning: evidence for behavioral mechanisms from experimental animal studies, *Neurotoxicology*, 14, 167, 1993.

80. Mirsky, A.F., Behavioral and psychophysiological markers of disordered attention, *Environ. Health Perspect.*, 8, 157, 1987.

81. Mirsky, A.F., Anthony, B., Duncan, C., Ahearn, M., and Kellam, S., Analysis of the elements of attention: a neuropsychological approach, *Neuropsychol. Rev.*, 2, 109, 1991.

82. Bellinger, D., Hu, H., Titlebaum, L., and Needleman, H.L., Attentional correlates of dentin and bone lead levels in adolescents, *Arch. Environ. Health*, 49, 98, 1994.

83. Needleman, H.L., Introduction: biomarkers in neurodevelopmental toxicology, *Environ. Health Perspect.*, 74, 149, 1987.

84. Hatzakis, A. et al., Psychometric intelligence and attentional performance deficits in lead-exposed children, in *Proc. 6th Intl. Conf. on Heavy Metals in the Environment*, Lindberg, S.E. and Hutchinson, T.C., Eds., CEP Consultants, Edinburgh, 1987, p. 204.

85. Stokes, L., Letz, R., Gerr, F., Kolczak, M., McNeill, F.E., Chettle, D.R., and Kaye, W.E., Neurotoxicity in young adults 20 years after childhood exposure to lead: the Bunker Hill experience, *Occup. Environ. Med.*, 55, 507, 1998.

86. Rice, D.C., Chronic low-level lead exposure in monkeys does not affect simple reaction time, *Neurotoxicology*, 9, 105, 1988.

87. Winneke, G. et al., Modulation of lead-induced performance deficit in children by varying signal rate in a serial choice reaction task, *Neurotoxicol. Teratol.*, 11, 587, 1989.

88. Winneke, G. et al., Neuropsychologic studies in children with elevated tooth-lead concentrations, II: extended study, *Int. Arch. Occup. Environ. Health*, 51, 231, 1983.

89. Winneke, G. and Kraemer, V., Neuropsychological effects of lead in children: interaction with social background variables, *Neuropsychobiology*, 11, 195, 1984.

90. Winneke, G., Hrdina, K., and Brockhaus, A., Neuropsychological studies in children with elevated tooth-lead concentration, *Int. Arch. Occup. Environ. Health*, 51, 169, 1982.

91. Walkowiak, J., Altmann, L., Krämer, U., Sveinsson, K., Turfeld, M., Weishoff-Houben, M., and Winneke, G., Cognitive and sensorimotor functions in 6-year-old children in relation to lead and mercury levels: adjustment for intelligence and contrast sensitivity in computerized testing, *Neurotoxicol. Teratol.*, 20, 511, 1998.

92. Hansen, O.M. et al., A neuropsychological study of children with elevated dentine lead level: assessment of the effect of lead in different socio-economic groups, *Neurotoxicol. Teratol.*, 11, 205, 1989.

93. Brockel, B.J. and Cory-Slechta, D.A., The effects of postweaning low-level Pb exposure on sustained attention: a study of target densities, stimulus presentation rate, and stimulus predictability, *Neurotoxicology*, 20, 921, 1999.

94. Morgan, R.E., Garavan, H., Smith, E.G., Driscoll, L.L., Levitsky, D.A., and Strupp, B.J., Early lead exposure produces lasting changes in sustained attention, response initiation, and relativity to errors, *Neurotoxicol. Teratol.*, 23, 519, 2001.

95. Brockel, B.J. and Cory-Slechta, D.A., Lead, attention, and impulsive behavior: changes in a fixed-ratio waiting-for-reward paradigm, *Pharmacol. Biochem. Behav.*, 60, 545, 1998.

96. Schwartz, J. and Otto, D., Blood lead, hearing thresholds, and neurobehavioral development in children and youth, *Arch. Environ. Health*, 42, 153, 1987.

97. Osman, K., Pawlas, K., Schütz, A., Gazdzik, M., Sokal, J.A., and Vahter, M., Lead exposure and hearing effects in children in Katowice, Poland, *Environ. Res.*, 80, 1, 1999.

98. Rice, D.C., Effects of lifetime lead exposure in monkeys on detection of pure tones, *Fundam. Appl. Toxicol.*, 36, 112, 1997.

99. Lasky, R.E., Maier, M.M., Snodgrass, E.B., Hecox, K.E., and Laughlin, N.K., The effects of lead on otoacoustic emissions and auditory evoked potentials in monkeys, *Neurotoxicol. Teratol.*, 17, 633, 1995.

100. Lilienthal, H., Winneke, G., and Ewert, T., Effects of lead on neurophysiological and performance measures: animal and human data, *Environ. Health Perspect.*, 89, 21, 1990.

101. Yamamura, K., Terayama, K., Yamamoto, N., Kohyama, A., and Kishi, R., Effects of acute lead acetate exposure on adult guinea pigs: electrophysiological study of the inner ear, *Fundam. Appl. Toxicol.*, 13, 509, 1989.

102. Lilienthal, H. and Winneke, G., Lead effects on the brain stem auditory evoked potential in monkeys during and after the treatment phase, *Neurotoxicol. Teratol.*, 18, 17, 1996.

103. Otto, D.A. and Fox, D.A., Auditory and visual dysfunction following lead exposure, *Neurotoxicology*, 14, 191, 1993.

104. Lasky, R.E., Luck, M.L., Torre, P., III, and Laughlin, N., The effects of early lead exposure on auditory function in rhesus monkeys, *Neurotoxicol. Teratol.*, 23, 639, 2001.

105. Stebbins, W.C., Clark, W.W., Pearson, R.D., and Weiland, N.G., Noise and drug-induced hearing loss in monkeys, *Advances Otorhinolaryngol.*, 20, 42, 1973.

106. Molfese, D.L. et al., Neuroelectrical correlates of categorical perception for place of articulatin in normal and lead-treated rhesus monkeys, *J. Clin. Experimental Neuropsychol.*, 8, 680, 1986.

107. Dietrich, K.N., Succop, P.A., Berger, O.G., and Keith, R.W., Lead exposure and the central auditory processing abilities and cognitive development of urban children: the Cincinnati lead study cohort at age 5 years, *Neurotoxicol. Teratol.*, 14, 51, 1992.

108. Campbell, T.F., Needleman, H.L., Reiss, J.A., and Tobin, M.J., Bone lead levels and language processing performance, *Dev. Neuropsychol.*, 18, 171, 2000.

109. Fergusson, D.M. and Horwood, L.J., The effects of lead levels on the growth of word recognition in middle childhood, *Int. J. Epidemiology*, 22, 891, 1993.

110. Bellinger, D., Needleman, H.L., Bromfield, R., and Mintz, M., A followup study of the academic attainment and classroom behavior of children with elevated dentine lead levels, *Biol. Trace. Elem. Res.*, 6, 207, 1984.

111. Altmann, L., Sveinsson, K., Krämer, U., Weishoff-Houben, M., Turfeld, M., Winneke, G., and Wiegand, H., Visual functions in 6-year-old-children in relation to lead and mercury levels, *Neurotoxicol. Teratol.*, 20, 9, 1998.

112. Allen, J.R., McWey, P.J., and Suomi, S.J., Pathobiological and behavioral effects of lead intoxication in the infant Rhesus monkey, *Environ. Health Perspect.*, 7, 239, 1974.

113. Bushnell, P., Scotopic vision deficits in young monkeys exposed to lead, *Science*, 196, 333, 1977.

114. Rice, D.C., Effects of lifetime lead exposure on spatial and temporal visual function in monkeys, *Neurotoxicology*, 19, 893, 1998.

115. Lilienthal, H. et al., Alteration of the visual evoked potential and the electroretinogram in lead-treated monkeys, *Neurotoxicol. Teratol.*, 10, 417, 1988.

116. Lilienthal, H. et al., Persistent increases in scotopic B-wave amplitudes after lead exposure in monkeys, *Exp. Eye Res.*, 59, 203, 1994.

7 Developmental Behavioral Toxicity of Methylmercury: Consequences, Conditioning, and Cortex

M. Christopher Newland
Auburn University

Wendy D. Donlin
Auburn University
and
Johns Hopkins School of Medicine

Elliott M. Paletz
University of Wisconsin — Madison

Kelly M. Banna
Auburn University

CONTENTS

101

Concern regarding human exposure to methylmercury (MeHg) has increased over the last several decades. Much of this concern is motivated by a growing public awareness of the deleterious and sometimes irreversible effects of MeHg that have been noted in human populations and in studies of laboratory animals. MeHg poses significant neurotoxic risks that span sensory and motor domains, associated especially with adult-onset exposures, and cognitive domains, associated with developmental exposure. MeHg's neurotoxic profile, therefore, is influenced not just by the dose, but also by the developmental period of exposure.

In this chapter, we describe the pattern of effects associated with MeHg exposure during development. Data from animal models are emphasized because these models provide control over diet, dose, and developmental windows. Animal studies also permit linkages to be drawn between the behavioral and neural mechanisms underlying MeHg's behavioral toxicity. It is argued here that such exposure presents a specific pattern of effects that can be characterized at the behavioral level as altered sensitivity to reinforcing consequences of responding, that is, as a deficit in the response-reinforcer relationship. In contrast, discrimination and memorial processes are relatively immune to developmental exposure. This pattern of behavioral effects corresponds to the pattern of effects seen in neural development after MeHg exposure. The development of cortical structures, as well as of monoamine and gamma-aminobutyric acid (GABA) neurotransmitter pathways, is especially sensitive to MeHg exposure.

The overall pattern of neural and behavioral effects points to developmental MeHg exposure as a model of how damage to the development of frontal cortical areas can have long-lasting behavioral consequences, especially on behavioral tasks that emphasize the acquisition of a response-reinforcer relationship, choice, or perseveration.

MERCURY FORMS AND SOURCES

The primary source of human mercury exposure is through the consumption of MeHg in contaminated fish. Fish are exposed to mercury that is introduced into the atmosphere by the burning of fossil fuels and municipal or medical waste, as well as volcanic activity. Mercury eventually deposits into lakes, rivers, and oceans, where microorganisms bioconvert elemental mercury to MeHg. At this point, it is introduced into the food web. Large, long-lived predatory fish, certain marine mammals, and terrestrial piscivores bioaccumulate mercury 100,000-fold or more over water concentrations [49, 112]. Large marine (shark, swordfish, tilefish, and king mackerel) and freshwater (bass, trout, and pike) predators contain the highest levels of MeHg.

Various organizations have estimated the daily intake of mercury (as MeHg) that is unlikely to be harmful. Naturally, these estimates depend on the assumptions that enter the risk assessment process. The U.S. Environmental Protection Agency has estimated that a daily intake of 0.1 µg/kg/day over the course of a lifetime is unlikely to be harmful, a risk assessment that was directed at pregnant women with the goal of protecting the fetus from undue exposure [49]. The World Health Organization has recently revised its estimate of how much mercury (as MeHg) is unlikely to be harmful downward from about 0.5 µg/kg/day to 0.22 µg/kg/day, with pregnant women identified for concern [183]. By way of context, canned albacore tuna contains approximately 0.25 ppm of mercury, so the EPA recommendation means that about one can of albacore tuna per week can be consumed by a pregnant woman without posing a risk to the fetus. The top predators, such as shark and swordfish, contain about ten times as much mercury, so intake of these species should be curtailed accordingly, if not eliminated altogether.

Two other forms of organic mercury, ethyl- and phenylmercury, have received less scientific attention than MeHg. The limited information available suggests that ethylmercury's neurotoxicity resembles that of MeHg, but their kinetics differ [95, 113]. Ethylmercury penetrates the adult blood-brain barrier more poorly and is eliminated more rapidly than MeHg. Phenylmercury shows even poorer blood-brain barrier penetration and more rapid elimination.

Ethylmercury has received public attention because of claims that its presence in thimerosal, a preservative once used in childhood vaccines, contributed to a rise in autism [84, 114]. Experimentation on the relationship between mercury and autism is lacking, so the data pertaining to this issue are largely correlational. A relationship between vaccination and autism has been hypothesized based on the coincidence between increased vaccination and rises in the rate of autism [12, 62, 69]. However, the absence to date of a plausible common mechanism the limited bioavailability of ethylmercury, and the absence of a link in large epidemiological studies all suggest that the causes of autism lie elsewhere [3, 79, 114].

Inorganic mercury presents yet a different spectrum of effects. Mercury vapor is especially neurotoxic when inhaled, and the primary source of mercury vapor is occupational or, to a lesser extent, dental amalgam. Because of space limitations, however, inorganic mercury will not be considered further.

The extent of mercury toxicity is influenced by many factors: form (mercury, MeHg, ethylmercury, dimethylmercury), dose, and duration of mercury exposure; route of exposure (inhaled, ingested); age or developmental stage (gestation, development, or adulthood); nutrition; exposure to other toxicants; and any other brain insults. For example, gestational exposure may result in attentional and sensory dysfunction, while adult exposure primarily results in sensory and motor deficits. Organic forms of mercury are generally more problematic than inorganic forms. The present chapter focuses on developmental methylmercury in animal models that have been designed to reflect human exposure.

HUMAN EXPOSURES

DEVELOPMENTAL EXPOSURE

During the mid-20th century, inhabitants around Minamata Bay, Japan, were exposed to MeHg by unwittingly consuming contaminated fish. In children born to exposed mothers, athetosis and chorea, physical malformations resembling cerebral palsy, and signs of developmental retardation were common [99]. Between 1955 and 1958, 29% of children were born with varying degrees of mental retardation, and up to 18% of secondary-school students exhibited motor dysfunction in the form of general clumsiness and poor dexterity.

Since the Minamata episode, documented cases of MeHg exposure have occurred in Canada, Peru, the Faroe Islands, the Seychelles Islands, New Zealand, the Amazon Basin, and Iraq. With the exception of the Iraqi poisoning incident, in which exposure was through the consumption of bread made with grain treated with MeHg to prevent the growth of fungus, exposure occurred through the consumption of contaminated seafood. During some of these outbreaks, data were collected on blood, hair, and cord serum mercury levels, and longitudinal investigations into the potential effects of exposure on measures of intelligence, scholastic aptitude, and cognitive abilities were assessed (e.g., performance on the Boston Naming Test, Wechsler Intelligence Test for Children, California Verbal Learning Test). In Iraq, prenatal MeHg exposure was linked to delays in reaching developmental milestones [110]. The Faroe Islands [73, 74] and New Zealand cohorts [36] indicated that prenatal MeHg exposure was associated with difficulties on some measures of cognitive function, including the Boston Naming Test, the Wechsler Intelligence Scale for Children (WISC), and the California Verbal Learning Test. However, an intensive study of the Seychelles Islands cohort [5, 37, 38, 109, 111] failed to establish a relationship between any variables.

ADULT-ONSET EXPOSURE

Adult-onset exposure has been associated with sensory and motor impairments, constriction of the visual field, numbness or paresthesia, ataxia or other gait disorders, auditory disturbances, motor disturbances, tremor, chorea, athetosis, contracture, tendon reflex, pathologic reflexes, hemiplegia, hypersalivation, sweating, focal cramps, pain in limbs, and mental disturbances in Minamata [78] and the Amazon Basin [93]. The "early onset" type of Minamata disease sometimes resolved with a reduction in the symptoms following cessation of exposure.

The Minamata cohort is being tracked as it ages, and the resulting studies have revealed that subtle, perhaps subclinical, cases of adult-onset MeHg toxicity can be unmasked by aging. Many mild cases of MeHg poisoning in Minamata were overlooked initially, especially if the symptoms were atypical. Much later, however, it was discovered that signs and symptoms could appear after a delay and are exacerbated by aging. Thus the delayed-onset Minamata disease sometimes showed progressively worsening symptoms well after exposure ceased [23, 87].

NUTRITION

In addition to being the primary source of MeHg exposure in human populations, fish are also a major source of important nutrients, including n-3 (omega-3) polyunsaturated fatty acids (especially docosahexaenoic acid [DHA]) and selenium [26]. This is an important consideration when devising fish advisories based on the presence of MeHg. These warnings could backfire by frightening people away from eating fish, thereby diminishing consumption of important nutrients [48]. Nutritional status may also contribute to the variability in behavioral effects associated with MeHg exposure in human populations. For example, the failure to detect effects of MeHg exposure in the Seychelles Island population has led to hypotheses that improved nutritional status may ameliorate the effects of MeHg exposure [38, 48, 96]. Stated differently, poor nutrition may unmask MeHg neurotoxicity that would otherwise remain undetected.

SELENIUM

Selenium is an element present in rock, soil, water, air, and food. Animals are exposed to selenium primarily through food intake. In the U.S. water supply, undetectable levels are often noted, but selenium is allowed to reach levels up to 50 ppb [4]. Both deficiencies and excesses in selenium intake can lead to adverse effects. The U.S. Food and Drug Administration's recommended daily allowance (RDA) for both men and women is 55 µg/day, or approximately 0.8 µg/kg/day (Rayman, 2000). The highest intake level should not exceed 5.7 µg/kg/day [4].

In the early 1970s, it was reported that selenium taken concurrently with methylmercury can reduce its toxicity suggesting differential MeHg toxicity associated with consumption of various fish species depending on selenium content [68, 188]. The reduced toxicity may be because mercury and selenium form a complex, bis(methylmercuric) selenide (BMS) [81]. The joint administration of selenium and MeHg results in lower selenium concentrations in many organs [89], but higher mercury concentrations in blood, liver, testes, and brain [89, 141, 166, 178].

In addition, there is evidence that selenium ameliorates some neurobehavioral effects of MeHg. It reduces MeHg-related retardation of weight gain and delays the onset of signs of neurotoxicity [82]. The behavioral effects of combinations of MeHg and selenium were examined in young mice exposed prenatally to zero, two, or five doses of 3 mg/kg Hg and a diet either sufficient (0.4 µg Se/kg) or deficient (<0.02 µg Se/kg) in selenium [175]. The mice exposed to high doses of MeHg showed selenium-related deficiencies in the righting reflex, walking ability, and open-field locomotion, as well as increased thermal preference. These differences diminished by the time the animals were weaned.

DOCOSAHEXAENOIC ACID (DHA)

Fatty fish such as sardines, menhaden, and anchovies are rich in polyunsaturated fatty acids (PUFAs), substances that are highly biologically relevant to mammals. For instance, the mammalian brain is a lipid-rich organ, with 60% of its dry weight consisting of various lipids, including very high concentrations of PUFAs such as docosahexaenoic acid (DHA) and arachidonic acid (AA). Consumption of these PUFAs has been related to a reduced risk of up to 50% in coronary and ischemic heart disease.

Learning and behavior are also affected by PUFA deficits, especially DHA. Rats fed a diet deficient in n-3 fatty acids perform poorly on Y-maze learning tasks [91], brightness discrimination tasks [185, 186], and passive-avoidance tasks [16], and demonstrate longer latencies on Morris water mazes [35]. Mice exposed to dietary manipulations of n-3 fatty acids both pre- and postnatally exhibit deficits in passive-avoidance tests [21], require more time to learn the Morris water maze, and show retarded performance on rotorod tasks [67, 172]. Slower acquisition of the water maze task appears to have been related to swimming speed rather than memory.

NEUROTOXICANTS AND BEHAVIORAL PLASTICITY

Nervous system toxicity is revealed by overt neurological signs that are patently observable, as well as by more subtle effects on behavioral adaptation to ever-changing environments. The latter requires refined testing, but is worth the effort, as the impairments that are revealed can be highly informative about the subtle effects of low-level exposure. These effects fall under the umbrella of cognitive effects and are typically reflected in operant or respondent conditioning phenomena as well as "simpler" processes such as habituation and sensitization. They entail manipulation of plastic behavior that has been trained or established during the course of a lifetime. Impairment is reflected in the course of acquisition or in the expression of behavior during a stable baseline. The return for the investment in the extensive testing that is required is the revelation of effects on memory, learning, perception, sensory-motor function, or other subtle effects that significantly impair functioning in an industrialized society [177].

A number of different behavioral domains can be affected by exposure to MeHg. Sensory function, motor function, sensitivity to certain behaviorally active drugs, and learning have all been identified as toxic endpoints. Behavioral procedures can be designed to emphasize these influences over behavior, but with the caveat that these domains cannot be completely separated from one another or examined in isolation. To characterize the effects of MeHg on the acquisition of behavior, for example, one must necessarily rely on the perception of stimuli, sensitivity to reinforcement, and motor abilities that permit an animal to respond.

While behavioral analyses of MeHg toxicity typically involve the use of laboratory models, they address, and can even predict, subtle effects in humans, some of which have economic consequences. For example, estimates of the economic consequences of productivity loss associated with MeHg exposure in the U.S. begin with decrements of scores on IQ tests [169], a highly refined form of testing. Animal models of human IQ tests do not exist, but correlations between exposure levels that impair operant behavior in nonhuman species and those that affect scores on IQ tests [33, 137, 138] support the use of such testing as a crucial component in the characterization of MeHg neurotoxicity.

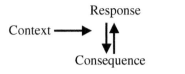

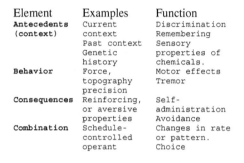

FIGURE 7.1 Three-term contingency describing the control of operant behavior. Simply put, consequences act on responses by changing their probability of occurrence. This response-reinforcer relationship always takes place in some context. The application of these different elements to the study of the behavioral effects of drugs and toxicants is described in the columns on the right.

Operant behavior is any behavior that is sensitive to its consequences, thus underlying a broad range of behavioral phenomena. While operants (i.e., voluntary responses) can be extraordinary complex and extensive over both time and space [98, 158], they often are studied with simple responses such as lever presses or traversal through a maze. The complexity can be reduced by identifying three important terms in the "three-term contingency" of operant behavior (Figure 7.1). These are responses, consequences, and the context in which behavior occurs. A consequence (food, drug, shock, or irritant) may follow a response. If that contingency between the response and the consequence increases the response rate over baseline levels, then it is called a reinforcer and the response is called an operant. Positive and negative reinforcement and punishment are defined according to how response rate changes as a function of the consequences [24]. All of this takes place in a context. Thus, for example, the stimulus context in which a left lever press is reinforced can differ from one in which a right-lever press is reinforced. Discrimination describes the extent to which response rates in the presence of these stimuli differ.

As noted in Figure 7.1, the experimental arrangement can be structured so as to emphasize the role of consequences and thereby examine the reinforcing efficacy of a drug or food [75], the irritancy of a chemical [184], or the sensitivity to the richness of different sources of reinforcers [39]. Patterns of lever pressing or the response's physical characteristics (force, duration, displacement) can be emphasized to study motor function [63, 116, 117]. The context can be exploited to examine sensory function [100, 146] as well as memory, discrimination, or generalization [105]. Schedule-controlled operant behavior, generally reflected in behavior supported by complex arrangements of contingencies, often involves complex behavioral patterns that can be quite sensitive to chemical exposure [34, 143, 176]. One important advantage of operant testing is that it reveals the extent to which impairment occurs in an intact, behaving organism. Behavior can be quite sensitive to toxicant exposure, and behavior analyses can test hypotheses generated in *in vitro* experiments as well as generate hypotheses for pursuit in further studies [90, 177].

IDENTIFYING EFFECTS OF MEHG EXPOSURE ON LEARNING

Newland and Paletz [121] proposed that MeHg's effects are dominated by disruptions in the relationship between a response and its consequences and that the role of stimulus control processes (discrimination, memory) are relatively minimal. These determinants of operant responses, their context and their consequences, are inseparable influences over the behavior of the whole animal, making it impossible to eliminate any single term in a behavioral analysis of a neurotoxicant [41]. It is possible, however, to design procedures that minimize the influence of a particular element.

If MeHg's effects are primarily on the response-reinforcer relationship, then it will be necessary to use procedures that exploit this relationship in order to identify these effects. These procedures should also minimize the role of memorial or discrimination processes. This runs counter to many approaches that are commonly used in behavioral toxicology because they emphasize discrimination or, especially, memory processes. For example, procedures such as radial arm mazes, delayed match to sample, and delayed discrimination procedures, which are commonly used in identifying "cognitive" effects of toxicant exposure, should be less sensitive to developmental MeHg if such exposure does not disrupt discrimination or contextual control processes.

CHOICE

At some level, all studies of operant behavior entail the study of choice, because the subject can either engage in the target response or do something else. Formal studies of choice arrange a situation in which two or more response options are made available and monitored explicitly. Choice is expressed as the allocation of behavior between two (or more) activities that are simultaneously available.

In the animal literature, choice is viewed as the allocation of behavior between two or more response alternatives (e.g., the distribution of responses between two concurrently available levers). Behavioral measures of preference include deliberate acts such as lever pressing or more passive measures such as time allocation [10]. The consequences available on the two alternatives typically vary in the rate of reinforcer delivery or reinforcement density, but they can also vary in magnitude, quality, or even type of activity. For example, an animal may choose between food and drug delivery or between drug delivery and an activity such as wheel running.

The definition of choice as time or response allocation is deliberately broad and can include many commonly used procedures. Behavior in a maze study, for example, is clearly an expression of choice: an animal chooses one arm and places itself in the position where a reinforcer may be delivered, as in a radial arm maze, or where a shock may be avoided, as in the shuttle-box or passive-avoidance task. The concurrent-schedule arrangement, used extensively in the study of choice, offers many advantages over these procedures (as will be addressed below), deriving both from its ability to measure choice in a continuous and repeatable fashion and from the impressively precise quantification of preference that this procedure affords [39]. These arrangements produce impressively precise measures of behavioral allocation in humans [88] and animals [171]. While the study of choice using concurrent schedules has received relatively little attention in behavioral toxicology, it has much potential for characterizing this important and complex behavior.

The sensitivity of choice procedures to subtle differences in the characteristics of the available consequences has been demonstrated many times [39]. For example, reinforcement magnitude by itself or reinforcement rate over a range of 30 to 120 reinforcers/h, say, are relatively weak variables until a choice is offered [25, 92, 161, 187]. When a choice is available, however, these variables become powerful determents of where responses occur [88, 171]. We experience this phenomenon almost daily. For example, a sizeable price of a necessary commodity such as gasoline has little effect on the rate at which it is purchased, but a relatively small difference in price between two service stations can greatly influence which station gets one's business.

To be concrete about the experimental setting in which choice is studied, imagine a situation in which a rat is responding under a concurrent schedule as described above. Pressing the left lever is reinforced under a variant internal (VI) 30s schedule (a reinforcer is delivered following the first response to occur after 30 sec, on average, has elapsed), and pressing the right lever is reinforced under a VI 60s schedule. This would be referred to as a concurrent (*conc*) VI 30s VI 60s schedule, and the left-to-right reinforcement ratio would be 2:1 (i.e., for each available reinforcer on the right lever, two are available on the left lever).

Simply examining behavior allocation using a single pair of VI schedules explains little about the effects of reinforcement rate on behavior. It is therefore customary to study how behavior changes under a variety of variant internal (VI) schedule pairs. This type of parametric manipulation allows for the assessment of many behavioral phenomena, such as choice, sensitivity to discrepancies in rates of reinforcement, sensory and motor function, and choice in transition. This last phenomenon, choice in transition after a change in the relative density of two sources of reinforcement, is especially sensitive to toxicant exposure [119, 123].

Behavior under concurrent schedules of reinforcement can be roughly divided into two categories: (1) that which occurs immediately after a change in schedule values and (2) that which is observed after a specific pair of schedules has been in effect for some period of time. The former category can be referred to as *transitional behavior* and the latter as *steady-state behavior*, since it is "a pattern of responding that exhibits little variation in its measured dimensional quantities over a period of time" [83]. It is characterized by a lack of trends, meaning that there are no consistent unidirectional changes or regular cyclical changes in a particular response.

Transitional behavior, on the other hand, is that which occurs during the change from one steady state to the next [83], and is characterized by a significant change (usually an increase or decrease) in the behavior ratio. Newland and Reile [119] define the beginning of the transition period as "the point at which programmed reinforcement contingencies change" (p. 323). A return to steady-state responding marks the end of the transitional period. It is important to stress that the data themselves dictate what type of behavior is being observed at any particular time, not the current condition of the experiment. For example, data that contain variation characterized by trends (e.g., a steady increase in response ratio across sessions) are indicative of transitional behavior whether it is observed following a change in the reinforcement schedules or after one pair of schedules has been effective for 20 consecutive sessions.

CHOICE IN STEADY-STATE CONDITIONS

Steady-state behavior is most commonly modeled using a formulation called the generalized matching relation, which takes the form

$$\frac{B_1}{B_2} = c \left[\frac{R_1}{R_2} \right]^a \tag{7.1}$$

This equation is usually represented in the logarithmic form

$$\log \frac{B_1}{B_2} = a \log \frac{R_1}{R_2} + \log c \tag{7.2}$$

Here, B_1 and B_2 (B for behavior) represent the number of responses or amount of time allocated to response alternatives 1 and 2, respectively. R_1 and R_2 refer to the number of reinforcers obtained by engaging in responses 1 and 2, respectively. The parameter a describes the slope of the linear function, and the parameter c describes the intercept of this function. The independent variable is the reinforcement ratio, R_1/R_2, and the dependent variable is the response (or time) ratio, B_1/B_2. The plots of behavior ratios vs. reinforcer ratios are called "matching functions" and are illustrated in Figure 7.2.

The structure of Equation 7.1 and Equation 7.2 is relatively simple, but these equations typically account for more than 80% of the variance in behavior. Equation 7.1 resembles Steven's Power Law used to describe sensory phenomena, and implies a psychophysical interpretation of choice similar to the detection of sound, light, or touch. The form of equation 7.2 is more tractable, since it describes a linear function between $\log (B_1/B_2)$ and $\log (R_1/R_2)$ that can be fit with linear regression. It also presents the equation in the same way that the relationship between behavior and reinforcer ratios is usually plotted, since both the horizontal and vertical axes are always scaled logarithmically (Figure 7.2).

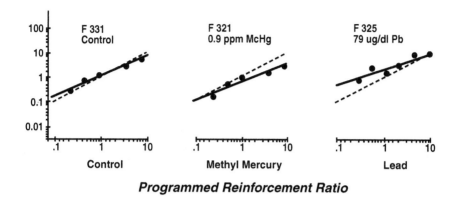

Programmed Reinforcement Ratio

FIGURE 7.2 Matching functions. Steady-state performance showing the allocation of lever pressing by monkeys between two response alternatives. The horizontal axis shows the ratio of reinforcement rates programmed to derive from the left and right levers, respectively. That is, it is the term R_1/R_2 in Equation 7.1 and Equation 7.2. The vertical axis shows the allocation of lever pressing to the left and right levers. The solid line shows the best-fit function (equation), and the dotted line shows perfect matching, i.e., response ratio equals reinforcer ratio. (From Newland, M.C. et al., *Toxicol. Appl. Pharmacol.*, 139, 374–386, 1996. With permission.)

The slope term, a, describes the sensitivity of the behavior monitored to reinforcement under the conditions employed. A value of zero describes a horizontal line and indicates that response allocation is completely insensitive to the reinforcer ratios. A value of 1.0 is called "strict matching" because it describes response ratios that match reinforcer ratios (assuming no bias, or $c = 1$). This is a mathematically convenient benchmark but not a behavioral standard. Slopes less than 1.0 ($\approx$0.8) referred to as "undermatching," are typical for laboratory animals as well as humans [39, 88, 171]. Thus, slightly less responding occurs on the rich alternative than would obtain under conditions of matching, and slightly more responding occurs on the lean alternative. The solid line shown in the left panel of Figure 7.2 has a slope of approximately 0.95. Slope can be influenced by some experimental conditions, such as the changeover cost. If there is a high cost for changing levers, then the animal leaves the rich side less frequently, and overmatching occurs [59, 168].

Bias, c, is an indicator of unaccounted-for preference for one of the two response devices, independent of the reinforcement ratio. This preference may be due to any number of factors. For example, one device may be easier to operate than the other. The characteristics that most distinguish bias are that (a) it is not driven by changes in the reinforcer ratio and (b) it may remain constant for each individual in a specific context. Responding is said to be unbiased when $c = 1$ ($\log c = 0$). When c is greater than 1, a preference exists for responding on the manipulandum represented in the numerator of the behavior ratio, and when c is less than 1, a preference exists for responding on the manipulandum represented in the denominator of the behavior ratio.

Neurotoxicant Exposure and Choice

Figure 7.2 shows matching functions from squirrel monkeys exposed gestationally to MeHg or to high lead levels [126]. These exposures were sufficiently high that they disrupted the expression of choice in steady state (i.e., after it had been acquired). They also demonstrate a permanent impairment in the expression of choice under the conditions of testing. Compared with unexposed animals, exposed monkeys' allocation of behavior to the rich and lean sources of reinforcement displayed diminished sensitivity to variations in reinforcement rates associated with these two alternatives.

Lower exposure levels had no such effect. In fact, at maternal lead levels below 40 μg/dl there was no detectable effect on steady-state responding in offspring. Prenatal exposure to high levels of lead, greater than 40 μg/dl, significantly impaired choice in young squirrel monkeys but, at least after acquisition, the expression in monkeys exposed to lower levels resembled that of controls.

Choice in Transition

Toxicant effects on learning can be assessed using concurrent-schedule behavior in transition. This approach has been used to assess the effects of developmental exposure to MeHg or lead on choice and learning in squirrel monkeys [126] and of MeHg on rats [125]. This sort of analysis can be combined with analyses of steady-state performance to form a complete characterization of choice and learning under concurrent schedules of reinforcement.

To produce a matching function such as that indicated in Figure 7.2, it is necessary to present a range of reinforcer ratios to a subject. Each new ratio affords an opportunity to investigate the acquisition of choice. Thus, there is a baseline of response allocation, a change in the reinforcement ratios, and acquisition of a new baseline. This study of the repeated acquisition of choice within a single subject has many advantages [119, 163]. The transition migrates from one baseline to another baseline, making it is possible to study repeatedly the sensitivity of behavior to changing conditions. However, other aspects of the experimental situation are similar to that of the initial acquisition: the same levers are pressed, their location does not change, the chamber remains the same, the reinforcer is the same, and the overall contingency between lever pressing and its consequences are similar. Therefore, the uncontrolled variability present in the initial acquisition of a response when all aspects of the environment are novel is minimized. This reduces experimental noise and isolates the variable of interest, that is, the allocation of behavior to different sources of reinforcement. Another strength of this approach is that parametric investigations of the acquisition of choice can be conducted repeatedly in a single subject.

An evaluation of transitional responding in monkeys exposed prenatally to either lead or MeHg demonstrates how this procedure can be used to identify toxicant effects on learning in a preparation designed to study choice [126]. Analyses of transitional behavior demonstrated slower transitions among exposed animals. The behavior of control monkeys tracked the schedule changes, and transitions were completed within three to six sessions. In contrast, monkeys exposed to MeHg during gestation demonstrated almost no changes in behavior during the first session. The transitions of exposed subjects progressed more slowly, ended before response ratios reached programmed reinforcement ratios, and were more inconsistent and variable than those of control subjects [124]. These impairments were observed at exposure levels lower than those that affected steady-state responding, demonstrating the strength of transition analyses to uncover untoward effects of exposure not otherwise detected by traditional measures of concurrent-schedule behavior.

Figure 7.3 shows representative transitions for a control (top panel), MeHg-exposed (middle), and lead-exposed (bottom) monkey. The exposure levels that produced these examples were relatively high, but they illustrate some important aspects of the acquisition of choice. Here, in this early approach to studying choice, its expression was measured using the proportion of responses on the left lever, rather than the ratio of left- to right-lever responses, which we use now [125], but for present purposes the distinction between response ratio and response proportion is unimportant.

Figure 7.3 also shows an intervention termed "behavior therapy." Monkeys whose behavior did not change after the reinforcer ratio was altered experienced an extreme discrepancy in the reinforcer ratios. In order to compensate for this, the programmed reinforcer rate on the rich lever for these monkeys was changed to 99.9% (rather than the originally programmed 80%). Behavior changes of an appropriate magnitude and in the correct direction were observed following instatement of therapy. This provided further evidence that sensitivity to reinforcement was the behavioral mechanism by which neurotoxicant exposure acted to disrupt acquisition. That is, prenatal exposure to MeHg resulted in a relative insensitivity to reinforcement or to the discrepancy between the reinforcement rates experienced on the two response alternatives.

Equation 7.3 shows the form of the logistic equation best suited to model the transitional data produced in our experiments [119]. When plotted, it generates an S-shaped function that describes how behavior changes as a function of the combined number of reinforcers delivered from both levers. This is often characterized by an initial period of slow change immediately following a change in the programmed reinforcement ratio, followed by a period of rapid behavior change that gradually culminates in an asymptote to the new steady state.

$$\log \frac{B_1}{B_2} = \log Y_{\text{init}} + \frac{\log(Y_M)}{1 + e^{k(X_{\text{half}} - X)}} \tag{7.3}$$

The shape of this function is seen in Figure 7.4. The independent variable, X, is some measure of the passage of the transition. While time might appear to be a natural dimension to use, we have found that the cumulative number of reinforcers delivered performs much better. This measure carries more behavioral relevance than elapsed time, since the transition is driven by reinforcer deliveries. This independent measure carries other advantages, too, not the least being that it permits more-direct comparisons across conditions and even across species using a relevant measure of

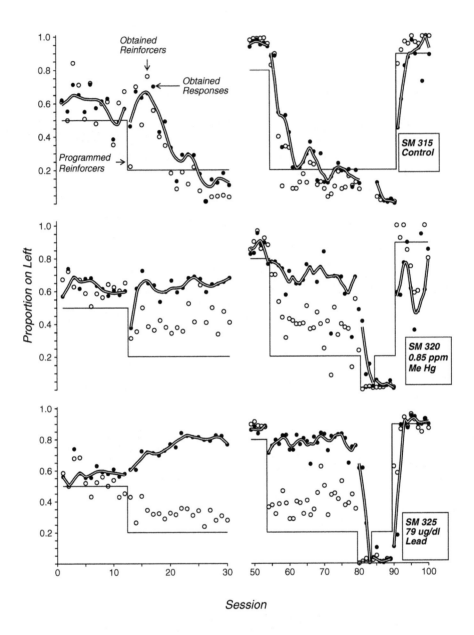

FIGURE 7.3 Lead- and MeHg-exposed monkeys, behavior in transition, showing the acquisition of choice in three squirrel monkeys. The horizontal axis shows session number, and the vertical axis shows the proportion of responses (filled symbols, thick line representing a Lowess smoothing of the filled symbols), obtained reinforcers (unfilled symbols), and programmed reinforcers (thin line). Three transitions plus a "behavior therapy" intervention are represented.

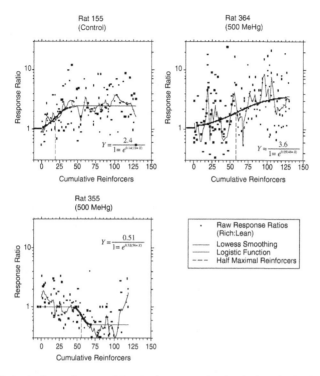

FIGURE 7.4 A single-session transition in the expression in choice. At the beginning of a 2.5 h session, each of two levers produces the same rate of reinforcement. After 30 min into the session (0 on the horizontal axis), the reinforcer rates may change so that one lever becomes richer than the other. The horizontal axis represents the number of reinforcers delivered since the transition's beginning. The vertical axis represents the ratio of rich to lean responses at the end of each pair of visits, normalized to the ratio obtained at the end of a 30-min baseline period. Thus the vertical axis represents the magnitude of change in choice since baseline. Individual data points represent the instantaneous response ratio after a pair of visits. The thin line is a Lowess smoothing of the data points. The fitted line is a logistic function, whose equation is shown in the inset. Other details in text. (From Newland, M.C. et al., *Neurotoxicol. Teratol.*, 26, 179–194, 2004. With permission.)

behavior change. In addition, one practical advantage to using cumulative reinforcers as the independent variable is that it yields model estimates that better fit the observed data.

The dependent variable is the logarithm of the behavior ratio $\log B_1/B_2$. The initial value Y_{init} represents the behavior ratio seen at the end of the baseline. The upper asymptote of the function, Y_M, describes the new steady state. We usually normalize the transition to the baseline and describe the magnitude of the transition, $\log B_1/B_2 - \log (Y_{init})$, relative to the baseline value. Thus, Y_M can be viewed as $Y_{ss} - Y_{init}$, where Y_{ss} represents the new steady state of choice. The parameter k describes the slope of the steep portion of the S-shaped curve. There are two ways of interpreting this parameter. First, it is the slope at the inflection point of the curve (where the curve changes from curving upward to curving downward), and it is the steepest portion of the function. Second, the value $1/k$ represents the number of reinforcers required to get from $1/e$ to $2/e$ of the entire transition. Since e is about 2.73, this describes the number of reinforcers that span approximately the middle third of the transition.

The value X_{half} is a "location parameter" that describes the number of reinforcers required to complete one-half of the transition. A high value locates the S toward the right (many reinforcers), and a low value locates it to the left (few reinforcers). The upper asymptote describes the behavior ratio (actually, the log of this value) after the transition is complete, so it represents the magnitude of the transition.

Choice in transition was used to evaluate the effects of low-to-moderate levels of gestational exposure to MeHg on single-session transitions in adult and aging rats [125]. The single-session transition offers advantages over the approach used in the earlier study, which examined transitions occurring across session [119]. In the earlier study, a transition required about 15 to 20 30-min sessions, or approximately three to four weeks, for completion. In the single-session procedure, transitions occur over the course of one extended session. Under these conditions, the upper asymptote (Y_M) was not as large as in the multiple-session transition. Most of the transition occurs within several dozen reinforcers, but a small amount of additional change in the behavior ratio occurs over the course of several weeks [42]. This results in slightly shallower slopes generated by matching functions fit to steady-state behavior from single-session transitions [119]. The fact that some of the transition is not modeled is slightly problematic, but this is offset by the ability to conduct multiple transitions. In addition, it provides a means by which the effects of drug challenges on choice in transition can be investigated.

In the single-session preparation, experimental sessions consisted of a 30-min baseline phase, during which the reinforcement ratio was 1:1, followed by a 120-min transition phase (described below, see also Figure 7.4). With earlier implementation, a random number generator was used to determine when a reinforcer was scheduled for delivery during the baseline phase, and the probability of reinforcer delivery following a response was equal across levers. Inspection of the data revealed the occasional presence of long runs of reinforcers on one lever, resulting in biased responding during the baseline phase. In later implementations, sequences of reinforcers were generated with a computer program to ensure that 40 to 60% of the reinforcers came from each lever during the last 20 min of the baseline phase, leading to a significant reduction in the development of position preferences during baseline.

The baseline phase was followed by one of three conditions: (1) the left lever became rich (the left-to-right reinforcement ratio was >1), (2) both levers produced reinforcement at the same rate (the left-to-right reinforcement ratio = 1), or (3) the right lever became rich (i.e., a left-to-right reinforcement ratio <1). The condition occurring on a particular day was pseudorandomly determined such that the same condition could not occur in three consecutive sessions.

To analyze transitions in a single session, it was important to track behavior at a molecular level. To accomplish this, behavior ratios were analyzed on a visit-by-visit basis. A series of consecutive responses on the left lever comprised a left visit. This visit was terminated by a response on the right lever, which initiated a right visit (i.e., a series of consecutive responses on the right lever). When the next visit on the left lever commenced, the behavior ratio on the previous pair of visits (left:right) was calculated. Both time and response ratios were inspected, and response ratios were chosen as the dependent variable of interest. A Lowess smoothing algorithm was used to smooth over nine visits. For example, to estimate the response ratio on the 10th visit, ratios from the 6th to the 14th visit were averaged using a procedure that assigned weights to visits based on their temporal distance from the visit of interest (i.e., for the 10th visit, response ratios at the 9th and 11th were the most heavily weighted, while those at the 6th and 14th were the least heavily weighted). This ensures that the more distal visits were the least influential. The resulting function was relatively smooth and maintained the important dynamics that become lost when response ratios are averaged within 30-min sessions.

Steady-state responding in single-session transitions was analyzed by applying Equation 7.2 to the last ten pairs of visits in a session, and transitions were analyzed by fitting Equation 7.3 to all data following a transition (i.e., that which occurred during the last 2 h of each session). Analyses of variance were carried out on all parameter estimates generated by Equation 7.2 and Equation 7.3 to determine the effects of MeHg on responding.

Prenatal exposure to MeHg did not affect steady-state responding in either young or aging rats, and slope estimates were slightly lower than what occurs when a behavior is allowed to stabilize over several weeks. There were no dose-related effects of MeHg on slope or intercept (c and a in Equation 7.1).

In Figure 7.4, for the figure on the top left, the magnitude of the transition was 2.4; it was half completed after 19 reinforcers, and the maximum slope was 0.14 (log percentage points/reinforcer). For the transition on the top right, representing a MeHg-exposed rat, the corresponding values are 3.6, 40, and 0.05. Thus the transition was larger (in this example) and slower. The magnitude of the transitions for MeHg-exposed rats was indistinguishable from controls, but they were slower. The bottom left figure shows a rat that transitioned in the wrong direction: most behavior was allocated to the lean alternative. This was sometimes seen in exposed rats.

The overall pace of behavior change (i.e., learning) was slower for MeHg-exposed rats, especially as the rats aged, as indicated by a greater number of reinforcers required to complete half of a transition in exposed rats (Figure 7.5). Since this effect appeared in the half-maximal reinforcers, it further indicates that the effect occurs because more reinforcer deliveries are required to effect the beginning of a transition. These exposed animals appeared relatively insensitive to the change in reinforcement. Once choice began to change, it proceeded at a pace that was indistinguishable from that of unexposed subjects, and it ended at the same value. That is, the magnitude of the transition and the maximum slope were indistinguishable across exposure groups.

The concurrent-schedule procedure emphasizes the acquisition of choice and its expression in steady state. The role of memory is minimized because all relevant stimuli are present at the same time that the animal responds, whereas in memory-type procedures the response at present must be influenced by a stimulus that was presented and removed in the past. Even the role of discrimination has been minimized, since there is no specific stimulus that is correlated with a "correct" or "incorrect" response; there is not even a "correct" or "incorrect" response.

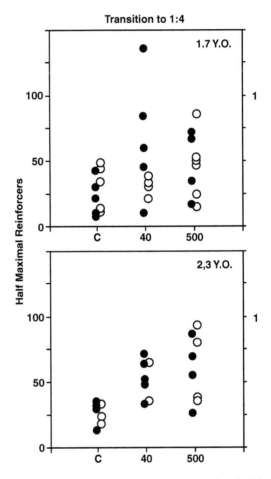

FIGURE 7.5 Mercury on choice in transition in middle-aged and old rats. Half-maximal reinforcers in a choice-in-transition procedure for middle-aged and old rats exposed to MeHg (0, 40, or 400 μg/kg/day via maternal water) during gestation. When the rats were over 2 years of age, they displayed significant retardation in allocating behavior to the rich lever. Filled and unfilled symbols represent males and females, respectively, but there was no gender difference in the effect.

ACQUISITION OF FIXED-RATIO (FR) RESPONDING

In acquiring lever pressing, a laboratory animal presses the lever and a reinforcing consequence such as food, sucrose, or water is delivered. If the animal is sufficiently food- or water-deprived, then this simple contingency increases the rate of lever pressing, which eventually occurs at a high rate and to the exclusion of other activities. There is some opportunity for discrimination and memorial processes to occur here — the position of the lever and the recent delivery of a reinforcer must be remembered — but their role is minimal. No stimulus is uniquely paired with reinforcer delivery either at the time that the animal responds, as in discrimination tasks, or prior to the opportunity to respond, as in memory tasks. The consequence is that the primary influence over behavior and the context is relatively unchanging.

In the FR-acquisition procedure, this contingency of one response/reinforcer is made more demanding by requiring more responses for each reinforcer [97, 127]. Thus, for example, after lever pressing under an FR-1 schedule (each lever press produces a reinforcer), an FR-5 schedule is imposed (five lever presses are required), then an FR-25 and finally an FR-75, with only a few sessions at each of the FR values. With each increase in the ratio between responses and reinforcers, the overall reinforcement rate becomes leaner. Thus, there is a risk that lever pressing could extinguish.

It might be tempting to assume that neurotoxicant exposures would always result in diminished responding under this procedure, but this analysis is incorrect. Certainly, motor deficits that increase the difficulty of lever pressing or, alternatively, an effect that reduces the efficacy of the reinforcer would result in more rapid extinction and diminished responding. A tendency to perseverate, however, would result in persistent responding as the reinforcement rate decreases, and could result in "improved" performance under the larger fixed-ratio value. In fact, as described below, this has been seen [132].

Another effect that could result in enhanced responding is enhanced reinforcing efficacy of the consequence, which could arise if neurotoxicant exposure altered reinforcement pathways. Incremental fixed-ratio schedules are well-established procedures for determining the reinforcing efficacy of abused drugs [30, 75, 151, 152]. There are also indirect ways by which increased responding might occur. For example, if exposure led to an increase in the caloric intake required to sustain an animal's body weight, then that could also enhance responding under such a procedure either by increasing local response rates or, more likely, by decreasing preratio pausing. Such pausing is powerfully influenced by the relative reinforcing value of the upcoming ratio and the consequences that it leads to [7, 44, 140]. In fact, these different mechanisms may all contribute to the elevated response rates, a situational outcome that one might interpret as perseveration. This is not to say that such an interpretation would be mistaken. Indeed, it could point to a behavioral mechanism by which "perseveration" occurs.

Developmental exposure to cadmium resulted in higher response rates as the fixed-ratio (FR) schedule became larger [126]. At the FR-75 schedule, the cadmium-exposed rats had much higher response rates than controls, and this effect persisted even with steady-state exposure to this schedule value. Cumulative records showing the responding of individual rats under this arrangement confirmed that, for controls, a pattern characteristic of "ratio strain" was evident.

The sensitivity of this procedure for detecting neurotoxicity has been seen also with MeHg exposure [133]. Figure 7.6 shows the rapid acquisition of FR responding in rats exposed during gestation, via maternal drinking water, to 0, 0.5, or 5 ppm of MeHg, resulting in about 0, 40, or 400 μg/kg/day exposure [132]. Some rats consumed a fish-oil diet and others consumed a coconut oil diet so as to produce diets rich or low in n-3 polyunsaturated fatty acids. Diet did not interact with MeHg, however, so only MeHg effects will be described.

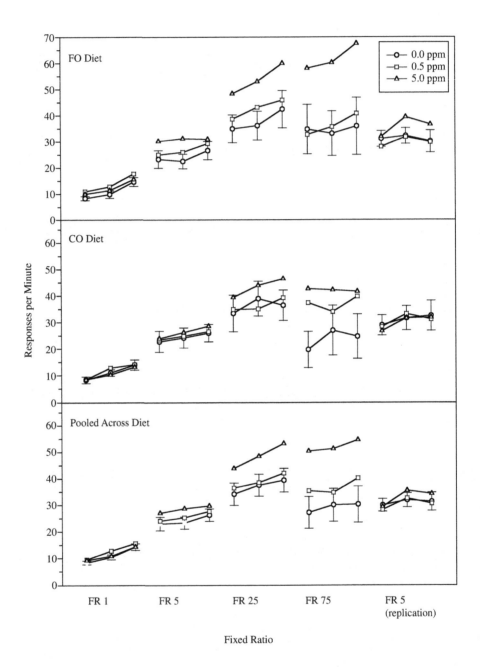

FIGURE 7.6 Rapid acquisition of fixed-ratio performance in rats exposed to MeHg (about 0, 40, or 400 µg/kg/day via 0, 0.5, or 5 ppm in maternal drinking water) during gestation. FO (fish oil) rats consumed a fish-oil diet rich in n-3 polyunsaturated fatty acids, and CO (coconut oil) rats consumed a coconut oil diet lean in these fatty acids. There was an effect of MeHg but not of diet on responding as the ratio value increased.

Here, as in the cadmium study, all exposure groups had similar overall response rates at the lower fixed-ratio values. As the fixed-ratio requirement increased, the groups became distinct. Rats exposed gestationally to MeHg showed higher response rates than controls. It may be that this effect is limited to ratio-type contingencies, which can produce high response rates. When the schedule was changed to a differential reinforcement of low-rate schedule, the three exposure groups were similar after the first day of exposure (not shown in figure).

This effect was systematically replicated with a progressive-ratio (PR) schedule. Under this arrangement, the response requirement increased by 5, 10, or 20% with each reinforcer delivery. For example, when the progressive ratio schedule increased by 5%, the first ratio requirement was FR 10, and it progressed using the following sequence: 10, 11, 11, 12, 12, 13, 13, 14, 15, 16, etc., responses per reinforcer. The repeated values (11, 11, for example) were due to rounding. The PR 20% schedule produced a more rapidly increasing sequence: 10, 12, 14, 17, 21, 25, 30, 36, 43, 52. The offspring of rats exposed to 5.0 ppm of MeHg had higher response rates under the progressive ratio schedule, and the appearance of this effect was most evident when the progressive ratio increased relatively slowly. As with the rapidly increasing fixed-ratio schedule, prenatal exposure to MeHg resulted in an increased response rate under the progressive-ratio schedule, too (Figure 7.7). The progressive ratio shows that the rate of increase is also an important determinant and that effects are more likely to be discernable when the response ratio increases by 5% with each reinforcer.

Prenatal exposure to MeHg resulted in elevated response rates under rapidly increasing fixed-ratio schedules of sucrose reinforcement and under progressive-ratio schedules. The effects do not reflect motor deficits, since those would appear as diminished response rates. They are consistent with an effect that might be described as perseveration. They are also consistent with an interpretation that emphasizes an alteration in the efficacy of the reinforcing value of sucrose. These are not inconsistent interpretations, as enhanced reinforcing efficacy could produce higher response rates and reduced pausing, effects that would appear as perseveration.

These data demonstrate the importance not only of parametric explorations of environment-behavior interactions but, more importantly, of the value of conducting detailed analyses of behavior. There was no effect of MeHg at the lower FR requirements (Figure 7.6) of FR-1 or FR-5. The absence of an effect on behavior under an FR-5 is not due to lack of experience with the schedule, since a replication of this condition after the FR-75 condition resulted in nearly identical responding as seen during the first implementation. It is likely that the more demanding schedules provide opportunity for perseveration to be manifested.

DIFFERENTIAL REINFORCEMENT
OF HIGH-RATE BEHAVIOR

Many reinforcement schedules produce high rates of behavior, indirectly, due to the response patterns that they select. Two characteristics of the fixed-ratio schedule result in the appearance of high response rates after extended exposure. This

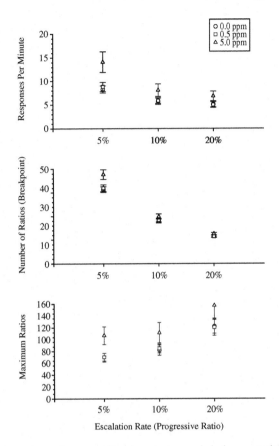

FIGURE 7.7 Progressive ratio responding in rats exposed during gestation to MeHg. Rats exposed to the high concentration of MeHg responded at higher rates, especially when the ratio increased at 5% after each reinforcer.

schedule, in which a set number of responses is required for reinforcement delivery, reinforces short interresponse times because reinforcers usually arrive in the middle of response bursts. In addition, there is a direct relationship between overall response rates and reinforcement rates [92, 187]. Because of these characteristics, FR schedules have been used to study motor function and its impairment by drugs and toxicants [64–66, 116–118].

The differential reinforcement of high rate (DRH) schedule reinforces high-rate behavior directly. Here, a sequence of lever presses is reinforced if it meets a time requirement. For example, using a nomenclature introduced by Newland and Rasmussen [122], a DRH 9:4 sec means that nine responses must be completed within 4 sec, or that each pair of responses must be separated by a 0.5 sec interresponse time (on average). Such a schedule was used to study the developmental neurotoxicity of MeHg, and behavior under this schedule was said to be an extraordinarily sensitive marker of effect [15]. While that particular sensitivity has not been replicated [122], behavior under a DRH schedule is a sensitive marker of MeHg's developmental neurotoxicity, especially as it deteriorates with aging [123].

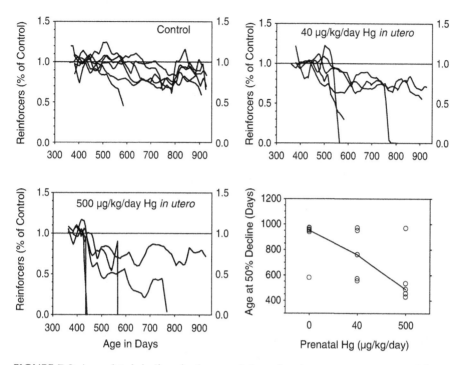

FIGURE 7.8 Age-related declines in the completion of a nine-response sequence of lever pressing within 4 sec, as required under a DRH 9:4 schedule of reinforcement. Control rats showed about an 80% decline as they aged. Exposed rats had sufficient difficulty that they stopped responding before they reached 900 days of age. Methylmercury-exposed rats showed dose-related declines in the age at which they experienced difficulty meeting the DRH 9:4 reinforcement contingency. (Adapted from Newland, M.C. and Rasmussen, E.B., *Neurotoxicol. Teratol.*, 22, 819–828, 2000. With permission.)

Rats were trained to respond under a DRH 9:4 schedule of reinforcement at 120 days of age, and the experiment continued until the oldest survivors were 900 days old [122]. This schedule produces response bursts of nine responses separated by pauses. The performance of each rat was compared against its performance in the early sessions by normalizing response rates against rates in the baseline established at the beginning of the study. The primary dependent variable was the number of reinforcers earned during a session, which matched the number of nine-response sequences that occurred within the 4-sec criterions.

Among control rats, all but one continued responding to 900 days of age, although there was an overall age-related decline in rate to about 80% of baseline at this old age (Figure 7.8). Only one of the rats exposed to 6.4 ppm continued responding this long. For the others, responding declined before they reached this age. For those rats that failed to continue responding to the end of the experiment, the age at which reinforcement rates fell to below 50% of baseline rate was determined.

There was a dose-related decline in the age at which reinforcement rate (and hence the number of criterion nine-response sequences) declined to 50% of baseline. Rats at higher MeHg doses reached the 50% of baseline at about 1.5 to 2 years of

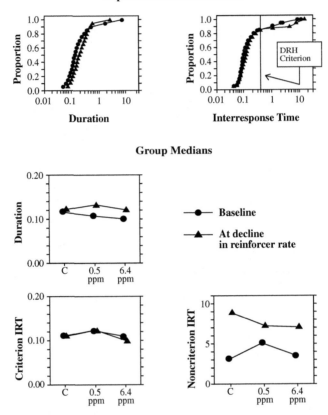

FIGURE 7.9 How DRH performance declined. The top panels show a cumulative distribution of lever-press duration (top left) and interresponse times (IRTs), exclusive of duration (top right), for a control rat averaged over ten sessions taken from baseline and the end of the experiment. The 95% confidence bands are obscured by the data points. Lever-press duration increased from 0.13 sec during baseline to 0.19 sec at the end of the experiment for this representative animal. The IRT was further analyzed into criterion and noncriterion IRTs, indicated by the vertical line in the top right panel located at 0.38 sec. If all IRTs were 0.38 sec and durations were 0.12 sec (the overall median), then the nine-response burst would just meet criterion. The middle panel shows median lever-press durations across control groups during baseline sessions and under a condition called "aged," taken from the session at which reinforcement rates were close to 50% of baseline, or the last day of testing. The bottom panels show similar curves for criterion and noncriterion IRTs. (Adapted from Newland, M.C. and Rasmussen, E.B., *Neurotoxicol. Teratol.*, 22, 819–828, 2000. With permission.)

age, depending on dose, while most unexposed rats' behavior declined to only about 80% of control by 2.5 years old. Such a decline in the overall output of high-rate behavior does not necessarily arise from impaired abilities to execute high-rate response bursts [122], at least not in a simple way. Figure 7.9 shows interresponse times taken from response bursts that met the DRH 9:4 criterion and those that did not. The latter included times between response bursts as well as the time to collect the reinforcer and return to the lever. Lever-press duration increased slightly as response rates fell, and this was interpreted as an indicator of age-related deficits in the motor execution of lever pressing.

Despite the increase in lever-press duration, the nine-response sequences produced still occurred within the required 4 sec, implying that the short (criterion) interresponse times (IRTs) remained intact. This was confirmed by investigating the distribution of interresponse times taken from the session at which the number of reinforcers (and criterion sequences) reduced to 50% of baseline (Figure 7.9). The reduction in responding most likely reflected greater increases in longer (noncriterion) IRTs, indicating increased pausing and longer delays in initiating responding. Such pausing could be due to difficulty in ambulating to the feeder, fishing the food out, and consuming it. It could, however, also be attributed to motivational, and not motor, influences. Evidence for this comes from a number of recent studies showing that such pausing is sensitive to the relative reinforcing value of the events following the pause: the more reinforcing an event (relative to other aspects of the experimental setting), the shorter is the pause between bouts of responding [7, 44, 140, 162].

MOTOR FUNCTION

Behavior always entails a motor act, so it is necessary to eliminate motor deficits as confounds in experiments on cognitive function. Sensory-motor function has been shown to be disrupted by MeHg exposure in humans, nonhuman primates, and laboratory rodents and pigeons, so the possibility that motor deficits confound effects on behavioral plasticity must be addressed [13, 31, 174]. With MeHg, the pattern of sensory-motor deficit differs, depending on whether exposure is developmental or chronic and beginning in adulthood.

A direct comparison of the effects of developmental and chronic, adult-onset exposure on motor skills showed dose-related effects of MeHg only after extensive, chronic exposure. Rats were exposed either chronically or prenatally (via maternal drinking water) to 0, 0.5, or 5.0 ppm of MeHg in drinking water, approximating 0, 40, or 400 µg/kg/day [43]. Half of the rats were maintained on a coconut oil diet, while the other half ate a fish oil diet high in docosahexaenoic acid (DHA), but diet did not influence methylmercury's effects, so will not be considered further here. Chronic, adult-onset exposure produced dose-related increases in grip strength, hindlimb crossing when the rat is held by the tail (a marker of chronic, high-dose exposure), gait abnormalities, and diminished running wheel activity. Similar effects have been reported in other experiments, but with higher exposure levels and, consequently, more rapid onset of effects [155]. Pigeons are also susceptible to sensory-motor deficits after chronic exposure [50].

Developmental exposure produced none of these effects [43]. Since the effects on choice, fixed-ratio acquisition, or high-rate behavior were all associated with developmental exposure, it can be concluded that they did not reflect such pronounced motor deficits. There remains the possibility that subtler motor deficits associated with developmental exposure influence the effects seen, but this seems unlikely. The elimination of subtler deficits that might be associated with developmental exposure [144] requires the more fine-grained analyses described elsewhere.

This issue has been addressed in experiments described above by noting distinctions between MeHg's effects on choice and overall response rate, or between pauses that separate high-rate bursts of responding and the initiation of such bursts. For example, it is difficult to argue that an effect on the acquisition of fixed-ratio responding is a motor deficit when the effect is an increase in overall response rates.

SENSORY DEFICITS

Deficits in visual, auditory, and tactile function have been reported in animal models [121, 147] and in human exposures [2, 6, 31, 174]. As with motor deficits, MeHg affects sensory function, and some of these effects could reflect MeHg's disruption of cortical development [128]. A large portion of the cortex is devoted to the processing and integration of sensory information, and these cortical regions accumulate MeHg during developmental exposure. Cortical deficits would be reflected in higher-order sensory functions such as integration of the visual stimuli. These include clumsiness and deficient tactile sensitivity [150], impaired contrast sensitivity function [148], and high-frequency hearing [149]. While these effects could reflect cortical function, they could, and probably do, also result from toxicity in sensory pathways afferent to the cortex.

Cortical damage could produce agnosia, and the range of possible effects is broad. This is an area that has not been pursued aggressively in the animal literature, although the study of contrast-sensitivity contours in exposed monkeys does represent an examination of complex visual function [148].

Evidence that young MeHg-exposed monkeys have difficulty recognizing faces even after several presentations have been reported [76, 77]. In these experiments, young monkeys were shown faces of monkeys, and the duration of gaze was recorded. Exposed monkeys tended to treat previously viewed faces as novel ones. These could reflect short-term memory loss or deficient recognition of faces. The latter is a function of visual cortex efferent from the primary cortex.

Sensory effects could result in deficits on tests of human cognitive function. Language disorders in children, for example, have been linked to deficits in higher-order auditory processing [14, 61]. Dyslexia and other disorders of language development have been traced, in some cases, to the initial processing of auditory information in the primary auditory cortex. For example, difficulties in distinguishing stop consonants can delay language development because of resulting difficulties in distinguishing between, say, the phonemes /b/ and /t/, with the consequence that, say, the syllables "ba" and "ta" sound the same. It is not known at present whether sensory deficits associated with developmental exposure to MeHg have such consequences.

DISCRIMINATION AND MEMORY PROCESSES

Discrimination and memory processes are disrupted by exposure to numerous drugs and toxic substances [105]. It is natural, therefore, to expect that MeHg exposure would produce similar effects. However, several studies have failed to show effects of developmental MeHg exposure on commonly used tests of discrimination or memory.

Monkeys exposed during gestation to 50 to 90 µg/kg/day were tested on a delayed spatial-alternation procedure [70]. Under a nondelayed spatial-alternation procedure, the monkey is presented with two levers, and the lever that is paired with the reinforcer alternates after each reinforcer. Thus, a left response is reinforced, then a right response, then a left response, and so on throughout the session. The delayed spatial-alternation procedure is similar, except that a delay is imposed between the previous reinforcer delivery and the opportunity to respond.

In this procedure, the response-reinforcer relationship remains unchanged throughout the session (refer to Figure 7.1): a single response on the left or right lever is always reinforced. The context, however, changes with each response. If the previous correct response was on the left side, then the next response must be on the right side. This procedure therefore emphasizes discrimination processes and not the response-reinforcer relationship. No MeHg-induced impairment has been found on this task [70]. If anything, the MeHg-exposed monkeys may have performed slightly better than controls.

Another discrimination-based procedure is the visual discrimination reversal, which can be arranged in a T-maze or in an operant chamber. In the traditional operant chamber it is arranged as follows. When the keylight above the right lever is illuminated, responding on that lever is reinforced and responding on the left lever is on extinction (EXT, no reinforcement for lever pressing is available). When the light above the left lever is illuminated, the opposite is true; responding on the left lever is reinforced and responding on the right lever is on EXT. A trial begins when a light is illuminated and ends following a response or some specified period of time. The active lever varies pseudorandomly from trial to trial. This preparation arranges for a perfect correlation between the location of a visual stimulus and the location of the lever paired with reinforcement.

Monkeys exposed gestationally to 10 to 50 µg/kg/day showed no effects on nonspatial discrimination reversals in which geometric stimuli served as the discriminative cue indicating which lever to press [145]. Rats have also failed to show effects of developmental MeHg exposure on accuracy in visual-discrimination-based experiments, even after exposure to as much as 4 mg/kg/day of MeHg (p.o.) during gestation, on delayed or nondelayed spatial-alternation procedures using a standard T-maze [71]. Mice exposed to 8 mg/kg/day also do not show effects of MeHg on standard visual discrimination reversal procedures [46, 72], but in a modification of this procedure, females showed some deficits when required to choose the arm opposite to one selected in a reference-memory task [72].

In a recently published experiment, rats were exposed developmentally to 0.5 ppm of MeHg in drinking water and trained, as adults, on a spatial discrimination reversal task as well as several alternation tasks [182]. There was no effect on the spatial discrimination reversal procedure, but developmental MeHg exposure resulted in a small but statistically significant increase in the number of errors on both delayed and nondelayed (actually a very short delay) spatial-alternation tasks. MeHg's effects were unrelated to the duration of the delay, suggesting that memory processes were not affected by MeHg. The pattern of errors indicated a tendency to perseverate on a lever, and especially on the lever that did not produce a reinforcer, i.e., a "lose-stay" pattern. The authors concluded that the effects reflected a diminished sensitivity to the consequences of the animals' behavior, an interpretation consistent with the hypothesis that MeHg selectively targets the response-reinforcer relationship while sparing discrimination processes.

In an unpublished dissertation, Paletz identified an effect of developmental exposure to MeHg (0.5 and 5 ppm) on spatial and visual discrimination reversal performance. MeHg impaired the acquisition of the first reversal, a pattern often seen in such procedures. The pattern of errors was nonrandom and was consistent with perseveration on the lever associated with extinction.

DRUG CHALLENGES

Drug challenges in behaving animals can be used both to test hypotheses deriving from *in vivo* studies and to generate hypotheses for further examination in such studies. Thus, this represents a natural area for linking behavioral and neural mechanisms of neurotoxicant action. With *in vitro* approaches, tissues are taken from animals exposed to MeHg under some protocol, and the concentration of a neurotransmitter, of metabolites, or of receptors might be examined [173]. These approaches have identified monoamine, GABAergic, glutamatergic, and cholinergic pathways as being affected by some sort of exposure protocol [9, 11, 18, 103, 106, 129–131].

Both monoamine and GABA systems play a role in the developmental neurotoxicity of MeHg as reflected in behaving animals [18, 80, 142]. Fully adult rats exposed during gestation to MeHg showed enhanced sensitivity to amphetamine (a dopamine and noradrenergic agonist) and diminished sensitivity to pentobarbital (a GABA agonist) [142]. They did not display sensitivity to a cholinergic or glutamate antagonist, nor did they show altered sensitivity to a dopamine receptor blocker. These effects have been replicated and extended in unpublished studies in which sensitivity to cocaine has been identified but not to arecoline (a muscarinic cholinergic agonist) or clomipramine (a serotonergic agonist).

Taken together these data indicate that there are long-term, irreversible alterations in some aspect of dopaminergic, GABAergic, and perhaps noradrenergic neurotransmitter systems, but not cholinergic, serotonergic, or, perhaps, glutamatergic systems. The absence of an effect of haloperidol (a dopamine receptor blocker) suggests that the effects lie in the regulation of neurotransmitter levels (i.e., in the generation, release, or reuptake) and not in postsynaptic receptors. It should be noted that the absence of an effect on glutamate systems might be viewed cautiously because it derives from the absence of an effect of a glutamate antagonist. After all, effects of amphetamine and cocaine have been seen, but not for haloperidol, a dopamine receptor blocker. The glutamate agonists are difficult to work with because many produce their own toxicity, so it cannot be concluded that presynaptic effects might be present.

NEURAL AND BEHAVIORAL MECHANISMS

In the present review, an argument has been presented that MeHg affects a response-reinforcer relationship and leaves relatively intact contextual (stimulus)-control processes such as discrimination and memory processes. In this section, we summarize the approach used to define a behavioral mechanism and then link it to the neural consequences of MeHg exposure. Specifically, we argue that the behavioral consequences of developmental exposure to MeHg (adult-onset exposure is quite different) can be traced to MeHg's disruption of cortical development and, possibly, of dopamine and GABA pathways. To do this it is necessary to show (a) that a response-reinforcer relationship can be isolated experimentally and affected specifically in behavioral procedures used to examine MeHg's neurotoxicity, (b) that the neural regions affected have effects similar to what has been seen with MeHg exposure, and (c) that developmental MeHg exposure affects development of the appropriate neurochemical pathways as well as of the cortex.

MeHg and the Response-Reinforcer Relationship

Operant behavior, which is any behavior that is sensitive to its consequences, reflects both discrimination (also called stimulus control or, in the more recent literature, contextual control) and reinforcement processes. It is impossible to eliminate completely any category (context, consequence, or response) and still have behavior. Contextual control and reinforcement processes can be isolated, however, by arranging conditions such that one set of processes does not play a differential role, thereby allowing the other's effects to be seen in greater relief [41]. For example, in the fixed-ratio acquisition and progressive-ratio experiments, the number of responses required for a reinforcer changes frequently but the context remains the same. Therefore, contextual processes cannot exert differential control over behavior. On the other hand, if stimulus conditions change frequently but reinforcement rates remain constant, then contextual control processes are likely to play a major role relative to reinforcement processes.

Under a concurrent schedule, two levers are available simultaneously and lever pressing on each one is reinforced. However, responding on one lever is usually reinforced at a higher rate than the other. This presents a more complex case than the acquisition of FR responding. Evidence suggests that concurrent-schedule procedures nevertheless tap reinforcement rather than discrimination processes. The evidence derives from comparisons of this standard implementation with modifications that have been made to examine memory [1, 102], sensory [40, 179, 180], or discrimination [1, 115] processes. In those experiments, concurrent schedules were arranged to highlight contextual control processes and demonstrate the minimal degree to which these processes act in the standard arrangement.

To see this, compare a two-lever concurrent schedule with a two-lever discrimination procedure. For example, under the typical concurrent schedule, the animal is presented with a left and a right lever that, when pressed, produce reinforcers at rates of one and four reinforcers/min, respectively. Contextual stimuli are limited to the spatial location of the levers; after time, right becomes the context for richly reinforced lever presses, and left becomes the context for leanly reinforced presses. In a subsequent condition, the locations of the levers remain unchanged, but the left and right levers produce reinforcers at, say, four and one reinforcers/min respectively. The key here is that the context barely changes but the relationship between a response and a reinforcer does. Therefore, reinforcement processes should have relatively greater, and therefore primary, impact on behavior [41].

Now consider a common discrimination procedure. The animal faces the same two levers, but stimulus conditions are explicitly arranged to play a larger role in behavior. For example, the left lever may be rich when the houselight flickers, but the right lever is rich when the houselight is steady. As these two houselight conditions alternate, the established response-reinforcer relationship between lever pressing and food will depend on the houselight. Accordingly, discrimination or contextual control processes are allowed to play a larger role in response rates on the two levers as the two response-reinforcer relationships (left- and right-lever pressing) are brought under the control of the context. If the houselight stimulus is presented and then removed before the animal can respond, then the procedure has changed to a memory task, and time and discrimination both become factors, but still associated with contextual control [1, 102]. This approach may be too complicated to study discrimination or memory processes *per se*, but has been used successfully to examine competing control between discrimination and reinforcement processes [1, 40, 115, 179, 180].

Note that in a "spatial discrimination reversal" procedure, an analysis similar to the one used with simple concurrent schedules applies, except that the difference in reinforcement rates on the two levers is more extreme: one reinforcer/response and extinction for one set of sessions and the reverse for a second set. This is a large difference between reinforcement rates that could increase the role of discrimination processes, but the dominant influence still seems to be the relationship between pressing a lever and receiving a reinforcer. After all, the largest change is in the response-reinforcer relationship (left lever produces reinforcement, then the right lever produces reinforcement); the animal always faces the same two levers.

This stands in contrast to a "visual discrimination reversal" which, although it sounds similar to a spatial discrimination reversal, may functionally be quite different. In the visual discrimination procedure, a light is presented over the lever that will produce a reinforcer. As usually implemented, this will be the left lever for half of the trials. Thus, the stimulus conditions change from trial to trial: sometimes the light is over the left lever and sometimes it is over the right. The discrimination is more complex because it enlists multiple sensory systems and ignores spatial position. The animal still faces two levers, as in the spatial discrimination reversal, but left lever presses are reinforced only if the light is on over the left lever (and lights can be aversive to nocturnal rats), and they are not reinforced if the light is over the right lever. We have conducted visual and spatial discrimination reversals with Long-Evans rats, and they are quite different procedures. The spatial discrimination can be acquired in 3 to 10 sessions, depending on the rat. The visual discrimination can take as few as 10 and as many as 90 sessions to acquire. This could represent a case in which it might be assumed that behavior is under the control of something simple, like which lever the light is over, but instead the controlling context is more complex and could be the entire configuration of stimuli (see Carter and Werner [22]).

The idea that discrimination and reinforcer processes are distinct, independent, and experimentally separable has been examined rigorously and quantitatively in empirical and theoretical treatments [1, 41, 101]. Thus, while it is impossible to remove the role of context or the role of consequences from studies of operant behavior, it is possible to arrange conditions so that the role of one process is minimal across all opportunities to behave while the role of the other is exaggerated. This minimizes or eliminates its role in differentiating behavior and permits the other set of processes to play a large role.

DOPAMINE, THE CORTEX, AND RESPONSE-REINFORCER RELATIONSHIPS

The complexities involved in separating learning and memory in behavior are shared with attempts to distinguish neural pathways. In a sense, the distribution of behavioral processes that contribute to the origin and maintenance of behavior is reflected in the distribution of function throughout the intact central nervous system. The behavioral distinction between a response-reinforcer relationship and the context in which it occurs, i.e., between reinforcement and discrimination or contextual control processes, is consistent with broad distinctions made in the structure and pharmacology of certain neural pathways [45]. It is well established that reinforcement processes underlying the role played by consequences in voluntary behavior are mediated by dopamine pathways [156, 164, 183]. These pathways originate with dopamine-containing cell bodies in midbrain structures, especially the ventral tegmental area.

Monoamine systems, including dopamine and noradrenergic pathways, in frontal areas also play a role in perseveration, impulsivity, and choice [51, 86, 108] as well as in behavior under progressive ratios of reinforcement [85, 107]. The effects on progressive ratios are generally consistent with our observations that fixed-ratio acquisition and progressive ratios are affected by MeHg exposure. The results of these studies and ours may differ in some important specifics. For example, lesions of the orbital-frontal cortex increased response rates under progressive-ratio schedules at smaller ratio values, consistent with our observations. These lesions also decreased the maximum ratio that the animal would tolerate, but that is not consistent with our observations.

The activity of midbrain dopamine-containing neurons is powerfully influenced by the delivery of reinforcing consequences. Midbrain dopamine neurons are important to the monitoring of reinforcing consequences and, especially, to discrepancies between a current consequent and previously presented ones, i.e., to "prediction errors" [159]. A suggestion that these neurons' activity may preclude choice derives from the presence of (a) midbrain dopamine-containing neurons that fire selectively when the reinforcer is larger than anticipated and (b) separate neurons that fire when the reinforcer is smaller than anticipated [60, 167].

Further evidence for the involvement of dopamine pathways in mediating choice and the response-reinforcer relationship derives from studies in which dopamine antagonists alter the preference for a food reinforcer without changing overall food intake. For example, when presented with a choice between lever pressing under an FR-5 schedule for a preferred food pellet and a standard chow pellet placed on the floor of the operant chamber, a rat will press a lever for the preferred food pellet. Administration of D1 and D2 antagonists does not alter the total food intake, but they do dramatically distort choice by shifting preference away from the preferred one and toward the previously nonpreferred alternative [157].

Distortions of the role of consequences can appear as impulsivity or perseveration. While these are not single behaviors with a specific neural mechanism, there is evidence that dopamine pathways originating in the VTA participate in many of the behaviors that are described as impulsive or perseverative [52]. Amphetamine, a dopamine agonist, can result in rapid switching between two response alternatives in a choice procedure, foster perseveration of lever pressing after a reinforcer is delivered [54], or promote persistent responding on a lever that has not produced reinforcement [53].

Studies in which the activity of individual neurons has been tracked as animals engage in certain voluntary behaviors suggest that the nucleus accumbens, basal ganglia, and frontal cortical structures are involved with the sorts of behavior that MeHg affects. Cortical processes and dopamine pathways seem to be the most likely candidate at present. Neurons in the nucleus accumbens appear to keep track of the type of reinforcer that has been presented [19, 20, 156]. Striatal neurons evolve specific response patterns only after some experience with a response-reinforcer relationship, and would therefore be involved with steady-state, "automatic" responding [156, 160]. Only at higher exposure levels does MeHg affect steady-state behavior more than behavior in transition, and only at higher levels does it appear in the basal ganglia.

Dopamine cell bodies in the midbrain send fibers to, among other places, the nucleus accumbens, striatum, and cortex [32, 58]. Experiments with nonhuman primates implicate the involvement of frontal cortical regions, especially the orbital frontal cortex, in choice. Neurons in frontal cortical areas come into play when a novel response-reinforcer relation is encountered, at least with a spatial task [160, 170]. In addition, cortical neurons have been identified that fire specifically when the subject (a monkey) is presented with the preferred of two consequences in a choice setting, regardless of the nature of the consequences [171].

Where they have been characterized, frontal cortical regions play a crucial role in the orchestration and expression of choice. For example, the firing rates of neurons from the orbital frontal cortex of monkeys were monitored as they chose between reinforcers of differing value [170]. Monkeys were trained with three stimuli, each paired with separate reinforcers: stimulus A was preferred to B, which was preferred to C. Neurons were identified that consistently responded to B when it was presented with C (i.e., B was preferred), but that responded to A, and not B, when A and B were presented. Thus, in this experimental context, these neurons came to respond to the better of two stimuli, suggesting that this cortical region participants in choice.

The delay-discounting procedure links choice to impulsivity. In this procedure, an animal is presented with two alternatives: a small reinforcer delivered immediately or a larger one delivered after a time delay. The larger, later reinforcer is discounted by the delay to its delivery. If, for example, the choice is between a dollar now or two dollars in one minute, then one might wait a minute for two dollars. However, if the choice is between one dollar now and two dollars in a year, the subject is likely to choose the smaller, immediate reinforcer. This procedure entails the scaling of both the relative magnitude of reinforcement and the relative delays. In a rodent model, lesions to the orbital prefrontal cortex of the rat significantly distort the scaling of both the relative magnitude and the delay in such a way that longer delays sometimes are tolerated [86].

Other areas of the cortex have also been implicated in the expression of choice. Neurons in the parietal cortex fire differentially for preferred rewards [140, 165]. The premotor cortex has been shown to be involved with discrimination reversal tasks, reminiscent of differential effects reported with methylmercury [136]. All this is not meant to imply that cortical areas play no role in memory. Certainly they do, especially those involving working memory [94], but many of the procedures used in neurotoxicity testing tap subcortical structures, such as the hippocampus.

Taken together, the cited literature indicates that the role of consequences in selecting responses, as well as in choice, perseveration, and perhaps in impulsive behavior, can be separated behaviorally from contextual control processes present in discrimination and memory tasks. Neural mechanisms underlying this behavioral category involve dopamine pathways originating in the midbrain and spreading to other regions, including the cortex. Cortical regions are clearly involved in mediating the role of consequences and of differential consequences arising from choice situations.

NEURAL EFFECTS OF MEHG EXPOSURE

The profile of MeHg's developmental behavioral toxicity profile is consistent with the sensitivity of laminar neural structures, such as the cortex, to developmental MeHg exposure. Some of the earliest animal models of MeHg neurotoxicity as well as postmortem examination of exposed humans implicate disruptions in cell migration and associated derangement of the layers characteristic of the neocortex or cerebellum [29, 154]. In other experiments with nonhuman primates, MeHg induces the presence of reactive glia in the visual cortex [27, 28].

The sensitivity of the neocortex has been noted also with developmental exposure to low-to-moderate levels of MeHg [8]. In that study, the effects were especially pronounced in the occipital neocortex, but were also noted in basal forebrain nuclei, hippocampus, and brain-stem nuclei. In the cortex, MeHg exposure resulted in distortions of the morphology and size of cortical lamina of 10- and 21-day old rats. This exposure interfered with the nerve growth factor (NGF) signal cascade (i.e., NGF transduction), resulting in a dramatic loss of neurites on neurons as well as decreased cell density and cell size in the occipital neocortex of rats [133–135]. In *in vitro* models (PC12 and cortical cells), MeHg's effects occurred after differentiation had taken place and were specific to the important developmental processes of dendritic elaboration and axogenesis. These are the processes by which neurons contact and communicate with one another and by which the activity of one portion of the nervous system influences or modulates activity in another.

Neurite outgrowth was impaired at mercury concentrations as low as about 30 nM, or about 0.006 ppm, in the medium, concentrations that were 40-fold lower than those that resulted in cytotoxicity. At first, this concentration sounds extraordinarily low, close to background levels, but the concentration in tissue is substantially higher than that in medium [104]. In the studies showing impaired neurite formation at 0.006 ppm, the concentration of mercury inside cortical cells was on the order of 0.3 to about 3 ppm. This compares with brain concentration of 0.5 to

9.5 ppm in neonates using the exposure protocol reported by Newland and colleagues, a protocol that altered choice in transition, high-rate behavior, and sensitivity to amphetamine [120, 123, 125].

Neurochemically, methylmercury exposure targets catecholamines, including dopamine, as well as GABA systems. Both dopamine and GABA neurotransmitter systems are closely linked to reinforcement processes. In adult rats, the presence of MeHg in striatum produces dose-dependent increases in dopamine in this region, one of the regions receiving dopamine projections that originate in the midbrain [55–57]. In *in vitro* striatal punches, the presence of MeHg increases the concentration of dopamine in extracellular media and decreases its concentration in the neuron, an effect consistent with enhanced release of dopamine [11]. Postnatal developmental exposure produced dose-dependent increases in the turnover of catecholamines (dopamine and norepinephrine) at 20 to 40 days of age. With developmental exposure, GABA systems in the cortex are also especially sensitive [130], although norepinephrine systems seem to be more sensitive in the caudate-putamen [131]. The aforementioned neurochemical effects were seen in the tissue of adults after developmental MeHg exposure or from tissue exposed *in vitro*, so they are relevant to developmental exposure.

The behavioral significance of these effects has been demonstrated using drug challenges in behaving animals. Developmental exposure to methylmercury (relatively high levels compared with contemporary exposure protocols) affect sensitivity to amphetamine, a dopamine and noradrenergic agonist in relatively young animals [17, 18, 47, 80, 153]. Moreover, these effects are specific to certain drug classes, extend into fully adult animals, and arise with dosing protocols involving relatively low exposure levels, suggesting that they represent permanent changes in the neurochemical tone of the intact nervous system under physiologically relevant conditions [142]. Adult rats exposed during gestation to MeHg (0.5 and 6 ppm, resulting in 40 and 500 µg/kg/day) were especially sensitive to amphetamine when tested as adults [142]. They were also relatively insensitive to pentobarbital, a GABA agonist. No specific sensitivity was demonstrated to a cholinergic or glutamatergic antagonist. An unpublished study in our lab shows no sensitivity to cholinergic or serotonergic agonists.

CONCLUSION

Methylmercury is a neurotoxicant that is well known to disrupt sensory-motor function with adult-onset exposure and, at high exposure levels, to produce cerebral palsy-like signs and mental retardation after developmental exposure. It is becoming evident that alterations in cognitive or intellectual function appear with low-level exposures. While these effects can be subtle, they can carry a significant economic cost [169].

The pattern of behavioral effects seen in animal studies of developmental MeHg exposure suggests a disruption in the acquisition or maintenance of a response-reinforcer relationship, with minimal contribution from contextual (stimulus)-control processes such as discrimination or even memory. The effects include disruption of choice, the acquisition of choice, persistent or perseverative responding on changing fixed-ratio schedules, and effects on discrimination reversal procedures suggestive of perseveration or diminished sensitivity to the consequences of behavior. Animal studies even suggest behavioral interventions to ameliorate MeHg's effects, interventions that entail the exaggeration of the discrepancy in the relative rate of reinforcement from concurrently available response alternatives. Behavioral procedures that tap memory processes are relatively insensitive to MeHg exposure. Where memorial processes do apply, they seem to involve what has been called "working memory," which entails stimuli that the animal must remember but that can change from task to task.

Studies of cortical function indicate that these are the very behavioral processes that affect the response-reinforcer relationship, choice, and perseveration. Drugs that act on dopamine systems have behavioral effects that include alterations in choice, in delay discounting, and in perseveration. The activity of neurons in dopamine-rich midbrain areas and of neurons in selected cortical areas is sensitive not just to reinforcer deliveries, but also to the relative value of different reinforcing consequences as compared with others that are available concurrently. This pattern of effects is, therefore, consistent with the neuroanatomical and neurochemical effects of developmental MeHg exposure.

Finally, in postmortem examination of methylmercury-exposed people as well as in experimental methylmercury exposures in laboratory rodents, the size and interconnectivity of the neocortex appears to be especially sensitive to developmental exposure. Psychopharmacological challenges have implicated both dopamine and GABAergic systems as being permanently altered by developmental methylmercury exposure. Taken together, these raise the possibility that developmental methylmercury exposure is a model of abnormal cortical development with effects that implicate the sensitivity of behavior to reinforcing consequences.

REFERENCES

1. Alsop, B. and Davison, M., Effects of varying stimulus disparity and the reinforcer ratio in concurrent schedule and signal-detection procedures., *J. Exp. Anal. Behav.*, 56, 67–80, 1991.
2. Amin-Zaki, L. and Majeed, M.A., Methylmercury poisoning in the Iraqi suckling infant: a longitudinal study over five years, *J. Appl. Toxicol.*, 1, 210–214, 1981.
3. Andrews, N., Miller, E., Grant, A., Stowe, J., Osborne, V., and Taylor, B., Thimerosal exposure in infants and developmental disorders: a retrospective cohort study in the United Kingdom does not support a causal association, *Pediatrics*, 114, 584–591, 2004.
4. ATSDR, Toxicological Profile for Selenium, U.S. Department of Health and Human Services, Agency for Toxic Substances and Disease Registry, Washington, DC, 2003.
5. Axtell, C.D., Cox, C., Myers, G.J., Davidson, P.W., Choi, A.L., Cernichiari, E., Sloane-Reeves, J., Shamlaye, C.F., and Clarkson, T.W., Association between methylmercury exposure from fish consumption and child development at five and a half years of age in the Seychelles Child Development Study: an evaluation of nonlinear relationships, *Environ. Res.*, 84, 71–80, 2000.
6. Bakir, F., Rustam, H., Tikriti, S., Al-Damluji, S.F., and Shihristani, H., Clinical and epidemiological aspects of methylmercury poisoning, *Postgrad. Med. J.*, 56, 1–10, 1980.

7. Baron, A. and Derenne, A., Progressive-ratio schedules: effects of later schedule requirements on earlier performances, *J. Exp. Anal. Behav.*, 73, 291–304, 2000; see also erratum in *J. Exp. Anal. Behav.*, 78, 94, 2002.

8. Barone, S., Jr., Haykal-Coates, N., Parran, D.K., and Tilson, H.A., Gestational exposure to methylmercury alters the developmental pattern of trk-like immunoreactivity in the rat brain and results in cortical dysmorphology, *Brain Res. Dev. Brain Res.*, 109, 13–31, 1998.

9. Bartolome, J., Trepanier, P., Chait, E.A., Seidler, F.J., Deskin, R., and Slotkin, T.A., Neonatal methylmercury poisoning in the rat: effects on development of central catecholamine neurotransmitter systems, *Toxicol. Appl. Pharmacol.*, 65, 92–99, 1982.

10. Baum, W.M. and Rachlin, H.C., Choice as time allocation, *J. Exp. Anal. Behav.*, 12, 861–874, 1969.

11. Bemis, J.C. and Seegal, R.F., Polychlorinated biphenyls and methylmercury act synergistically to reduce rat brain dopamine content *in vitro*, *Environ. Health Perspect.*, 107, 879–885, 1999.

12. Bernard, S., Enayati, A., Roger, H., Binstock, T., and Redwood, L., The role of mercury in the pathogenesis of autism, *Mol. Psychiatry*, 7, s42–43, 2002.

13. Beuter, A., De Geoffroy, A., and Edwards, R., Analysis of rapid alternating movements in Cree subjects exposed to methylmercury and in subjects with neurological deficits, *Environ. Res.*, 80, 64–79, 1999.

14. Bishop, D.V., Bishop, S.J., Bright, P., James, C., Delaney, T., and Tallal, P., Different origin of auditory and phonological processing problems in children with language impairment: evidence from a twin study, *J. Speech Language Hearing Res.*, 42, 155–168, 1999.

15. Bornhausen, M., Musch, H.R., and Greim, H., Operant behavior performance changes in rats after prenatal methylmercury exposure, *Toxicol. Appl. Pharmacol.*, 56, 305–310, 1980.

16. Bourre, J.M., Francois, M., Youyou, A., Dumont, O., Piciotti, M., Pascal, G., and Durand, G., The effects of dietary alpha-linolenic acid on the composition of nerve membranes, enzymatic activity, amplitude of electrophysiological parameters, resistance to poisons and performance of learning tasks in rats, *J. Nutr.*, 119, 1880-1892, 1989.

17. Buelke-Sam, J., Kimmel, C.A., Adams, J., Nelson, C.J., Vorhees, C.V., Wright, D.C., St. Omer, V., Korol, B.A., Butcher, R.E., and Geyer, M.A., Collaborative behavioral teratology study: results, *Neurobehav. Toxicol. Teratol.*, 7, 591–624, 1985.

18. Cagiano, R., De Salvia, M.A., Renna, G., Tortella, E., Braghiroli, D., Parenti, C., Zanoli, P., Baraldi, M., Annau, Z., and Cuomo, V., Evidence that exposure to methyl mercury during gestation induces behavioral and neurochemical changes in offspring of rats, *Neurotoxicol. Teratol.*, 12, 23–28, 1990.

19. Carelli, R.M., Nucleus accumbens cell firing during goal-directed behaviors for cocaine vs. "natural" reinforcement, *Physiol. Behav.*, 76, 379–387, 2002.

20. Carelli, R.M., Ijames, S.G., and Crumling, A.J., Evidence that separate neural circuits in the nucleus accumbens encode cocaine versus "natural" (water and food) reward, *J. Neurosci.*, 20, 4255–4266, 2000.

21. Carrie, I., Clement, M., De Javel, D., Frances, H., and Bourre, J.M., Learning deficits in first generation of mice deficient in (n-3) polyunsaturated fatty acids do not result from visual alteration, *Neurosci. Lett.*, 266, 69–72, 1999.

22. Carter, D.E. and Werner, T.J., Complex learning and information processing by pigeons: a critical analysis, *J. Exp. Anal. Behav.*, 29, 565–601, 1978.

23. Castoldi, A., Coccini, T., and Manzo, L., Neurotoxic and molecular effects of methylmercury in humans, *Rev. Environ. Health*, 18, 19–31, 2003.

24. Catania, A.C., Glossary, in *Experimental Analysis of Behavior: Part 2*, Iversen, I.H. and Lattal, K.A., Eds., Elsevier, Amsterdam, 1991, pp. G1–G44.

25. Catania, A.C. and Reynolds, G.S., A quantitative analysis of the responding maintained by interval schedules of reinforcement, *J. Exp. Anal. Behav.*, 11, 327–383, 1968.

26. Chapman, L. and Chan, H.M., The influence of nutrition on methylmercury intoxication, *Environ. Health Perspect.*, 108 (Suppl. 1), 29–56, 2000.

27. Charleston, J.S., Body, R.L., Mottet, N.K., Vahter, M.E., and Burbacher, T.M., Autometallographic determination of inorganic mercury distribution in the cortex of the calcarine sulcus of the monkey *Macaca fascicularis* following long-term subclinical exposure to methylmercury and mercuric chloride, *Toxicol. Appl. Pharmacol.*, 132, 325–333, 1995.

28. Charleston, J.S., Bolender, R.P., Mottet, N.K., Body, R.L., Vahter, M.E., and Burbacher, T.M., Increases in the number of reactive glia in the visual cortex of *Macaca fascicularis* following subclinical long-term methyl mercury exposure, *Toxicol. Appl. Pharmacol.*, 129, 196–206, 1994.

29. Choi, B.H., Lapham, L.W., Amin-Zaki, L., and Saleem, T., Abnormal neuronal migration, deranged cerebral cortical organization, and diffuse white matter astrocytosis of human fetal brain: a major effect of methylmercury poisoning *in utero*, *J. Neuropathol. Experimental Neurol.*, 37, 719–733, 1978.

30. Collins, R.J., Weeks, J.R., Cooper, M.M., Good, P.I., and Russell, R.R., Prediction of abuse liability of drugs using IV self-administration by rats, *Psychopharmacology*, 82, 6–13, 1984.

31. National Research Council, Committee on the Toxicological Effects of Methylmercury, *Toxicological Effects of Methylmercury*, National Academy Press, Washington, DC, 2000.

32. Cooper, J.R., Bloom, F.E., and Roth, R.H., *The Biochemical Basis of Neuropharmacology*, Oxford Press, New York, 2003.

33. Cory-Slechta, D.A., Bridging experimental animal and human behavioral toxicity studies, in *Behavioral Measures of Neurotoxicity*, National Academy Press, Washington, DC, 1990, pp. 137–158.

34. Cory-Slechta, D.A., Implications of changes in schedule-controlled operant behavior, in *Neurobehavioral Toxicity: Analysis and Interpretation*, Weiss, B. and O'Donoghue, J., Eds., Raven Press, New York, 1994, pp. 195–214.

35. Coscina, D.V., Yehuda, S., Dixon, L.M., Kish, S.J., and Leprohon-Greenwood, C.E., Learning is improved by a soybean oil diet in rats, *Life Sci.*, 38, 1789–1794, 1986.

36. Crump, K.S., Kjellstrom, T., Shipp, A.M., Silvers, A., and Stewart, A., Influence of prenatal mercury exposure upon scholastic and psychological test performance: benchmark analysis of a New Zealand cohort, *Risk Anal.*, 18, 701–713, 1998.

37. Davidson, P.W., Palumbo, D., Myers, G.J., Cox, C., Shamlaye, C.F., Sloane-Reeves, J., Cernichiari, E., Wilding, G.E., and Clarkson, T.W., Neurodevelopmental outcomes of Seychellois children from the pilot cohort at 108 months following prenatal exposure to methylmercury from a maternal fish diet, *Environ. Res.*, 84, 1–11, 2000.

38. Davidson, P.W., Myers, G.J., Cox, C., Axtell, C., Shamlaye, C., Sloane-Reeves, J., Cernichiari, E., Needham, L., Choi, A., Wang, Y., Berlin, M., and Clarkson, T.W., Effects of prenatal and postnatal methylmercury exposure from fish consumption on neurodevelopment: outcomes at 66 months of age in the Seychelles Child Development Study, *JAMA*, 280, 701–707, 1998.

39. Davison, M. and McCarthy, D., *The Matching Law: A Research Review*, Erlbaum, Hillsdale, NJ, 1988.

40. Davison, M. and McCarthy, D., *Matching Models of Signal Detection: The Matching Law*, Erlbaum, Hillsdale, NJ, 1988, pp. 216–248.

41. Davison, M. and Nevin, J.A., Stimuli, reinforcers, and behavior: an integration, *J. Exp. Anal. Behav.*, 71, 439–482, 1999.

42. Davison, M.C. and Hunter, I.W., Concurrent schedules: undermatching and control by previous experimental conditions, *J. Exp. Anal. Behav.*, 32, 233–244, 1979.

43. Day, J., Reed, M., and Newland, M.C., Neuromotor deficits in aging rats exposed to methylmercury and n-3 fatty acids, *Neurotoxicol. Teratol.*, 27, 629–642, 2005.

44. Derenne, A. and Baron, A., Preratio pausing: effects of an alternative reinforcer on fixed- and variable-ratio responding, *J. Exp. Anal. Behav.*, 77, 273–282, 2002.

45. Donahoe, J.W., Burgos, J.E., and Palmer, D.C., A selectionist approach to reinforcement, *J. Exp. Anal. Behav.*, 60, 17–40, 1993.

46. Dore, F.Y., Goulet, S., Gallagher, A., Harvey, P.O., Cantin, J.F., D'aigle, T., and Mirault, M.E., Neurobehavioral changes in mice treated with methylmercury at two different stages of fetal development, *Neurotoxicol. Teratol.*, 23, 463–472, 2001.

47. Eccles, C.U. and Annau, Z., Prenatal methyl mercury exposure, II: alterations in learning and psychotropic drug sensitivity in adult offspring, *Neurobehav. Toxicol. Teratol.*, 4, 377–382, 1982.

48. Egeland, G.M. and Middaugh, J.P., Balancing fish consumption benefits with mercury exposure, *Science*, 278, 1904–1905, 1997.

49. EPA, Mercury Report to Congress: Health Effects of Mercury and Mercury Compounds, U.S. Environmental Protection Agency, Washington, DC, 1997, p. 349.

50. Evans, H.L., Garman, R.H., and Laties, V.G., Neurotoxicity of methylmercury in the pigeon, *Neurotoxicology*, 3, 21–36, 1982.

51. Evenden, J., The pharmacology of impulsive behavior in rats, V: the effects of drugs on responding under a discrimination task using unreliable visual stimuli, *Psychopharmacology*, 143, 111–122, 1999.

52. Evenden, J.L., Varieties of impulsivity, *Psychopharmacology*, 146, 348–361, 1999.

53. Evenden, J.L. and Robbins, T.W., Dissociable effects of d-amphetamine, chlordiazepoxide and alpha-flupenthixol on choice and rate measures of reinforcement in the rat, *Psychopharmacology*, 79, 180–186, 1983.

54. Evenden, J.L. and Robbins, T.W., Increased response switching, perseveration and perseverative switching following d-amphetamine in the rat, *Psychopharmacology*, 80, 67–73, 1983.

55. Faro, L.R., Do Nascimento, J.L., Alfonso, M., and Duran, R., Acute administration of methylmercury changes *in vivo* dopamine release from rat striatum, *Bull. Environ. Contamination Toxicol.*, 60, 632–638, 1998.

56. Faro, L.R., Do Nascimento, J.L., Alfonso, M., and Duran, R., Mechanism of action of methylmercury on *in vivo* striatal dopamine release: possible involvement of dopamine transporter, *Neurochem. Int.*, 40, 455–465, 2002.

57. Faro, L.R., Duran, R., Do Nascimento, J.L., Alfonso, M., and Picanco-Diniz, C.W., Effects of methyl mercury on the *in vivo* release of dopamine and its acidic metabolites dopac and hva from striatum of rats, *Ecotoxicol. Environmental Saf.*, 38, 95–98, 1997.

58. Feldman, R.S., Meyer, J.S., and Quenzer, L.F., *Principles of Neuropsychopharmacology*, Sinauer Associates, Sunderland, MA, 1997.

59. Findley, J.D., Preference and switching under concurrent scheduling, *J. Exp. Anal. Behav.*, 1, 123–144, 1958.

60. Fiorillo, C.D., Tobler, P.N., and Schultz, W., Discrete coding of reward probability and uncertainty by dopamine neurons, *Science*, 299, 1898–1902, 2003.

61. Fitch, R.H., Miller, S., and Tallal, P., Neurobiology of speech perception, *Ann. Rev. Neurosci.*, 20, 331–353, 1997.

62. Fombonne, E., Epidemiological trends in rates of autism, *Molecular Psychiatry*, 7 (Suppl. 2), S4–6, 2002.

63. Fowler, S.C., Force and duration of operant responses as dependent variables in behavioral pharmacology, in *Advances in Behavioral Pharmacology: Neurobehavioral Pharmacology*, Thompson, T., Dews, P.B., and Barrett, J.E., Eds., Erlbaum, Hillsdale, NJ, 1987, pp. 83–128.

64. Fowler, S.C., Filewich, R.J., and Leberer, M.R., Drug effects upon force and duration of response during fixed-ratio performance in rats, *Pharmacol. Biochem. Behav.*, 6, 421–426, 1977.

65. Fowler, S.C., Gramling, S.E., and Liao, R.M., Effects of pimozide on emitted force, duration and rate of operant response maintained at low and high levels of required force, *Pharmacol. Biochem. Behav.*, 25, 615–622, 1986.

66. Fowler, S.C., Lewis, R.M., Gramling, S.E., and Nail, G.L., Chlordiazepoxide increases the force of two topographically distinct operant responses in rats, *Pharmacol. Biochem. Behav.*, 19, 787–790, 1983.

67. Frances, H., Coudereau, J.P., Sandouk, P., Clement, M., Monier, C., and Bourre, J.M., Influence of a dietary alpha-linolenic acid deficiency on learning in the Morris water maze and on the effects of morphine, *Eur. J. Pharmacol.*, 298, 217–225, 1996.

68. Ganther, H., Goudie, C., Sunde, M., Kopecky, M., and Wagner, P., Selenium: relation to decreased toxicity of methylmercury added to diets containing tuna, *Science*, 175, 1122–1124, 1972.

69. Geier, D. and Geier, M., An assessment of the impact of thimerosal on childhood neurodevelopmental disorders, *Pediatr. Rehabil.*, 6, 97–102, 2003.

70. Gilbert, S.G., Burbacher, T.M., and Rice, D.C., Effects of *in utero* methylmercury exposure on a spatial delayed alternation task in monkeys, *Toxicol. Appl. Pharmacol.*, 123, 130–136, 1993.

71. Goldey, E.S., O'Callaghan, J.P., Stanton, M.E., Barone, S., Jr., and Crofton, K.M., Developmental neurotoxicity: evaluation of testing procedures with methylazoxymethanol and methylmercury, *Fundam. Appl. Toxicol.*, 23, 447–464, 1994.

72. Goulet, S., Dore, F.Y., and Mirault, M.E., Neurobehavioral changes in mice chronically exposed to methylmercury during fetal and early postnatal development, *Neurotoxicol. Teratol.*, 25, 335–347, 2003.

73. Grandjean, P., Weihe, P., White, R.F., and Debes, F., Cognitive performance of children prenatally exposed to "safe" levels of methylmercury, *Environ. Res.*, 77, 165–172, 1998.

74. Grandjean, P., Weihe, P., White, R.F., Debes, F., Araki, S., Yokoyama, K., Murata, K., Sorensen, N., Dahl, R., and Jorgensen, P.J., Cognitive deficit in 7-year-old children with prenatal exposure to methylmercury, *Neurotoxicol. Teratol.*, 19, 417–428, 1997.

75. Griffiths, R.R., Brady, J.V., and Bradford, L.D., Predicting the abuse liability of drugs with animal drug self-administration procedures: psychomotor stimulants and hallucinogens, in *Advances in Behavioral Pharmacology*, Academic Press, New York, 1977, pp. 164–208.

76. Gunderson, V., Grant, K., Burbacher, T., Fagan, J., and Mottet, N., The effect of low-level prenatal methylmercury exposure on visual recognition memory in infant crab-eating macaques, *Child Dev.*, 54, 1076–1083, 1986.

77. Gunderson, V.M., Grant-Webster, K.S., Burbacher, T.M., and Mottet, N.K., Visual recognition memory deficits in methylmercury-exposed *Macaca fascicularis* infants, *Neurotoxicol. Teratol.*, 10, 373–379, 1988.
78. Harada, M., Minamata disease: methylmercury poisoning in Japan caused by environmental pollution, *Crit. Rev. Toxicol.*, 25, 1–24, 1995.
79. Heron, J., Golding, J., and Team, A.S., Thimerosal exposure in infants and developmental disorders: a prospective cohort study in the United Kingdom does not support a causal association, *Pediatrics*, 114, 577–583, 2004.
80. Hughes, J.A. and Sparber, S.B., D-amphetamine unmasks postnatal consequences of exposure to methylmercury *in utero*: methods for studying behavioral teratogenesis, *Pharmacol. Biochem. Behav.*, 8, 365–375, 1978.
81. Imura, N. and Naganuma, A., Possible mechanisms of detoxifying effect of selenium on the toxicity of mercury compounds, in *Advances in Mercury Toxicity*, Suzuki, T., Imura, N., and Clarkson, T., Eds., Plenum Press, New York, 1991, p. 490.
82. Iwata, H., Okamoto, H., and Ohsawa, Y., Effect of selenium on methylmercury poisoning, *Res. Commn. Chemical Pathol. Pharmacol.*, 5, 673–680, 1973.
83. Johnston, J.M. and Pennypacker, H.S., *Strategies and Tactics of Behavioral Research*, Erlbaum, Hillsdale, NJ, 1993.
84. Kennedy, R.F.J., Deadly immunity, *Rolling Stone,* June 20, 2005.
85. Kheramin, S., Body, S., Herrera, F.M., Bradshaw, C.M., Szabadi, E., Deakin, J.F., and Anderson, I.M., The effect of orbital prefrontal cortex lesions on performance on a progressive ratio schedule: implications for models of inter-temporal choice, *Behav. Brain Res.*, 156, 145–152, 2005.
86. Kheramin, S., Body, S., Mobini, S., Ho, M.Y., Velazquez-Martinez, D.N., Bradshaw, C.M., Szabadi, E., Deakin, J.F., and Anderson, I.M., Effects of quinolinic acid-induced lesions of the orbital prefrontal cortex on inter-temporal choice: a quantitative analysis, *Psychopharmacology*, 165, 9–17, 2002.
87. Kinjo, Y., Higashi, H., Nakano, A., Sakamoto, M., and Sakai, R., Profile of subjective complaints and activities of daily living among current patients with Minamata disease after three decades, *Environ. Res.*, 63, 241–251, 1993.
88. Kollins, S.H., Newland, M.C., and Critchfield, T.S., Human sensitivity to reinforcement in operant choice: how much do consequences matter? *Psychonom. Bull. Rev.*, 4, 208–220, 1997.
89. Komsta-Szumska, E., Reuhl, K., and Miller, D., The effect of methylmercury on the distribution and excretion of selenium by the guinea pig, *Arch. Toxicol.*, 54, 303–310, 1983.
90. Kulig, B.M. and Jaspers, R.M.A., Assessment techniques for detecting neurobehavioral toxicity, in *Introduction to Neurobehavioral Toxicology*, Niesink, R.J.M., Jaspers, R.M.A., Kornet, L.M.W., van Ree, J.M., and Tilson, H.A., Eds., CRC Press, Boca Raton, FL, 1999, pp. 70–113.
91. Lamptey, M.S. and Walker, B.L., A possible role for dietary linolenic acid in the development of the young rat, *J. Nutr.*, 106, 86–93, 1976.
92. Lattal, A., Scheduling positive reinforcers, in *Techniques in the Behavioral and Neural Sciences: Experimental Analysis of Behavior*, Elsevier, Amsterdam, 1991, pp. 87–134.
93. Lebel, J., Mergler, D., Branches, F., Lucotte, M., Amorim, M., Larribe, F., and Dolbec, J., Neurotoxic effects of low-level methylmercury contamination in the Amazonian basin, *Environ. Res.*, 79, 20–32, 1998.
94. Lidow, M.S., Williams, G.V., and Goldman-Rakic, P.S., The cerebral cortex: a case for a common site of action of antipsychotics, *Trends Pharmacol. Sci.*, 19, 136–140, 1998.

95. Magos, L., Review on the toxicity of ethylmercury, including its presence as a preservative in biological and pharmaceutical products, *J. Appl. Toxicol.*, 21, 1–5, 2001.

96. Mahaffey, K.R., Fish and shellfish as dietary sources of methylmercury and the -3 fatty acids, eicosahexaenoic acid and docosahexaenoic acid: risks and benefits, *Environ. Res.*, 95, 414–428, 2004.

97. Markowski, V.P., Zareba, G., Stern, S., Cox, C., and Weiss, B., Altered operant responding for motor reinforcement and the determination of benchmark doses following perinatal exposure to low-level 2,3,7,8-tetrachlorodibenzo-p-dioxin, *Environ. Health Perspect.*, 109, 621–627, 2001.

98. Marr, M.J., Second-order schedules and the generation of unitary response sequences, in *Advances in the Analysis of Behavior*, Vol. 1, *Reinforcement and the Organization of Behavior*, Zeiler, M.D. and Harzem, P., Eds., Wiley, New York, 1979, pp. 223–260.

99. Marsh, D.O., Dose-response relationships in humans: methyl mercury epidemics in Japan and Iraq, in *The Toxicity of Methyl Mercury*, Johns Hopkins University Press, Baltimore, MD, 1987, pp. 45–53.

100. Maurissen, J.P.J., Neurobehavioral methods for the evaluation of sensory functions, in *Neurotoxicology: Approaches and Methods*, Academic Press, San Diego, 1995, pp. 239–264.

101. McCarthy, D. and Davison, M.C., Independence of sensitivity to relative reinforcement rate and discriminability in signal detection, *J. Exp. Anal. Behav.*, 34, 273–284, 1980.

102. McCarthy, D.C. and Davison, M.D., The interaction between stimulus and reinforcer control on remembering, *J. Exp. Anal. Behav.*, 86, 51–66, 1991.

103. McKay, S.J., Reynolds, J.N., and Racz, W.J., Differential effects of methylmercuric chloride and mercuric chloride on the l-glutamate and potassium evoked release of [3h]dopamine from mouse striatal slices, *Canadian J. Physiol. Pharmacol.*, 64, 656–660, 1986.

104. Meacham, C.A., Freudenrich, T.M., Anderson, W.L., Sui, L., Lyons-Darden, T., Barone, S., Jr., Gilbert, M.E., Mundy, W.R., and Shafer, T.J., Accumulation of methylmercury or polychlorinated biphenyls in *in vitro* models of rat neuronal tissue, *Toxicol. Appl. Pharmacol.*, 205, 177–187, 2005.

105. Miller, D.B. and Eckerman, D.A., Learning and memory measures, in *Neurobehavioral Toxicology*, Annau, Z., Ed., Johns Hopkins University Press, Baltimore, MD, 1986, pp. 94–149.

106. Minnema, D.J., Cooper, G.P., and Greenland, R.D., Effects of methylmercury on neurotransmitter release from rat brain synaptosomes, *Toxicol. Appl. Pharmacol.*, 99, 510–521, 1989.

107. Mobini, S., Chiang, T.J., Ho, M.Y., Bradshaw, C.M., and Szabadi, E., Comparison of the effects of clozapine, haloperidol, chlorpromazine and d-amphetamine on performance on a time-constrained progressive ratio schedule and on locomotor behaviour in the rat, *Psychopharmacology*, 152, 47–54, 2000.

108. Mobini, S., Body, S., Ho, M.Y., Bradshaw, C.M., Szabadi, E., Deakin, J.F., and Anderson, I.M., Effects of lesions of the orbitofrontal cortex on sensitivity to delayed and probabilistic reinforcement, *Psychopharmacology*, 160, 290–298, 2002.

109. Myers, G., Davidson, P., Cox, C., Shamlaye, C., Tanner, M., Marsh, D., Cernichiari, E., Lapham, L., Berlin, M., and Clarkson, T., Summary of the Seychelles Child Development Study on the relationship of fetal methylmercury exposure to neurodevelopment, *Neurotoxicology*, 16, 711–716, 1995.

110. Myers, G.J., Davidson, P.W., Cox, C., Shamlaye, C., Cernichiari, E., and Clarkson, T.W., Twenty-seven years studying the human neurotoxicity of methylmercury exposure, *Environ. Res.*, 83, 275–285, 2000.

111. Myers, G.J., Davidson, P.W., Shamlaye, C.F., Axtell, C.D., Cernichiari, E., Choisy, O., Choi, A., Cox, C., and Clarkson, T.W., Effects of prenatal methylmercury exposure from a high fish diet on developmental milestones in the Seychelles Child Development Study, *Neurotoxicology*, 18, 819–829, 1997.

112. National Research Council, Toxicological Effects of Methylmercury, National Academy Press, Washington, DC, 2000.

113. National Research Council, Immunization Safety Review: Thimerosal-Containing Vaccines and Neurodevelopmental Disorders, National Academy Press, Washington, DC, 2001.

114. National Research Council, Immunization Safety Review: Vaccines and Autism, National Academy Press, Washington, DC, 2004.

115. Nevin, J.A., Cate, H., and Alsop, B., Effects of differences between stimuli, responses, and reinforcer rates on conditional discrimination performance, *J. Exp. Anal. Behav.*, 59, 147–161, 1993.

116. Newland, M.C., Operant behavior and the measurement of motor dysfunction, in *Neurobehavioral Toxicity: Analysis and Interpretation,* Weiss, B. and O'Donoghue, J.L., Eds., Raven Press, New York, 1994, pp. 273–297.

117. Newland, M.C., Motor function and the physical properties of the operant: applications to screening and advanced techniques, in *Neurotoxicology: Approaches and Methods*, Chang, L.W. and Slikker, W., Eds., Academic Press, San Diego, 1995, pp. 265–299.

118. Newland, M.C. and Weiss, B., Persistent effects of manganese on effortful responding and their relationship to manganese accumulation in the primate globus pallidus, *Toxicol. Appl. Pharmacol.*, 113, 87–97, 1992.

119. Newland, M.C. and Reile, P.A., Learning and behavior change as neurotoxic endpoints., in *Target Organ Series: Neurotoxicology,* Tilson, H.A. and Harry, J., Eds., Raven Press, New York, 1999, pp. 311–338.

120. Newland, M.C. and Reile, P.A., Blood and brain mercury levels after chronic gestational exposure to methylmercury in rats, *Toxicological Sciences*, 50, 106–116, 1999.

121. Newland, M.C. and Paletz, E.M., Animal studies of methylmercury and PCBs: what do they tell us about expected effects in humans? *Neurotoxicology*, 21, 1003–1027, 2000.

122. Newland, M.C. and Rasmussen, E.B., Aging unmasks adverse effects of gestational exposure to methylmercury in rats, *Neurotoxicol. Teratol.*, 22, 819–828, 2000.

123. Newland, M.C. and Rasmussen, E.B., Behavior in adulthood and during aging is affected by contaminant exposure *in utero*, *Curr. Directions Psychological Sci.*, 12, 212–217, 2003.

124. Newland, M.C., Warfvinge, K., and Berlin, M., Behavioral consequences of *in utero* exposure to mercury vapor: alterations in lever-press durations and learning in squirrel monkeys, *Toxicol. Appl. Pharmacol.*, 139, 374–386, 1996.

125. Newland, M.C., Reile, P.A., and Langston, J.L., Gestational exposure to methylmercury retards choice in transition in aging rats, *Neurotoxicol. Teratol.*, 26, 179–194, 2004.

126. Newland, M.C., Yezhou, S., Logdberg, B., and Berlin, M., Prolonged behavioral effects of *in utero* exposure to lead or methyl mercury: reduced sensitivity to changes in reinforcement contingencies during behavioral transitions and in steady state, *Toxicol. Appl. Pharmacol.*, 126, 6–15, 1994.

127. Newland, M.C., Ng, W.W., Baggs, R.B., Gentry, G.D., Weiss, B., and Millar, R.K., Operant behavior in transition reflects neonatal exposure to cadmium, *Teratology*, 34, 231–241, 1986.

128. O'Kusky, J., Synaptic degeneration in rat visual cortex after neonatal administration of methylmercury, *Exp. Neurol.*, 89, 32–47, 1985.
129. O'Kusky, J.R. and McGeer, E.G., Methylmercury poisoning of the developing nervous system in the rat: decreased activity of glutamic acid decarboxylase in cerebral cortex and neostriatum, *Brain Res.*, 353, 299–306, 1985.
130. O'Kusky, J.R. and McGeer, E.G., Methylmercury-induced movements and postural disorders in developing rat: high-affinity uptake of choline, glutamate, and gamma-aminobutyric acid in the cerebral cortex and caudate-putamen, *Neurochemistry*, 53, 999–1006, 1989.
131. O'Kusky, J.R., Boyes, B.E., and McGeer, E.G., Methylmercury-induced movement and postural disorders in developing rat: regional analysis of brain catecholamines and indoleamines, *Brain Res.*, 439, 138–146, 1988.
132. Paletz, E.M., Craig-Schmidt, M.C., and Newland, M.C., Gestational exposure to methylmercury and n-3 fatty acids: effects on high- and low-rate operant behavior in adulthood, *Neurotoxicol. Teratol.*, 28, 59–73, 2006.
133. Parran, D.K., Mundy, W.R., and Barone, S., Jr., Effects of methylmercury and mercuric chloride on differentiation and cell viability in pc12 cells, *Toxicological Sciences*, 59, 278–290, 2001.
134. Parran, D.K., Barone, S., Jr., and Mundy, W.R., Methylmercury decreases ngf-induced trka autophosphorylation and neurite outgrowth in pc12 cells, *Brain Res. Dev. Brain Res.*, 141, 71–81, 2003.
135. Parran, D.K., Barone, S., Jr., and Mundy, W.R., Methylmercury inhibits trka signaling through the erk1/2 cascade after ngf stimulation of pc12 cells, *Brain Res. Dev. Brain Res.*, 149, 53–61, 2004.
136. Passingham, R.E., Myers, C., Rawlins, N., Lightfoot, V., and Fearn, S., Premotor cortex in the rat, *Behav. Neurosci.*, 102, 101–109, 1988.
137. Paule, M.G., Forrester, T.M., Maher, M.A., Cranmer, J.M., and Allen, R.R., Monkey versus human performance in the nctr operant test battery, *Neurotoxicol. Teratol.*, 12, 503–507, 1990.
138. Paule, M.G., Chelonis, J.J., Buffalo, E.A., Blake, D.J., and Casey, P.H., Operant test battery performance in children: correlation with IQ, *Neurotoxicol. Teratol.*, 21, 223–230, 1999.
139. Paulus, M.P., Hozack, N., Frank, L., Brown, G.G., and Schuckit, M.A., Decision making by methamphetamine-dependent subjects is associated with error-rate-independent decrease in prefrontal and parietal activation, *Biol. Psychiatry*, 53, 65–74, 2003.
140. Perone, M. and Courtney, K., Fixed-ratio pausing: joint effects of past reinforcer magnitude and stimuli correlated with upcoming magnitude, *J. Exp. Anal. Behav.*, 57, 33–46, 1992.
141. Prohaska, J. and Ganther, H., Interactions between selenium and methylmercury in rat brain, *Chem. Biol. Interact.*, 16, 155–167, 1977.
142. Rasmussen, E.B. and Newland, M.C., Developmental exposure to methylmercury alters behavioral sensitivity to *d* amphetamine and pentobarbital in adult rats, *Neurotoxicol. Teratol.*, 23, 45–55, 2001.
143. Rice, D.C., Quantification of operant behavior, *Toxicol. Lett.*, 43, 361–379, 1988.
144. Rice, D.C., Delayed neurotoxicity in monkeys exposed developmentally to methylmercury, *Neurotoxicology*, 10, 645–650, 1989.
145. Rice, D.C., Effects of pre- plus postnatal exposure to methylmercury in the monkey on fixed interval and discrimination reversal performance, *Neurotoxicology*, 13, 443–452, 1992.

146. Rice, D.C., Testing effects of toxicants on sensory system function by operant methodology, in *Neurobehavioral Toxicity: Analysis and Interpretation*, Raven Press, New York, 1994, pp. 299–318.

147. Rice, D.C., Sensory and cognitive effects of developmental methylmercury exposure in monkeys, and a comparison to effects in rodents, *Neurotoxicology*, 17, 139–154, 1996.

148. Rice, D.C. and Gilbert, S.G., Effects of developmental exposure to methyl mercury on spatial and temporal visual function in monkeys, *Toxicol. Appl. Pharmacol.*, 102, 151–163, 1990.

149. Rice, D.C. and Gilbert, S.G., Exposure to methyl mercury from birth to adulthood impairs high-frequency hearing in monkeys, *Toxicol. Appl. Pharmacol.*, 115, 6–10, 1992.

150. Rice, D.C. and Gilbert, S.G., Effects of developmental methylmercury exposure or lifetime lead exposure on vibration sensitivity function in monkeys, *Toxicol. Appl. Pharmacol.*, 134, 161–169, 1995.

151. Rodefer, J.S. and Carroll, M.E., Progressive ratio and behavioral economic evaluation of the reinforcing efficacy of orally delivered phencyclidine and ethanol in monkeys: effects of feeding conditions, *Psychopharmacology*, 128, 265–273, 1996.

152. Rodefer, J.S. and Carroll, M.E., A comparison of progressive ratio schedules versus behavioral economic measures: effect of an alternative reinforcer on the reinforcing efficacy of phencyclidine, *Psychopharmacology*, 132, 95–103, 1997.

153. Rossi, A.D., Ahlbom, E., Ogren, S.O., Nicotera, P., and Ceccatelli, S., Prenatal exposure to methylmercury alters locomotor activity of male but not female rats, *Exp. Brain Res.*, 117, 428–436, 1997.

154. Sager, P.R., Doherty, R.A., and Rodier, P.M., Effects of methylmercury on developing mouse cerebellar cortex, *Exp. Neurol.*, 77, 179–193, 1982.

155. Sakamoto, M., Wakabayashi, K., Kakita, A., Hitoshi, T., Adachi, T., and Nakano, A., Widespread neuronal degeneration in rats following oral administration of methylmercury during the postnatal developing phase: a model of fetal-type Minamata disease, *Brain Res.*, 784, 351–354, 1998.

156. Salamone, J.D., Correa, M., Mingote, S., and Weber, S.M., Nucleus accumbens dopamine and the regulation of effort in food-seeking behavior: implications for studies of natural motivation, psychiatry, and drug abuse, *J. Pharmacol. Exp. Ther.*, 305, 1–8, 2003.

157. Salamone, J.D., Arizzi, M.N., Sandoval, M.D., Cervone, K.M., and Aberman, J.E., Dopamine antagonists alter response allocation but do not suppress appetite for food in rats: contrast between the effects of skf 83566, raclopride, and fenfluramine on a concurrent choice task, *Psychopharmacology*, 160, 371–380, 2002.

158. Schick, K., Operants, *J. Exp. Anal. Behav.*, 15, 413–423, 1971.

159. Schultz, W., Dayan, P., and Montague, P.R., A neural substrate of prediction and reward, *Science*, 275, 1593–1599, 1997.

160. Schultz, W., Tremblay, L., and Hollerman, J.R., Reward processing in primate orbitofrontal cortex and basal ganglia, *Cereb. Cortex*, 10, 272–284, 2000.

161. Shull, R.L., The sensitivity of response rate to rate of variable-interval reinforcement for pigeons and rats: a review, *J. Exp. Anal. Behav.*, 84, 99–109, 2005.

162. Shull, R.L.G., Scott, T., and Grimes, J.A., Response rate viewed as engagement bouts: effects of relative reinforcement and schedule type, *J. Exp. Anal. Behav.*, 75, 247–274, 2001.

163. Sidman, M., *Tactics of Scientific Research*, Basic, New York, 1960.

164. Spanagel, R. and Weiss, F., The dopamine hypothesis of reward: past and current status, *Trends Neurosciences*, 22, 521–527, 1999.

165. Sugrue, L.P., Corrado, G.S., and Newsome, W.T., Matching behavior and the representation of value in the parietal cortex, *Science*, 304, 1782–1787, 2004.
166. Thomas, D. and Smith, J.C., Effects of coadministered sodium selenite on short-term distribution of methyl mercury in the rat, *Environ. Res.*, 34, 287–294, 1984.
167. Tobler, P.N., Fiorillo, C.D., and Schultz, W., Adaptive coding of reward value by dopamine neurons, *Science*, 307, 1642–1645, 2005.
168. Todorov, J.C., Concurrent performances: effect of punishment contingent on the switching response, *J. Exp. Anal. Behav.*, 16, 51–62, 1971.
169. Trasande, L., Landrigan, P.J., and Schecter, C., Public health and economic consequences of methylmercury toxicity to the developing brain, *Environ. Health Perspect.*, 113, 590–596, 2005.
170. Tremblay, L. and Schultz, W., Relative reward preference in primate orbitofrontal cortex, *Nature*, 398, 704–708, 1999.
171. Villiers, P.A.D., Choice in concurrent schedules and a quantitative formulation of the law of effect, in *Handbook of Operant Behavior*, Prentice-Hall, Englewood Cliffs, NJ, 1977, pp. 233–287.
172. Wainwright, P., Huang, Y., Bulman-Fleming, B., Levesque, S., and McCutcheon, D., The effects of dietary fatty acid composition combined with environmental enrichment on brain and behavior in mice, *Behavioural Brain Res.*, 60, 125–136, 1994.
173. Walsh, T.J. and Tilson, H.A., The use of pharmacological challenges, in *Neurobehavioral Toxicology*, Johns Hopkins University Press, Baltimore, MD, 1986, pp. 244–267.
174. Watanabe, C. and Satoh, H., Evolution of our understanding of methylmercury as a health threat, *Environ. Health Perspect.*, 104 (Suppl. 2), 367–379, 1996.
175. Watanabe, C., Yin, K., Kasanuma, Y., and Satoh, H., *In utero* exposure to methylmercury and se deficiency converge on the neurobehavioral outcome in mice, *Neurotoxicol. Teratol.*, 21, 83–88, 1999.
176. Weiss, B., *Microproperties of Operant Behavior as Aspects of Toxicity: Quantification of Steady-State Operant Behavior*, Elsevier, Amsterdam, 1981, pp. 249–265.
177. Weiss, B. and Cory-Slechta, D.A., Assessment of behavioral toxicity, in *Principles and Methods of Toxicology*, Raven Press, New York, 1994, pp. 1091–1155.
178. Whanger, P., Selenium in the treatment of heavy metal poisoning and chemical carcinogenesis, *J. Trace Elements Electrolytes Health Disease*, 6, 209–221, 1992.
179. White, K.G. and McKenzie, J., Delayed stimulus control: recall for single and relational stimuli, *J. Exp. Anal. Behav.*, 38, 305–312, 1982.
180. White, K.G. and Wixted, J.T., Psychophysics of remembering, *J. Exp. Anal. Behav.*, 71, 91–113, 1999.
181. WHO, Evaluation of certain food additives and contaminants, in *WHO Technical Report Series*, World Health Organization, Geneva, 2004.
182. Widholm, J.J., Villareal, S., Seegal, R.F., and Schantz, S.L., Spatial alternation deficits following developmental exposure to Aroclor 1254 and/or methylmercury in rats, *Toxicological Sciences*, 82, 577–589, 2004.
183. Wise, R.A., Dopamine, learning and motivation, *Nat. Rev. Neurosci.*, 5, 483–494, 2004.
184. Wood, R.W., Determinants of irritant termination behavior, *Toxicol. Appl. Pharmacol.*, 61, 260–268, 1981.
185. Yamamoto, N., Saitoh, M., Moriuchi, A., Nomura, M., and Okuyama, H., Effect of dietary alpha-linolenate/linoleate balance on brain lipid compositions and learning ability of rats, *J. Lipid Res.*, 28, 144–151, 1987.

186. Yamamoto, N., Hashimoto, A., Takemoto, Y., Okuyama, H., Nomura, M., Kitajima, R., Togashi, T., and Tamai, Y., Effect of the dietary alpha-linolenate/linoleate balance on lipid compositions and learning ability of rats, II: discrimination process, extinction process, and glycolipid compositions, *J. Lipid Res.*, 29, 1013–1021, 1988.
187. Zeiler, M., Schedules of reinforcement: the controlling variables, in *Handbook of Operant Behavior*, Honig, W.K. and Staddon, J.E.R., Eds., Prentice-Hall, Englewood Cliffs, NJ, 1977, pp. 201–232.
188. Rayman, M.P., The importance of selenium to human health, *Lancet*, 356(9225), 223–241, 2000.

8 Executive Function following Developmental Exposure to Polychlorinated Biphenyls (PCBs): What Animal Models Have Told Us

Helen J.K. Sable and Susan L. Schantz
University of Illinois at Urbana–Champaign

CONTENTS

INTRODUCTION

Polychlorinated biphenyls (PCBs) were first manufactured in the 1930s for industrial use. While they were eventually banned in the 1970s, their extreme stability allowed them to bioaccumulate over time, such that they remain one of the most prevalent environmental contaminants. A large number of research studies have focused on the neurotoxic properties of PCBs. Interpreting and synthesizing the results of these studies is challenging because special consideration must be given to (a) the class of PCB congeners examined, (b) the age at the time of PCB exposure, and (c) the specific neurological or behavioral function assessed. Nevertheless, animal models of developmental PCB exposure have contributed significantly to our understanding of the specific cognitive domains affected by exposure. In particular, a number of well-controlled experiments using animal models have assessed cognitive function, specifically executive function, following developmental PCB exposure in monkeys and rodents. Generally speaking, executive functions are higher-order cognitive processes involved in the planning, sequencing, and control of goal-directed behaviors. Specifically, these processes include formulating strategies as well as initiating, inhibiting, and shifting responding when necessary.

This chapter reviews the experiments examining executive function in animals following PCB exposure. The results of the animal studies are then compared and contrasted with the results of relevant epidemiological studies using cognitive tasks that tap similar functional domains. To provide some background, the different classes of PCB congeners (coplanar vs. *ortho*-substituted) are reviewed first. A brief summary of the literature on exposure to coplanar PCB congeners in animals during development is included, but coplanar PCBs will not be discussed in detail because they appear to have relatively minor effects on cognitive function. The remainder of the chapter focuses on research examining executive function in animals following exposure to single *ortho*-substituted PCB congeners as well as PCB mixtures.

CLASSES OF PCB CONGENERS

The PCB molecule consists of a biphenyl ring structure that has a carbon-carbon bond between carbon 1 on one ring and carbon $1'$ on the other ring (Figure 8.1). The remaining ten carbon positions are available for chlorine substitution. Thus, there are 209 different PCB congeners which, depending on the number and position of chlorine atoms on the molecule, differ in their chemical and physical properties as well as their toxicity. Chlorines bonded to a carbon immediately adjacent to the biphenyl ring bond (positions 2 or $2'$ and 6 or $6'$) are in an *ortho* position. *Meta*-bonded chlorines are attached on carbons 3 or $3'$ or 5 or $5'$, and *para*-bonded chlorines are attached to carbons 4 or $4'$. In general, the congeners are classified into two main

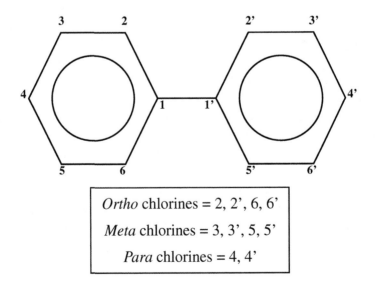

Ortho chlorines = 2, 2', 6, 6'
Meta chlorines = 3, 3', 5, 5'
Para chlorines = 4, 4'

FIGURE 8.1 Structural formula of an unsubstituted PCB molecule.

classes: coplanar and *ortho*-substituted. Coplanar congeners have zero or one *ortho* chlorines, two *meta* chlorines, and at least one *para* chlorine. This configuration allows the molecule to assume a planar configuration. The *ortho*-substituted PCB congeners have two or more *ortho* chlorines that repel each other and keep the molecule from assuming a planar conformation.

Coplanar PCB congeners including PCB 77 (3,3′,4,4′-tetrachlorobiphenyl), PCB 126 (3,3′,4,4′,5-pentachlorobiphenyl), and PCB 169 (3,3′,4,4′,5,5′-haxachloro-biphenyl) have chemical structures and properties that resemble dioxins (PCDDs) and dibenzofurans (PCDFs). Coplanar PCBs, PCDDs, and PCDFs interact with the aryl hydrocarbon (Ah) receptor protein to produce their toxic effects and can do so at very low doses (Rowlands and Gustafsson, 1997). *Ortho*-substituted PCB congeners have a much lower affinity for the Ah receptor, require higher doses to produce acute toxicity, and produce a pattern of toxic effects different from that of coplanar PCBs, PCDDs, and PCDFs.

Even though they have greater acute toxicity, developmental exposure to coplanar PCBs produces few deficits on tests of executive function in rodents or monkeys (Schantz and Widholm, 2001). For example, rats exposed to coplanar PCBs during gestation or lactation have shown no deficits on visual discrimination learning (Bernhoft et al., 1994), spatial learning (Schantz et al., 1996; Rice, 1999), two different tests of attention (Bushnell and Rice, 1999), or operant schedules including fixed ratio, progressive ratio, and differential reinforcement of low rates (Rice and Hayward, 1998; Rice and Hayward, 1999).

Deficits in performance on a fixed-interval (FI) operant schedule have been reported (Holene et al., 1995; 1998). However, reinterpretation of the data from Holene et al. (1995) suggested that no decrement was present (see Schantz and Widholm, 2001), and a similar study assessing FI found no clear-cut deficits attributable to coplanar PCB exposure (Rice and Hayward, 1998). A deficit attributable to coplanar PCB exposure was observed when the task included a series of three concurrent random interval–random interval (RI-RI) schedules, but the deficit was limited to the first schedule in the series (Rice and Hayward, 1999).

ORTHO-SUBSTITUTED PCB CONGENERS

Unlike the coplanar PCBs, deficits on a variety of behavioral tasks have been observed following exposure to individual *ortho*-substituted PCB congeners and PCB mixtures. The discussion in this chapter is limited to a discussion of tests of executive functions that rely on cognitive processing within the prefrontal cortex (PFC), including tests of cognitive flexibility, working memory, and inhibitory control (IC).

COGNITIVE FLEXIBILITY

Cognitive flexibility is an assessment of the ability to transfer knowledge and skills beyond initial learning to new situations with different demands. Reversal learning (RL) is a type of discrimination learning that measures cognitive flexibility. Subjects are required to discriminate between two possible actions and choose the correct action to obtain a reward. Once the action-outcome association is well learned, the response requirement is shifted such that the previously incorrect action is now required to obtain the reward. The response "reversals" are done to evaluate how flexible the subject is at adjusting goal-directed behavior as a result of changes to the action-outcome association.

In a spatial RL problem, a response on either the left or the right side is the correct option for each trial within the particular phase. The subject must figure out which side is the correct one and respond on that side to obtain the reward. For visual RL, a specific characteristic such as the color or shape of a discriminative cue varies between the response options and is independent of location. The subject is required to figure out which characteristic is the relevant one and then respond consistently to that visual cue. In both spatial and visual RL, the initial phase is known as original learning (OL). Once a predetermined criterion is attained on OL, the animal is often overtrained for a few more sessions before a reversal is made. A series of reversals is usually given, and the first reversal is typically the most difficult for the subject to learn. With progressive reversals, a "learning set" (Harlow, 1949) develops whereby the animal formulates a strategy for dealing with the new response requirement. Establishment of the learning set allows the response shift to be learned in fewer trials, with fewer errors being made in the process.

A number of studies have examined RL in monkeys exposed to PCBs during gestation or lactation (Table 8.1). In one such study, female rhesus monkeys were fed 2.5 ppm or 5.0 ppm of the commercial PCB mixture Aroclor 1248 (primarily *tetra*-chlorinated congeners, 48% chlorine by weight). Daily exposure occurred via the monkeys' chow for 16 to 21 months beginning approximately 8.5 to 13.5 months prior to conception and continuing until 3 months of lactation were completed (Bowman et al., 1978). Offspring of the dosed mothers and age-matched controls were weaned at 4 months and later tested on spatial, color, and shape reversal learning at 6, 7, and 8 months of age, respectively. Offspring of the treated animals showed deficits on the spatial and color reversal tasks, but shape discrimination reversal learning was not affected. Offspring born 18 months after the dosing of the female rhesus monkeys with 2.5 ppm Aroclor 1248 showed no deficits on spatial, color, or shape RL when compared with age-matched controls (Schantz et al., 1989). Chemical analyses revealed that the PCB concentration was significantly lower in the milk fat during the postexposure study, suggesting a lesser degree of exposure.

In a second follow-up experiment, female monkeys were dosed with Aroclor 1016 (primarily di- and trichlorinated congeners, 42% chlorine by weight). Mothers were given 0.25 ppm or 1.0 ppm Aroclor 1016 (7 days/week) in their chow beginning 7 months prior to breeding and continuing until weaning at 4 months of age. In addition, the mothers given 2.5 ppm Aroclor 1248 were bred again 32 months after PCB exposure had ended. The offspring were trained on a simple spatial RL problem and then on modified spatial, color, and shape RLs in which additional but irrelevant cues were present. Compared with age-matched controls, infants whose mothers were exposed to 1.0 ppm Aroclor 1016 required more trials to learn the simple spatial RL problem. The offspring born to mothers who were given 0.25 ppm Aroclor 1016 or to mothers previously exposed to 2.5 ppm Aroclor 1248 were not impaired on simple spatial RL (Schantz et al., 1989).

TABLE 8.1
Effects of PCB Exposure on Cognitive Flexibility

Exposure/Cohort	Species	Dose and Timing	Results of Exposure	Reference
Aroclor 1248	Monkey	2.5 or 5.0 ppm; preconception, gestation, and lactation	$\downarrow$ Spatial and color RL $-$ Shape RL	Bowman et al., 1978
		2.5 ppm; preconception (dosing ended 18 months before births)	$-$ Spatial, shape, or color RL	Schantz et al., 1989
		2.5 ppm; preconception (dosing ended 32 months before breeding)	$-$ Simple spatial, modified spatial, or color RL $\uparrow$ Shape RL	Schantz et al., 1989
Aroclor 1016	Monkey	0.25 or 1.0 ppm; preconception, gestation, and lactation	$\downarrow$ Simple spatial RL (1.0 ppm only) $-$ Modified spatial or color RL $\uparrow$ Shape RL	Schantz et al., 1989
Mixture of congeners in breast milk	Monkey	7.5 µg/kg/day; birth to 20 weeks of age	$-$ Color, shape, or spatial RL	Rice and Hayward, 1997; Rice, 1998
Aroclor 1254	Rat	6 mg/kg/day; gestation and lactation	$\downarrow$ Spatial RL by males on first reversal and females on fourth and fifth reversal	Widholm et al., 2001
Michigan	Human	Exposure varied; gestation and lactation	$\downarrow$ Performance on WCST	Jacobson and Jacobson, 2003

Note: $\downarrow$ = impaired performance; $\uparrow$ = enhanced performance; $-$ = no effect; RL = reversal learning; WCST = Wisconsin Card Sorting Task.

No deficits on modified spatial or color RL were observed in any of the treatment groups. However, PCB exposure produced an interesting change in performance on shape RL. Offspring of mothers exposed to 1.0 ppm Aroclor 1016 and 2.5 ppm Aroclor 1248 required significantly fewer trials than controls to reach criterion when shape was the relevant stimulus cue and spatial location and color cues were present but irrelevant to solving the problem. A similar pattern was observed in offspring of animals exposed to 0.25 ppm Aroclor 1016, although the difference from controls was not significant (Schantz et al., 1989). Controls appeared better able to dismiss the irrelevant color and shape cues during the modified spatial RL task given first, but then had a harder time recognizing the relevance of the shape cue during the shape RL task given later. The PCB-treated monkeys were apparently less focused on spatial location during the spatial RL tasks given first, allowing them to more efficiently focus on shape when it became relevant later. In fact, when irrelevant shape and color cues were not included during the spatial RL problem given first in previous experiments, no improvement by PCB-exposed monkeys was observed on shape RL.

The effects of postnatal PCB exposure on reversal learning have been examined in a study modeling mammary transmission of PCBs. Male monkeys were dosed from birth until 20 weeks of age with 7.5 μg/kg/day of a PCB mixture that included 15 PCB congeners regularly detected in human breast milk in Canada (Mes and Marchand, 1987). The exposed monkeys were compared with age-matched controls on color and shape RL. PCB exposure did not significantly affect performance on either the shape, color, or spatial RL tasks (Rice and Hayward, 1997; Rice, 1998). While previous studies examining RL in monkeys did find deficits (Bowman et al., 1978; Schantz et al., 1989), the dosing regimen in those studies included both pre- and postnatal exposure. In contrast, the Rice studies only examined postnatal exposure. The PCB mixture was also quite different from those used in the earlier studies, which could explain the disparity in results.

Spatial RL has also been assessed in rats following PCB exposure during gestation and lactation (Widholm et al., 2001) (Table 8.1). Long-Evans rats were given 0 or 6 mg/kg/day Aroclor 1254 (primarily penta- and hexachlorinated congeners, 54% chlorine by weight) from gestational day 6 to postnatal day 21. One male and one female from each litter were tested on spatial RL using standard two-lever operant testing chambers. OL and five spatial reversals were completed. There were no differences between the exposed and control rats on OL. However, Aroclor 1254-exposed males made more errors than control males on the first reversal. The exposed females, on the other hand, made more errors during the fourth and fifth reversals.

Analysis of the response patterns indicated that the increased errors seen for the male and female rats were due to different underlying deficits. For the males, the increased error rate on the first reversal was due to their tendency to perseverate on the previously correct, now incorrect, lever. In other words, the exposed males took longer to adopt a new response strategy, suggesting a lack of inhibitory control. The males were able to recover and perform the task equal to controls on later reversals. The increased errors committed by the females during reversal four and five were not due to perseverative responding. Instead, the increase in errors seemed to be due to impaired acquisition of a learning set. While the exposed females did require fewer trials with each additional reversal, they required significantly more trials than the controls, suggesting they were not able to learn the response strategy as efficiently.

Overall, reversal learning deficits have been reported in most, but not all, PCB studies using animal models. These deficits are present whether gestational and lactational exposure is to PCB mixtures containing mostly lightly chlorinated congeners (Schantz et al., 1989) or more highly chlorinated congeners (Bowman et al., 1978; Widholm et al., 2001). Postnatal exposure to a PCB congener mixture representative of that found in human breast milk did not produce measurable spatial RL deficits (Rice, 1998).

The experiments examining RL in animals exposed to PCBs relate well to what has been observed in exposed children (Table 8.1). In particular, performance deficits were observed on the Wisconsin Card Sorting Task (WCST) in 11-year-old children of the Michigan cohort, whose mothers had eaten PCB-contaminated Lake Michigan fish during gestation and lactation (Jacobson and Jacobson, 2003). The WCST is a neuropsychological task similar to reversal learning, which requires the participant to sort cards that vary on three dimensions: color, shape, and number. Only the experimenter knows the correct stimulus dimension, and the participant is told after each card if he/she has sorted correctly. If incorrect, the participant must sort according to another stimulus dimension. Once correct, the participant is required to continue until ten consecutive cards have been sorted correctly. At that point, the experimenter changes the stimulus dimension without warning, and the ability of the participant to change his/her response strategy is observed. In this sense, the task is similar to RL because the response requirement is changed after a particular action-outcome association is well learned. The participant must adjust his/her goal-directed behavior to succeed at the task. The ability to make the adjustment is indicative of the participant's cognitive flexibility. Children prenatally exposed to PCBs had difficulty changing response strategy when the stimulus dimension changed. Instead, they failed to make the switch and perseverated on the previously correct stimulus dimension, even when told it was no longer correct (Jacobson and Jacobson, 2003).

WORKING MEMORY

Deficits in working memory can be assessed on alternation tasks that require the subject to alternate between two responses from one trial to the next. Alternation tasks can be based on visual cues (e.g., presence versus absence of cue lights) or on spatial location (e.g., left versus right). When a delay is interposed between trials, a working-memory component is added to the task. During the delay, the subject must maintain a memory trace of the previous response, such that it can correctly respond to the alternative cue after the delay.

The effects of prenatal PCB exposure on delayed spatial alternation (DSA) have been examined by a number of investigators in both monkeys and rodents (Table 8.2). The earliest experiment by Levin et al. (1988) compared DSA performance in control monkeys with age-matched monkeys chronically exposed to Aroclor 1248 (2.5 ppm in diet) or Aroclor 1016 (0.25 or 1.0 ppm in diet) during gestation and while nursing until 4 months of age. They were tested at 4 to 6 years of age and had previously been tested on a series of RL problems (Schantz et al., 1989). Variable delays of 5, 10, 20, and 40 sec were used in counterbalanced order.

When compared with controls, Aroclor 1248 exposure resulted in a reduction in the percent correct at the 5-, 10-, and 20-sec delays. There was no difference between the control and PCB-exposed rats at the 40-sec delay, primarily due to a drop in performance by the control monkeys. Given that impaired performance in the PCB-exposed monkeys was observed at even the shortest delay, the observed deficit does not appear to be attributable to a decay in the memory trace. Instead, it seems that PCB-exposed monkeys are unable to adequately *establish* the memory trace at the offset of the preceding trial, suggesting an attentional or associational impairment. This type of deficit on the DSA task is characteristic of damage to the prefrontal cortex (Miller and Orbach, 1972; Skeen and Masterton, 1976; Treichler

TABLE 8.2
Effects of PCB Exposure on Working Memory

Exposure/Cohort	Species	Dose and Timing	Results of Exposure	Reference
Aroclor 1248	Monkey	2.5 ppm; preconception, gestation, and lactation	$\downarrow$ DSA	Levin et al., 1988
Aroclor 1016	Monkey	0.25 or 1.0 ppm; preconception, gestation, and lactation	$\downarrow$ DSA (1.0 ppm); $\uparrow$ DSA (0.25 ppm)	Levin et al., 1988
	Rat	10 mg/kg/day; gestation and lactation	– DSA	Zahalka et al., 2001
Mixture of congeners in breast milk	Monkey	7.5 µg/kg/day; birth to 20 weeks of age	$\downarrow$ DSA	Rice and Hayward, 1997
Aroclor 1254	Rat	8 mg/kg/day; gestation and lactation	– DSA	Zahalka et al., 2001
		6 mg/kg/day; gestation and lactation	$\downarrow$ DSA	Widholm et al., 2004
PCB 28	Rat	8 or 32 mg/kg/day; gestation	$\downarrow$ DSA in females at highest dose	Schantz et al., 1995
PCB 118	Rat	4 or 16 mg/kg/day; gestation	$\downarrow$ DSA in females at highest dose	Schantz et al., 1995
PCB 153	Rat	16 or 64 mg/kg/day; gestation	$\downarrow$ DSA in females at highest dose	Schantz et al., 1995
Michigan	Human	Exposure varied; gestation and lactation	$\downarrow$ Fagan test and McCarthy memory scale	Jacobson et al., 1985
			$\downarrow$ Sternberg test and WISC-R digit-span test	Jacobson et al., 1992; Jacobson and Jacobson, 1996; Jacobson and Jacobson, 2003
Oswego	Human	Exposure varied; gestation and lactation	$\downarrow$ Fagan test – McCarthy memory scale	Darvill et al., 2000; Stewart et al., 2003b
Dutch	Human	Exposure varied; gestation and lactation	$\downarrow$ GCI and McCarthy memory scale in children with low-IQ parents or young mothers	Vreugdenhil et al., 2002; Vreugdenhil et al., 2004
German	Human	Exposure varied; gestation and lactation	– Fagan test	Winneke et al., 1998
Faroe Islands	Human	Exposure varied; gestation and lactation	– PCB-related effects on WISC-R digit-span test, Bender visual-motor test, and CVLT	Grandjean et al., 2001

TABLE 8.2
Effects of PCB Exposure on Working Memory (continued)

Exposure/Cohort	Species	Dose and Timing	Results of Exposure	Reference
North Carolina	Human	Exposure varied; gestation and lactation	– McCarthy GCI or memory scale	Gladen and Rogan, 1991

Note: $\downarrow$ = impaired performance; $\uparrow$ = enhanced performance; – = no effect; DSA = delayed spatial alternation; WISC-R = Wechsler Intelligence Scale for Children — Revised; GCI = general cognitive index (of McCarthy); CVLT = California Verbal Learning Test.

et al., 1971) and, in particular, of damage to the mesocortical dopamine pathway (Murphy et al., 1996). A perseveration problem became evident when the types of errors made by the PCB-exposed monkeys were examined more closely. Specifically, the PCB-exposed monkeys committed a significantly higher number of win-stay errors. Win-stay errors occur when a monkey makes a correct response (win) followed by an incorrect response on the same side (stay) instead of switching to the opposite side. There were no differences in response latency between the PCB-exposed and control monkeys, indicating that there were no differences in motor control or motivation to perform the task.

The effects of Aroclor 1016 were more subtle and difficult to interpret. No significant effects of PCB exposure were found at any of the individual delays. Averaged across delays, neither exposure group was significantly different from controls, but the 0.25-ppm and 1.0-ppm groups were significantly different from each other. With increasing sessions, the low-dose group showed facilitated performance on the DSA task relative to controls, while the high-dose group looked impaired. Similar biphasic dose effects have been observed on the DSA task following postnatal exposure to lead (Levin and Bowman, 1986; 1988). The authors hypothesized that low PCB and lead exposure may have filtered out peripheral stimuli and ultimately narrowed the monkeys' focus of attention to the relevant stimuli, while high levels of exposure produced a global reduction in attention to all cues (relevant or not). Gestational and lactational exposure to 10 mg/kg/day Aroclor 1016 in rats produced no notable spatial alternation deficits, although performance at different delays was not reported (Zahalka et al., 2001). Aroclor 1248 contains a more substantial percentage of tetra-substituted congeners in comparison to Aroclor 1016, which has more di- and trichlorine substitutions. Differences in the congener profile of the two mixtures may underlie the disparity of DSA results.

Rice and Hayward (1997) examined early postnatal exposure of monkeys to a PCB mixture representative of the PCBs typically found in human breast milk and found DSA deficits similar to those following Aroclor 1248 exposure. Specifically, monkeys were given 7.5 µg/kg/day of the mixture seven days per week from birth until 20 weeks of age. PCB-exposed monkeys required more sessions to learn the alternation task and made more incorrect responses across the acquisition sessions. When the delays were a fixed length within a session, increased errors were observed in the exposed animals at short delays, which were done at the beginning of the experiment. There were no differences between exposed and control animals at long delay values, which were tested later. When variable delays were presented within a session during the last portion of the experiment, there was no difference between the groups on the total number of incorrect responses, but perseverative-type errors were increased in the exposed animals relative to controls at all of the delays.

In rats, research examining DSA performance after gestational and lactational exposure to a mixture containing a higher number of penta- and hexa-chlorine substitutions, Aroclor1254, has produced mixed results. Zahalka et al. (2001) administered Aroclor 1254 to pregnant females via oral gavage (8 mg/kg/day) starting on gestational day 6 and terminating on postnatal day (PND) 21. On PND 23, delayed spatial alternation was assessed across five 12-trial blocks using a T-maze. Each trial consisted of a forced run in which only one arm was available and a choice run where the rat had to choose the arm opposite that visited during the forced run to be rewarded. The time that elapsed between the end of the forced run and the commencement of the choice run constituted the delay and was not systematically timed. The authors reported no deficits present on the DSA task in the PCB-exposed rats based on the fact that there was no difference between the PCB-exposed and control animals on the percent-correct trials within any of the five 12-trial blocks.

Widholm et al. (2004) also examined the effects of gestational and lactational exposure to Aroclor 1254, and they did observe DSA deficits attributable to PCB exposure. Specifically, 6 mg/kg Aroclor 1254 or corn oil (control) was administered to female rats beginning 28 days before breeding and continued until PND 16. DSA testing of the offspring began when the rats were approximately 85 days old and was conducted in operant chambers with variable delays of 0, 3, 6, 9, or 18 sec interposed between trials. Animals exposed to PCBs were impaired relative to controls at the 0-, 6-, and 9-sec delay but not at the 3- and 18-sec delay. Analyses of the win-stay and lose-stay errors were conducted to determine if PCB exposure influenced the degree of perseverative responding. No significant differences between the PCB-exposed animals and controls were found on the win-stay errors. This result indicated that both exposure groups were equally likely to be influenced by recent reinforcement history, returning to the same lever instead of correctly alternating to the other side. Lose-stay errors, on the other hand, were significantly affected by PCB exposure. Animals exposed to Aroclor 1254 perseverated on an incorrect, nonreinforced lever instead of alternating to the other side more often than controls, particularly during the final two five-session blocks of testing. Overall, the results of Widholm et al. (2004) were similar to the results of Levin et al. (1988) and Rice and Hayward (1997). The impairment present at even the shortest delay suggested that the exposed rats had difficulty establishing an accurate memory trace at the offset of the previous trial, again suggesting a prefrontal site of action. Analyses of lose-stay errors confirmed that increased perseverative errors in the PCB-exposed animals across all delays were significantly contributing to the deficit.

The different results obtained by Zahalka et al. (2001) and Widholm et al. (2004) were likely due to two important procedural differences between the studies. The first related to the substantial difference in the nature of the DSA tasks used between the two studies. Zahalka et al. used a T-maze DSA procedure that did not employ systematically timed delays between trials, while Widholm et al. used an operant setup with timed variable delays between trials. When using the T-maze for DSA, testing is interrupted at the conclusion of every trial in order to place the rat in the start box for the next trial. This is not the case when DSA testing is done using an operant setup because the house lights signal the start of the next trial, making handling or a change in location unnecessary.

The second major difference was the age of the rats at the time of DSA testing. Zahalka et al. conducted the five 12-trial session blocks on a single day while the rats were adolescents. Widholm et al. tested the rats during early adulthood, with the 25 DSA sessions occurring over the course of 25 days. The timing of testing is a critical component because spatial alternation tasks such as DSA are known to be highly dependent on the prefrontal cortex (Murphy et al., 1996), which is not fully mature until the end of adolescence (Bourgeois et al., 1994). Therefore, the results of Zahalka et al. must be interpreted with caution, because testing the same animals on the same DSA task as adults might have yielded very different results.

It is unclear exactly how the *ortho*-substituted PCB congeners in the various mixtures discussed above produce neurobehavioral deficits on tasks such as DSA. Some research has focused on the effects of perinatal exposure to individual *ortho*-substituted PCB congeners on DSA. For example, Schantz et al. (1995) exposed female rats to *ortho*-substituted PCB congeners PCB 28 (2,4,4′-trichlorobiphenyl) at doses of 8 or 32 mg/kg/day, PCB 118 (2,3′,4,4′,5-pentachlorobiphenyl) at doses of 4 or 16 mg/kg/day, or PCB 153 (2,2′,4,4′,5,5′-hexachlorobiphenyl) at doses of 16 or 64 mg/kg/day from gestational days 10 to 16. DSA was assessed in offspring of the control and exposed females using a T-maze DSA that incorporated systematic variable delays of 15, 25, or 40 sec between trials. The higher dose of all three congeners impaired the ability of female rats to learn the DSA task, while males were unaffected. Specifically, the females were impaired at all delays, again suggesting an attentional or associational impairment related to the establishment of the memory trace.

A number of epidemiological studies have examined working memory in children following developmental exposure to PCBs (Table 8.2). A rather extensive assessment of working memory has been conducted in the children of the Michigan cohort at 7 months and at 4 and 11 years of age. As previously mentioned, the children were exposed to PCBs during gestation and lactation as a result of contaminated fish consumption by their mothers. Prenatal exposure was determined by measuring PCB levels in umbilical cord serum. Postnatal exposure was determined by measuring breast milk PCB levels at birth and at 5 months of age in women who breast fed.

At 7 months of age, exposed infants were administered the Fagan test of visual recognition memory (Jacobson et al., 1985). Stimuli consisted of pairs of faces. One of the photos, known as the target, was presented first in both the left and right positions, and the baby was required to fixate on the target photo for 20 sec. The target was then removed and subsequently presented along with a novel target for two 5-sec recognition periods. The positions of the target and novel stimuli were reversed between the two recognition periods. The normal response is for the baby to spend more time viewing the novel stimulus. As cord serum PCB level increased, the amount of time fixating on the novel stimulus decreased. In fact, the mean fixation to novelty score in the most highly exposed infants was 50%, indicating that these infants showed no preference for novelty. The failure to show a preference suggests a working-memory impairment.

Tests at age 4 included a simplified version of the Sternberg memory paradigm (Jacobson et al., 1992) as well as the McCarthy scales of children's abilities, which includes a memory scale (Jacobson et al., 1990). Rather than using digits or letters for the Sternberg task, the researchers first showed the children a memory set consisting of one or three drawings of familiar objects on a computer screen. A set of 16 test stimuli (half from the previous memory set) was then presented one at a time on the computer screen, and the child was instructed to press a button as quickly as possible if the test stimulus was from the original memory set. Response time was longer and errors were greater when the memory set on the Sternberg was three items rather than one. This finding indicated that the task was a valid assessment of working memory. Higher cord serum PCB levels were associated with more recall errors on the Sternberg.

The McCarthy test provides standardized test scores for a general cognitive index (GCI) and five scales including verbal, perceptual-performance, quantitative, memory, and motor. Higher umbilical cord PCB levels were associated with significant impairment on the memory scale. Two subtests constituting three-fourths of the memory-scale score were examined in more detail. The verbal-memory subtest assesses recall for strings of words, sentences, and a story. The numerical-memory subtest assesses the child's ability to repeat strings of numbers (both forward and backward) as the strings get progressively longer. Both subtests are working-memory dependent, and deficits associated with cord PCB levels have been seen on both of these subtests (Jacobson et al., 1990). Overall, the results at 4 years suggested that the working-memory deficit observed on the Fagan test at 7 months of age (Jacobson et al., 1985) was persistent.

Working memory was assessed again at age 11 (Jacobson and Jacobson, 1996; Jacobson and Jacobson, 2003). The traditional Sternberg memory paradigm (using memory sets of one, three, or five digits) was administered, as was the Wechsler Intelligence Scale for Children — Revised (WISC-R). Of particular relevance to working memory was the digit-span subtest of the WISC-R, which required the child to repeat progressively longer strings of numbers in both forward and reverse direction. Overall, PCB exposure was associated with lower accuracy on the Sternberg (replicating the results of Jacobson et al. [1992] from age 4). Breast-feeding history influenced the magnitude of working-memory deficit. In particular, among children breast fed for less than 6 weeks, PCB exposure was negatively correlated with number correct on the Sternberg (-0.37 vs. 0.01 for ≥ 6 weeks) and WISC-R digit-span accuracy (-0.31 vs. -0.03) (Jacobson and Jacobson, 2003).

The Fagan Test of Infant Intelligence (Darvill et al., 2000) and the McCarthy Scales of Children's Abilities (Stewart et al., 2003b) were also administered to exposed children of the Oswego cohort. Participants of the Oswego cohort included pregnant women who had been exposed to PCBs as a result of eating contaminated Lake Ontario fish. Umbilical cord plasma and placenta samples were collected at birth to determine the degree of prenatal PCB exposure. Some women also provided breast milk samples for determination of postnatal exposure. Results on the Fagan test were similar to those observed in the Michigan cohort (Jacobson et al., 1985), even though cord PCB concentrations were lower (Darvill et al., 2000) than those measured in the Michigan cohort (Schwartz et al., 1983). Higher total PCBs in umbilical cord serum were associated with lower fixation times to the novel stimulus at both 6 and 12 months of age (Darvill et al., 2000). There was no relationship between cord PCB level and performance on the memory subscale of the McCarthy at 38 or 54 months of age (Stewart et al., 2003b).

A Dutch cohort consisting of pregnant women who had either been living in the heavily industrialized city of Rotterdam (or surrounding areas) for at least 5 years or pregnant women who had lived in Groningen (or surrounding rural areas) for at least 5 years has also been evaluated. Developmental exposure was determined by measuring PCBs in maternal blood during the last month of pregnancy, in umbilical cord serum, and in breast milk.

The Rotterdam children were reassessed on the McCarthy Scales of Children's Abilities at 84 months (7 years) of age (Vreugdenhil et al., 2002) and then again on the Rey Complex Figure Test and Auditory-Verbal Learning Test at 9 years of age (Vreugdenhil et al., 2004). For the Rey Complex Figure Test, the child was shown the complex figure and asked to copy it. After a distractor task, the child was asked to draw the complex figure from memory. In the Auditory-Verbal Learning Test, a list of 15 words was orally presented to the child five times. Immediately after each presentation and later after a distractor task, the child was asked to recall the words.

PCB exposure appeared to have little effect on the McCarthy GCI and memory scale scores after adjustment for covariables. Likewise, separate analysis of breast-fed and formula-fed infants did not show a relationship between PCB exposure and scores on the McCarthy for either subgroup. However, additional analyses determined that prenatal PCB exposure did negatively affect the GCI in children of parents with less education as well as the GCI and memory-scale scores of children born to parents with lower verbal IQ scores and to younger mothers (Vreugdenhil et al., 2002). Children exposed to high levels of PCBs were no different from children exposed to low PCB levels on the Rey Complex Figures Test or the Auditory-Verbal Learning Test (Vreugdenhil et al., 2004).

In 1993, the European Union, hoping to expand the Dutch PCB study into a transnational study and improve risk assessment, provided additional funding to create a German cohort in Dusseldorf and a Danish cohort in the Faroe Islands (Winneke et al., 1998). In the German cohort, prenatal PCB exposure was determined by measuring three marker congeners (PCBs 138, 153, and 180) in cord blood. Breast milk taken at the time the infants were 2 and 4 weeks of age was screened for the same three congeners. The German children were given the Fagan test at 7 months. Unlike the children of the Michigan (Jacobson et al., 1985) and Oswego (Darvill et al., 2000) cohorts, there was no relationship between the PCB exposure and Fagan test scores (Winneke et al., 1998). However, the authors suggest that the lack of a relationship may have been due to low test-retest and inter-rater reliability of the version of the Fagan test they used.

Two PCB cohorts are being studied in the Faroe Islands. The first cohort was assembled in 1986–1987 and the other was recruited in 1994–1995 as part of the European Union transnational project. The primary source of PCB exposure for both samples was consumption of pilot whale meat and blubber, which also contained significant amounts of methylmercury. PCB exposure for the older cohort was estimated by summing the amount of PCBs 138, 153, and 180 in umbilical cord. In the newer cohort, maternal serum, breast milk, and a subset of cord blood were analyzed for 28 individual PCB congeners. Only a limited number of neuropsychological tests have been done and, for the most part, it has been difficult to attribute deficits to PCB exposure independent of methylmercury exposure. For example, at 7 years of age, children from the earlier cohort were assessed on a number of different neuropsychological tests and exhibited deficits on working-memory tasks, including the digit-span subtest of the WISC-R, the Bender Visual-Motor Gestalt Test (drawing figures based on immediate recall), and the California Verbal Learning Test (recalling lists of semantically related words). Reanalysis of the data was later done in an attempt to isolate potential PCB-related effects (as opposed to methylmercury-related effects). After adjusting for mercury exposure, none of the working-memory tasks was significantly influenced by PCB exposure, indicating that methylmercury exposure seems to be a greater hazard in this population (Grandjean et al., 2001).

Unlike the other PCB-exposed cohorts, children from a North Carolina cohort have not exhibited any impairment in cognitive function. The North Carolina cohort included pregnant women from the general population who were exposed to PCBs via the general food supply. The McCarthy was administered when the children were 3, 4, and 5 years of age. Unlike the results of Jacobson et al. (1990), Stewart et al. (2003b), and Vreugdenhil et al. (2002), no significant associations between transplacental or breast-feeding PCB exposure and scores on the McCarthy GCI or memory scale were found at any age (Gladen and Rogan, 1991).

In summary, developmental PCB exposure has been shown to produce impairments on working-memory tests in children from some, but not all, of the PCB-exposed cohorts that have been evaluated. These effects include poorer performance on the Fagan test of visual recognition memory (Darvill et al., 2000; Jacobson et al., 1985), the Sternberg test (Jacobson et al., 1992; Jacobson and Jacobson, 2003), and memory subtests of the McCarthy (Jacobson et al., 1990) and the WISC-R (Jacobson and Jacobson, 2003). The discrepancies between studies could be due to differences in the congener profiles and degree of exposure among the cohorts or to differences in the tasks used to assess working memory and the age at which they were administered. Although not consistently reported across all studies, the working-memory effects reported in the epidemiological studies are consistent with the findings from studies conducted in laboratory animals, highlighting the utility of animal models for predicting the specific cognitive domains that are impacted by developmental PCB exposure.

INHIBITORY CONTROL

The PFC has been implicated in the process of behavioral inhibition (Neill, 1976; Neill et al., 1974; Wilcott, 1982; 1984). As previously discussed, perseverative errors on the PFC-dependent tasks of spatial RL (Widholm et al., 2001) and DSA (Rice and Hayward, 1997; Widholm et al., 2004) are increased following developmental PCB exposure (Table 8.3). Perseverative errors occur when there is a failure to inhibit

TABLE 8.3
Effects of PCB Exposure on Inhibitory Control

Exposure/Cohort	Species	Dose and Timing	Results of Exposure	Reference
Aroclor 1254	Rat	6 mg/kg/day; gestation and lactation	↓ RL and DSA tasks (increased perseverative errors)	Widholm et al., 2001; Widholm et al., 2004
Mixture of congeners in breast milk	Monkey	7.5 µg/kg/day; birth to 20 weeks of age	↓ DSA task (increased perseverative errors)	Rice and Hayward, 1997
			↓ FI schedule	Rice, 1997
			↓ DRL schedule	Rice, 1998
Aroclor 1248	Monkey	0.5 or 2.5 ppm; preconception, gestation, and lactation	↓ ICs on longer FI trials (0.5 ppm only)	Mele et al., 1986
		2.5 ppm; preconception (dosing ended 20 months before breeding)	↓ ICs on longer FI trials	Mele et al., 1986
PCB 153	Rat	5 mg/kg; lactation	↓ FI schedule in males but not females ↓ DRL performance in females	Holene et al., 1998; Holene et al., 1999
Michigan	Human	Exposure varied; gestation and lactation	↓ Sternberg test and CPT (increased errors of commission) ↓ WCST (increased perseverative errors)	Jacobson et al., 1992; Jacobson and Jacobson, 2003 Jacobson and Jacobson, 2003
Oswego	Human	Exposure varied; gestation and lactation	↓ CPT (increased errors of commission)	Stewart et al., 2003a

Note: ↓ = impaired performance; ↑ = enhanced performance; – = no effect; IC = inhibitory control; RL = reversal learning; DSA = delayed spatial alternation; FI = fixed interval; DRL = differential reinforcement of low rates; CPT = continuous performance test; WCST = Wisconsin Card Sorting Task.

an incorrect response and represent a problem with inhibitory control. Temporal discrimination is also regulated by the PFC (Manning, 1973). Fixed interval (FI) and differential reinforcement of low-response rate (DRL) operant schedules have been used to examine the effects of developmental PCB exposure in monkeys and rodents because they incorporate a time-keeping component and a response-inhibition component (Table 8.3). A fixed-interval schedule requires a single response after a specified period of time has elapsed in order to obtain the reward. Responses occurring before the time period is over do not have any consequence, but they require effort and reduce efficiency. DRL is similar to FI in that only a single response is required at the end of a specified time interval. However, unlike FI, responding prior to the end of the interval resets the timer, thereby delaying access to the reinforcer.

Mele et al. (1986) examined the effects of perinatal exposure to Aroclor 1248 in monkeys on a series of FI tasks. Four groups of offspring were tested in two separate experiments. In the first experiment, offspring of mothers that had been fed 2.5 ppm Aroclor 1248 daily in their chow prior to breeding and during gestation and lactation were compared with offspring of control monkeys who had no PCB exposure. In the second experiment, offspring born to the same mothers rebred 20 months after exposure had ended were tested. An additional lower-PCB-dose group was also included. These monkeys were given 0.5 ppm Aroclor1248 in their chow three times per week. A final group of offspring whose mothers did not have any PCB exposure was also tested. All the monkeys were given 10 sessions on FI 30 sec and 15 sessions each at FI 60, 300, and 600 sec. No consistent differences were observed between the PCB-exposed animals and the control animals in FI performance in the first experiment. Results were similar in the second experiment, except that a significant reduction in the index of curvature (IC) was observed in the 0.5-ppm PCB group under FI 300 and 600 and in the 2.5-ppm delayed-exposure group at FI 600. Index of curvature represents the relative distribution of responses as they occur throughout the FI interval. An IC of zero indicates that responding was relatively constant throughout the interval, while a positive value indicates that the responding increased toward the end of the interval. The low IC seen in the PCB-treated animals indicated that they had impaired efficiency. Only a response made at the end of the preset interval was rewarded, yet the PCB-treated animals responded at relatively high levels throughout the interval. Such responding was not cost effective in terms of effort made and payoff received.

Rice (1997) examined inhibitory control in monkeys exposed from birth until 20 weeks of age to 7.5 $\mu g/kg/day$ of a PCB mixture representative of the PCBs typically found in human breast milk. The specific dosing protocol has been described earlier in this chapter, as these monkeys were previously tested on reversal learning and DSA (Rice and Hayward, 1997). When the monkeys were 4 years old, performance under a multiple FI–fixed ratio (FR) schedule of reinforcement was assessed across 48 sessions. The FI was 6 min for all sessions. The FR schedule required a fixed number of responses in order to receive the reinforcer and was increased sequentially from FR 2 to FR 10 by session 16. The active schedule was indicated by illumination of a different color on the response button (FI = white, FR = yellow). PCB exposure resulted in impaired performance on the FI schedule; the FR component was not substantially affected. The PCB-exposed monkeys had shorter mean interresponse times (IRTs) than the control animals, especially during earlier sessions. The FI pause time, which was the time between the start of the schedule and the first response, developed more slowly over trials in the PCB-treated animals. The PCB-treated animals also had a greater number of short (<5 sec) interresponse times than controls. Similar to the results of Mele et al. (1986), monkeys were less efficient, making nonreinforced responses (as indicated by lower pause times and more short IRTs) on the FI schedule.

Additional rodent work, has focused on the effects of mammary transmission of PCBs on FI performance. Mothers were dosed with 5 mg/kg of the *ortho*-substituted PCB congener PCB 153 every second day from days 3 to 13 after delivery. Male (Holene et al., 1998) and female (Holene et al., 1999) offspring were tested on multiple FI 2-min extinction (EXT) schedules of reinforcement. The status of the house light determined which schedule was active (on = FI, off = EXT). A response after the 2-min interval delivered a water reinforcer. No reinforcer was ever delivered during EXT, regardless of response rate. Male PCB-exposed rats had an increased frequency of lever presses, including short IRTs, shortly before the next reinforcer was scheduled to be delivered. The male PCB-exposed rats also made significantly more unnecessary visits to the water tray. These results indicate that the male PCB-treated rats were less efficient and disinhibited. The female offspring showed no PCB-related impairments on the FI schedule relative to controls. Thus, in rats tested with the specific PCB congener PCB 153, the effect on FI appears to be sex specific.

As previously mentioned, DRL is also sensitive to PFC disruption. The female rats in the Holene et al. (1999) study were also examined on concurrent variable interval (VI) 120-sec–DRL 14-sec schedules. The 120-sec VI schedule delivered a reinforcer approximately every second minute following an appropriate response on the right lever. The DRL contingency was set up on the left lever. The contingency on both levers had to be satisfied before a reward would be delivered. The initial five sessions showed no effect of PCB exposure on IRTs. During the second group of five sessions, PCB-exposed rats had lower IRTs than the controls, and fewer long IRTs and reinforcers were present during the third block of five sessions.

Impairments in DRL have also been observed in the offspring of monkeys exposed to a 7.5-μg/kg/day PCB mixture representative of the PCB congeners present in breast milk (Rice, 1998). DRL testing in these animals occurred after nonspatial RL and DSA (Rice and Hayward, 1997). Treated monkeys had robust efficiency problems exhibiting shorter IRTs, obtaining fewer reinforcements, and making more nonreinforced responses. These results, in combination with the DRL results of Holene et al. (1999), suggest that postnatal exposure to PCBs can disrupt response inhibition.

Evidence of inhibitory control problems exists in the epidemiological literature as well (Table 8.3). As previously mentioned, exposed children of the Michigan cohort were tested at age 4 and 11 years. At age 4, PCB exposure was associated with increased errors of commission (making a response when one should be inhibited) on the modified Sternberg task (Jacobson et al., 1992); at age 11, exposure was associated with increased perseverative responses on the WCST (Jacobson and Jacobson, 2003). Inhibitory control was also assessed in the 11-year-old children via a continuous performance test (CPT), where a response was required when a predetermined stimulus appeared on a computer screen. The more highly exposed children were more impulsive and made more errors of commission on the CPT (Jacobson and Jacobson, 2003). Exposed children of the Oswego cohort tested at 4.5 years of age also exhibited a positive relationship between cord-blood PCB level and errors of commission on a CPT (Stewart et al., 2003a).

CONCLUSIONS

Neurodevelopmental exposure to PCBs has been shown to produce significant disruption of tasks that require executive function. Specifically, functional domains known to rely on the PFC — including cognitive flexibility, working memory, and inhibitory control — are adversely affected by developmental PCB exposure to mixtures containing mostly *ortho*-substituted PCBs as well as individual *ortho* PCB congeners. Extensively studied and well-understood animal behavioral paradigms such as RL, DSA, FI, and DRL tap these functional domains and allow for an accurate assessment of potential PCB-related decrements. As a result of the reliability and validity of these tasks, the PCB research using animal models has contributed a substantial amount to what is now understood about the neurobehavioral effects of developmental PCB exposure.

Animal research has also helped (a) to reveal the differential effects of coplanar versus *ortho*-substituted PCB congeners and complex PCB mixtures on tasks requiring executive function and (b) to highlight the specific aspects of executive function — working memory and inhibitory control — that are most severely impacted by early PCB exposure. As a testament to the usefulness of animal models, recent follow-up studies of several human PCB cohorts have employed domain-specific tests such as the Wisconsin Card Sorting Task (Jacobson and Jacobson, 2003) and continuous performance tasks (Jacobson and Jacobson, 2003; Stewart et al., 2003a) that assess specific aspects of executive control. These recent assessments have yielded results that are strikingly consistent with the findings in animal models, and they highlight the utility of animal models for predicting specific cognitive domains that are likely to be impacted in exposed human populations.

ACKNOWLEDGMENTS

The first author was supported by grants from the USEPA (R-82939001) and NIEHS (P01 ES11263) during the preparation of this manuscript.

REFERENCES

Bernhoft, A., Nafstad, I., Engen, P., and Skaare, J.U. (1994), Effects of pre- and postnatal exposure to 3,3′,4,4′,5-pentachlorobiphenyl on physical development, neurobehavior and xenobiotic metabolizing enzymes in rats, *Environmental Toxicol. Chem.*, 13, 1589–1597.

Bourgeois, J.P., Goldman-Rakic, P.S., and Rakic, P. (1994), Synaptogenesis in the prefrontal cortex of rhesus monkeys, *Cerebral Cortex*, 4, 78–96.

Bowman, R.E., Heironimus, M.P., and Allen, J.R. (1978), Correlation of PCB body burden with behavioral toxicology in monkeys, *Pharmacol. Biochem. Behav.*, 9, 49–56.

Bushnell, P.J. and Rice, D.C. (1999), Behavioral assessments of learning and attention in rats exposed perinatally to 3,3′,4,4′,5-pentachlorobiphenyl (PCB 126), *Neurotoxicol. Teratol.*, 21, 381–392.

Darvill, T., Lonky, E., Reihman, J., Stewart, P., and Pagano, J. (2000), Prenatal exposure to PCBs and infant performance on the Fagan test of infant intelligence, *Neurotoxicology*, 6, 1029–1038.

Fagan, J.F. and Singer, L.T. (1983), Infant recognition memory as a measure of intelligence, in *Advances in Infancy Research*, Vol. 2, Lipsitt, L.P., Ed., Ablex, Norwood, NJ, pp. 31–78.

Gladen, B. and Rogan, W. (1991), Effects of polychlorinated biphenyls and dichlorodiphenyl dichloroethane on later development, *J. Pediatrics*, 119, 58–63.

Grandjean, P., Weihe, P., Burse, V.W., Needham, L.L., Storr-Hansen, E., Heinzow, B., Debes, F., Murata, K., Simonsen, H., Ellefsen, P., Budtz-Jorgensen, E., Keiding, N., and White, R.F. (2001), Neurobehavioral deficits associated with PCB in 7-year-old children prenatally exposed to seafood neurotoxicants, *Neurotoxicol. Teratol.*, 23, 305–317.

Harlow, H.F. (1949), The formation of learning sets, *Psychological Rev.*, 56, 51–65.

Holene, E., Nafstad, I., Skaare, J.U., Bernhoft, A., Engen, P., and Sagvolden, T. (1995), Behavioral effects of pre- and postnatal exposure to individual polychlorinated biphenyl congeners in rats, *Environmental Toxicol. Chem.*, 14, 967–976.

Holene, E., Nafstad, I., Skaare, J.U., Krogh, H., and Sagvolden, T. (1999), Behavioural effects in female rats of postnatal exposure to sub-toxic doses of polychlorinated biphenyl congener 153, *Acta Paediatric Suppl.*, 429, 55–63.

Holene, E., Nafstad, I., Skaare, J.U., and Sagvolden, T. (1998), Behavioral hyperactivity in rats following postnatal exposure to sub-toxic doses of polychlorinated biphenyl congeners 153 and 126, *Behavioral Brain Res.*, 94, 213–224.

Jacobson, J.L. and Jacobson S.W. (1996), Intellectual impairment in children exposed to polychlorinated biphenyls *in utero*, *New England J. Med.*, 335, 783–789.

Jacobson, J.L. and Jacobson S.W. (2003), Prenatal exposure to polychlorinated biphenyls and attention at school age, *J. Pediatrics*, 143, 780–788.

Jacobson, J.L., Jacobson, S.W., and Humphrey, H.E.B. (1990), Effects of *in utero* exposure to polychlorinated biphenyls and related contaminants on cognitive functioning in young children, *J. Pediatrics*, 116, 38–45.

Jacobson, J.L., Jacobson, S.W., Padgett, R.J., Brumitt, G.A., and Billings, R.L. (1992), Effects of prenatal PCB exposure on cognitive processing efficiency and sustained attention, *Developmental Psychol.*, 28, 297–306.

Jacobson, S.W., Fein, G.G., Jacobson, J.L., Schwartz, P.M., and Dowler, J.K. (1985), The effect of intrauterine PCB exposure on visual recognition memory, *Child Development*, 56, 853–860.

Levin, E.D. and Bowman, R.E. (1986), Long-term lead effects on the Hamilton search task and delayed alternation in monkeys, *Neurobehavioral Toxicol. Teratol.*, 8, 219–224.

Levin, E.D. and Bowman, R.E. (1988), Long-term effects of chronic developmental lead exposure on delayed spatial alternation in monkeys, *Neurotoxicol. Teratol.*, 10, 505–510.

Levin, E.D., Schantz, S.L., and Bowman, R.E. (1988), Delayed spatial alternation deficits resulting from perinatal PCB exposure in monkeys, *Arch. Toxicol.*, 62, 267–273.

Manning, F.J. (1973), Performance under temporal schedules by monkeys with partial ablations of prefrontal cortex, *Physiol. Behav.*, 11, 563–569.

Mele, P.C., Bowman, R.E., and Levin, E.D. (1986), Behavioral evaluation of perinatal PCB exposure in rhesus monkeys: fixed interval performance and reinforcement-omission, *Neurobehavioral Toxicol. Teratol.*, 8, 131–138.

Mes, J. and Marchand, L. (1987), Comparison of some specific polychlorinated biphenyl isomers in human and monkey milk, *Bull. Environmental Contamination Toxicol.*, 39, 736–742.

Miller, M.H. and Orbach, J. (1972), Retention of spatial alternation following frontal lobe resections in stump-tailed macaques, *Neuropsychologia*, 10, 291–298.

Murphy, B.L., Arnsten, A.F., Goldman-Rakic, P.S., and Roth, R.H. (1996), Increased dopamine turnover in the prefrontal cortex impairs spatial working memory performance in rats and monkeys, *Proc. Natl. Acad. Sci. U.S.A.*, 93, 1325–1329.

Neill, D.B., Ross, J.F., and Grossman, S.P. (1974), Comparison of the effects of frontal, striatal, and septal lesions in paradigms thought to measure incentive motivation or behavioral inhibition, *Physiol. Behav.*, 13, 297–305.

Neill, D.B. (1976), Frontal-striatal control of behavioral inhibition in the rat, *Brain Res.*, 105, 89–103.

Rice, D.C. (1997), Effect of postnatal exposure to a PCB mixture in monkeys on multiple fixed interval-fixed ratio performance, *Neurotoxicol. Teratol.*, 19, 429–434.

Rice, D.C. (1998), Effect of postnatal exposure of monkeys to a PCB mixture on spatial discrimination reversal and DRL performance, *Neurotoxicol. Teratol.*, 20, 391–400.

Rice, D.C. (1999), Effect of exposure to 3,3′,4,4′,5-pentachlorobiphenyl (PCB 126) throughout gestation and lactation on development and spatial delayed alternation performance in rats, *Neurotoxicol. Teratol.*, 21, 59–69.

Rice, D.C. and Hayward, S. (1997), Effects of postnatal exposure to a PCB mixture in monkeys on nonspatial discrimination reversal and delayed alternation performance, *Neurotoxicology*, 18, 479–494.

Rice, D.C. and Hayward, S. (1998), Lack of effect of 3,3′,4,4′,5-pentachlorobiphenyl (PCB 126) throughout gestation and lactation on multiple fixed interval-fixed ratio and DRL performance in rats, *Neurotoxicol. Teratol.*, 20, 645–650.

Rice, D.C. and Hayward, S. (1999), Effects of exposure to 3,3′,4,4′,5-pentachlorobiphenyl (PCB 126) throughout gestation and lactation on behavior (concurrent random interval-random interval and progressive ratio) in rats, *Neurotoxicol. Teratol.*, 21, 679–687.

Rowlands, J.C. and Gustafsson, J.A. (1997), Aryl hydrocarbon receptor-mediated signal transduction, *Crit. Rev. Toxicol.*, 27, 109–134.

Schantz, S.L., Levin, E.D., Bowman, R.E., Heironimus, M.P., and Laughlin, N.K. (1989), PCB effects on reversal learning in rhesus monkeys, *Neurotoxicol. Teratol.*, 11, 243–250.

Schantz, S.L., Moshtaghian, J., and Ness, D.K. (1995), Spatial learning deficits in adult rats exposed to *ortho*-substituted PCB congeners during gestation and lactation, *Fundam. Appl. Toxicol.*, 26, 117–126.

Schantz, S.L., Moshtaghian, J., Peterson, R.E., and Moore, R.W. (1996), Effects of gestational and lactational exposure to TCDD or coplanar PCBs on spatial learning, *Neurotoxicol. Teratol.*, 18, 305–313.

Schantz, S.L. and Widholm, J.J. (2001), Effects of PCB exposure on neurobehavioral function in animal models, in *PCBs: Recent Advances in Environmental Toxicology and Health Effects*, Robertson, L. and Hansen, L., Eds., University Press of Kentucky, Lexington, pp. 221–240.

Schwartz, P.M., Jacobson, S.W., Fein, G.G., Jacobson, J.L., and Price, H. (1983), Lake Michigan fish consumption as a source of polychlorinated biphenyls in human cord serum, maternal serum, and milk, *Public Health Briefs*, 73, 293–296.

Skeen, L.C. and Masterton, R.B. (1976), Origins of anthropoid intelligence, III: role of prefrontal system in delayed-alternation and spatial-reversal learning in a prosimian (*Galago senegalensis*), *Brain Behav. Evol.*, 13, 179–195.

Stewart, P., Fitzgerald, S., Reihman, J., Gump, B., Lonky, E., Darvill, T., Pagano, J., and Hauser, P. (2003a), Prenatal PCB exposure, the corpus callosum, and response inhibition, *Environmental Health Perspect.*, 111, 1670–1677.

Stewart, P.W., Reihman, J., Lonky, E.I., Darvill, T.J., and Pagano, J. (2003b), Cognitive development in preschool children prenatally exposed to PCBs and MeHg, *Neurotoxicol. Teratol.*, 25, 11–22.

Treichler, R.F., Hamilton, D.M., and Halay, M.A. (1971), The influence of delay interval on severity of the spatial alternation deficit in frontal monkeys, *Cortex*, 7, 143–151.

Vreugdenhil, H.J.I., Lanting, C.I., Mulder, P.G.H., Boersma, E.R., and Weisglas-Kuperus, N. (2002), Effects of prenatal PCB and dioxing background exposure on cognitive and motor abilities in Dutch children at school age, *J. Pediatrics*, 140, 48–56.

Vreugdenhil, H.J.I., Mulder, P.G.H., Emmen, H.H., and Weisglas-Kuperus, N. (2004), Effects of perinatal PCBs on neuropsychological functions in the Rotterdam cohort at 9 years of age, *Neuropsychology*, 18, 185–193.

Widholm, J.J., Clarkson, G.B., Strupp, B.J., Crofton, K.M., Seegal, R.F., and Schantz, S.L. (2001), Spatial reversal learning in Aroclor 1254-exposed rats: sex-specific deficits in associative ability and inhibitory control, *Toxicol. Appl. Pharmacol.*, 174, 188–198.

Widholm, J.J., Villareal, S., Seegal, R.F., and Schantz, S.L. (2004), Spatial alternation deficits following developmental exposure to Aroclor 1254 and/or methylmercury in rats, *Toxicological Sci.*, 82, 577–589.

Wilcott, R.C. (1982), Frontal lesions and the rate of operant behavior in the rat, *Physiological Psychol.*, 10, 371–375.

Wilcott, R.C. (1984), Prefrontal cortex and bulbar reticular formation and behavioral inhibition in the rat, *Brain Res. Bull.*, 12, 63–69.

Winneke, G., Bucholski, A., Heinzow, B., Krämer, U., Schmidt, E., Walkowiak, J., Wiener, J.-A., and Steingrüber, H.-J. (1998), Developmental neurotoxicity of polychlorinated biphenyls (PCBS): cognitive and psychomotor functions in 7-month old children, *Toxicol. Lett.*, 102–103, 423–428.

Zahalka, E.A., Ellis, D.H., Goldey, E.S., Stanton, M.E., and Lau, C. (2001), Perinatal exposure to polychlorinated biphenyls Aroclor 1016 or 1254 did not alter brain catecholamines nor delay alternation performance in Long-Evans rats, *Brain Res. Bull.*, 55, 487–500.

9 Modeling Cognitive Deficits Associated with Parkinsonism in the Chronic-Low-Dose MPTP-Treated Monkey

Jay S. Schneider
Thomas Jefferson University

CONTENTS

COGNITIVE DEFICITS ASSOCIATED WITH PARKINSON'S DISEASE

Historically, Parkinson's disease (PD) has been considered to be distinctly a disorder of motor function. While motor dysfunction is clearly demonstrable in PD patients, nonmotor aspects of the disease, while less obvious, are no less real or troublesome. It is now generally recognized that cognitive impairment is a common feature of PD, even in the early stages of the disease [1–4]. Parkinson's disease patients may develop neuropsychological deficits across a range of cognitive functions [5–8], and many of these impairments, particularly during the early stages of the disease,

169

resemble those associated with frontal lobe dysfunction [9]. These deficits include: difficulty in attention set shifting and set formation [10, 11]; temporal ordering, sequencing, and planning [3]; impaired nonverbal and verbal short-term recall; impaired spatial short-term memory [12]; and impairments in focused attention [13, 14].

Although recent data suggest that the cognitive deficits of PD initially appear to involve frontostriatal circuits and are similar to those found in patients with frontal lobe dysfunction [9], as PD progresses, impairments are seen on tests that may involve more posterior cortical regions [4, 15, 16]. For example, early-stage PD patients seem to have more attentional/executive functioning problems, whereas later-stage patients seem to have a broader range of deficits that include certain types of memory and working-memory impairments [9]. Studies have shown that non-medicated early-PD patients are unimpaired on performance of spatial working-memory tasks [17], whereas medicated mild-PD patients and patients with moderate to severe symptoms are impaired on performance of spatial working-memory tasks [18–20]. More-severe patients are also impaired on verbal and visual memory tasks that may depend more on the functional integrity of the temporal lobe than the frontal cortex [9]. However, the apparent "progression" of cognitive deficits in parallel with progressively more severe motor disability is a complicated issue, and patients at different stages of PD can be differentiated in terms of their performance on tests of spatial memory that make different demands of executive processes [21].

The cognitive deficits in PD patients appear to arise from dysfunction of multiple cortical and subcortical neurotransmitter systems and functional circuits. Dopaminergic as well as nondopaminergic systems (perhaps cholinergic and noradrenergic) may play important roles in the development of cognitive disturbances in PD patients [22]. Although levodopa treatment can improve some cognitive functions, it can also induce or exacerbate deficits in performance of certain cognitive tasks in PD patients [23]. Administration of levodopa to PD patients may have different effects on different sectors of frontostriatal circuitry [21, 24]. The amount of levodopa required to restore function to more heavily denervated dorsal frontostriatal regions and striatal regions primarily involved in motor control may effectively overdose relatively less depleted ventral frontostriatal circuits and actually worsen certain aspects of cognitive performance.

MODELING COGNITIVE DEFICITS OF PARKINSONISM WITH CHRONIC ADMINISTRATION OF MPTP

Our initial idea was to produce cognitive deficits in monkeys pretrained to perform a variety of cognitive tasks by administering the dopaminergic neurotoxin MPTP (1-methyl-4-phenyl-1,2,3,6-tetrahydropyridine) in doses too low to induce gross parkinsonian motor deficits. Since the clinical literature indicated that cognitive deficits are present at the earliest stages of PD, our goal was try to model this early stage of PD in nonhuman primates. Over several years we developed and refined the chronic-low-dose MPTP administration protocol to produce cognitive deficits in monkeys with minimal or no motor impairments. With continued MPTP exposure, animals develop parkinsonian motor deficits superimposed on earlier appearing cognitive deficits.

In the chronic-low-dose MPTP model of "early" parkinsonism, animals develop cognitive deficits analogous to those described in early PD patients [25–28], including deficits in attention, attention set shifting, cognitive flexibility, planning, and problem solving, but not in working or reference memory *per se*. Although it has been suggested that there may be a visuospatial working-memory deficit in early *medicated* PD patients [9], it is possible that this deficit may not be a working-memory deficit *per se* but may reflect an impairment in attentional processes involved in visuospatial working-memory tasks [29]. This latter interpretation would be consistent with our data from MPTP-treated monkeys.

Continued administration of low doses of MPTP after cognitive deficits have appeared can result in the development of parkinsonian motor symptoms [30]. Patients with more advanced PD exhibit a broad range of cognitive impairments that include attentional and other "frontal-related" deficits but also impaired performance on at least some memory tasks [9]. Monkeys with moderate parkinsonism continue to have frontal-related cognitive deficits [30] that interestingly do not respond to the same pharmacotherapies that improve attentional and executive functioning in "early" parkinsonian monkeys, suggesting that pharmacological strategies to improve cognition in PD may need to change as symptoms evolve and as the disease progresses.

ATTENTIONAL AND EXECUTIVE FUNCTION DEFICITS CHARACTERIZE "EARLY" PARKINSONIAN MPTP-TREATED MONKEYS

In the first studies of chronic-low-dose MPTP-treated monkeys, deficits were observed in performance of spatial delayed response (both with fixed and variable delays), delayed alternation, delayed matching-to-sample, visual discrimination reversal, and object retrieval tasks [26, 27]. Deficits in variable delayed-response performance have proven to be particularly interesting and have led to further insights into the nature of the cognitive deficits in these animals. In the variable delayed-response task, the monkeys perform a spatial delayed-response task with different delay lengths randomly distributed over the trials that make up a daily test session. These delay conditions yield approximately chance performance at the longest delay, and normal monkeys perform this task with a delay-dependent increase in errors that reflects the limits of their short-term spatial memory. This task appears to place little load on executive processes, since there is no specific requirement for manipulation or reorganization of the information in working memory.

Chronic-low-dose MPTP-treated monkeys are significantly impaired on performance of this task, with deficits in performing short- and medium-duration delay trials, suggesting an attentional deficit [31]. Because of the nature of the variable delayed-response task, it is difficult to unequivocally separate attention from memory components of task performance. Thus, an additional study was conducted to try to assess more directly the degree to which deficiencies in variable delayed-response task performance could be attributed to attentional or working-memory disturbances. To do this, we used a modified variable delayed-response task with attentional cueing [32].

On the standard variable delayed-response task, an opaque screen is lifted, and cue presentation (baiting of one of two food wells) occurs over a fixed 2-sec interval. The opaque screen is then lowered for an intratrial delay, and the food wells are covered with two identical sliding plates. The screen is then raised, and the animal is presented with the two wells and is allowed to respond (i.e., uncover one of the two wells and retrieve the reward).

The protocol for the variable delayed-response task with attentional cueing was identical to the variable delayed-response procedures described above, except that the examiner alerted the animal to the baiting of the well to ensure that the monkey was attending to the cue presentation before the opaque screen was lowered. Cue presentation still occurred over a standard 2-sec interval. Test sessions with attentional cueing were given to each animal once per week between standard variable delayed-response (VDR) testing sessions. Attentional cueing significantly improved overall task performance in comparison with standard VDR performance in chronic-low-dose MPTP-treated monkeys, and restored a normal pattern of responding. In particular, performance at short- to medium-duration delays was significantly improved by attentional cueing, while cueing had no significant effect on performance of long-duration delay trials (Figure 9.1). Thus, directing the animal's attention to the target presentation significantly improved performance on a spatial working-memory task. Interestingly, attentional cueing, as described above, had no effect on performance of short-, medium-, or long-delay trials in intact animals. It appears then that the main contributor to impaired variable delayed-response performance of chronic-low-dose MPTP-treated monkeys is an attentional deficit that disrupts the encoding of behaviorally relevant information to be kept in short-term memory.

Other studies have shown additional attentional and executive functioning deficits in motor-asymptomatic chronic-low-dose MPTP-treated monkeys [28]. Animals were trained to perform the following attentional and executive function tasks that had low memory demands.

ATTENTIONAL/EXECUTIVE TASKS

Attention set-shifting ability: This test of cognitive flexibility and set-shifting ability is based on the principle of the Wisconsin Card Sorting Test (WCST), which is known to be impaired in patients with frontal lobe dysfunction and in PD patients, and was modeled after the intradimensional/extradimensional set-shift task from the Cambridge Neuropsychological Test Automated Battery [33]. In this task, monkeys were required to shift attention from one perceptual dimension of a complex stimulus to another, e.g., from line to shape. The task consisted of five subtests: simple visual discrimination, simple visual discrimination reversal, compound visual discrimination, and intra/extradimensional shifts (IDS/EDS).

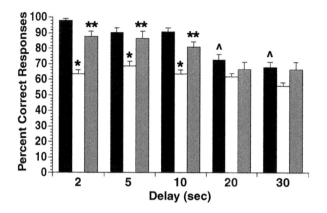

FIGURE 9.1 Variable delayed-response performance prior to (black bars) and following chronic-low-dose MPTP exposure (white bars) and the effect of attentional cueing (shaded bars) on task performance. In intact monkeys, there was a delay-dependent decrease in the number of correct responses, with performance at 20- and 30-sec delays significantly worse than performance at 2-, 5-, and 10-sec delays ($^\wedge p < 0.05$). After chronic-low-dose MPTP exposure, performance at short- and medium-duration delays (2, 5, and 10 sec) was significantly worse than when animals were normal (*$p < 0.01$). Task performance at the longest delays (20 and 30 sec) remained the same. With attentional cueing, the performance of chronic-low-dose MPTP-treated animals was significantly improved at 2-, 5-, and 10-sec delays (**$p < 0.01$ vs. post-MPTP performance without attentional cueing). Overall task performance with attentional cueing was not significantly different from baseline performance prior to MPTP exposure. (From Decamp, E. et al., *Behavioural Brain Res.*, 152, 260, 2004. With permission).

Visuospatial attention shifting: This task assessed the ability of the animal to shift attention once it had been directed away from the location of an impending target [34, 35].

Cued reaction time (focused attention): This test (a four-choice cued and noncued reaction time task) assessed the monkey's ability to focus attention and use advance information for successful task performance.

Motor readiness (impulse control) task: This test, which required time estimation, had a high attentional load and has been shown to depend on the integrity of the dopaminergic system to be performed correctly [36]. An animal must hold a lever for a variable amount of time and wait for a target to appear and then touch the target for a reward. As time elapses, the probability for appearance of the target increases. The relation between reaction time and each specific delay was taken as a measure of "motor readiness." Premature release of the lever reflected increased impulsivity, leading to poor task performance.

Effects of Chronic Low Dose MPTP Exposure

Animals received cumulative MPTP amounts of 10.64, 10.9, 12.6, and 15.7 mg over periods of 106, 158, 98, and 109 days, respectively. Sensorimotor deficits were not

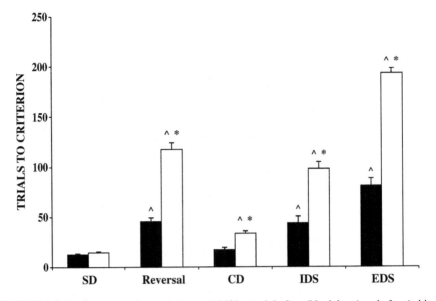

FIGURE 9.2 Performance of an attention set-shifting task before (black bars) and after (white bars) chronic-low-dose MPTP exposure. Results are expressed as mean trials to criterion (± SEM) for each subtest: SD (simple visual discrimination), CD (compound visual discrimination), reversal (discrimination reversal), IDS (intradimensional shift), and EDS (extradimensional shift). When normal, animals performed the SD and CD subtests easily but needed more trials to reach criterion at the reversal, IDS, and EDS subtests. After MPTP exposure, performance remained normal at the SD task but significantly worsened on the other subtests (with particular difficulty with the discrimination reversal and EDS subtests). $^\wedge p < .01$ compared with SD performance; $*p < .01$ compared with corresponding normal baseline performance. (From Decamp, E. and Schneider, J.S., *Eur. J. Neurosci.*, 20, 1374, 2004. With permission).

observed in any animal at the end of MPTP administration or after deficits were observed on cognitive tasks.

Attention set-shifting ability: When normal, animals performed all subtests of the task. The discrimination reversal and EDS subtests required significantly more trials to reach criterion (i.e., six consecutive correct responses) than did the other subtests. After chronic-low-dose MPTP administration, animals were still able to perform all stages of the task. While the ability to perform the simple discrimination task did not change, animals performed significantly worse on all other subtests after chronic MPTP exposure (indicating a disturbance in set stability), with particular difficulty with the discrimination reversal and EDS tasks (Figure 9.2). Response latencies were not significantly changed following MPTP exposure, indicating that deficits were not due to motor impairment. Difficulty in completing the discrimination reversal and EDS portions of this task in particular suggests deficits in cognitive shifting ability (mental flexibility). These data are consistent with reports of cognitive inflexibility in mild PD patients [17, 23] and impairment in mild and nonmedicated PD patients at the earlier stages of an attention set-shifting task that requires the formation and maintenance of a response set for the relevant stimulus dimension [17].

Visuospatial attention shifting: When normal, animals responded to valid cue trials with shorter reaction times (460 ± 37 msec) than to invalid cue trials (815 ± 103 msec). After chronic MPTP exposure, reaction time for valid cue trials was significantly increased, resulting in equivalent times to respond to valid (825 ± 72 msec) or invalid cues (852 ± 106 msec). Animals also made significantly more early-release errors after MPTP exposure compared with when they were normal. These data suggest an impaired ability to sustain spatial attention rather than a specific impairment in visuospatial attention shifting per se. Even mild PD patients have been reported to have visuospatial attention deficits [37, 38], particularly when sustained attention is required.

Cued reaction time (focused attention): Before MPTP exposure, animals performed this task easily, making no or few early bar-release errors. Animals also had significantly shorter reaction times on trials for which there was advance information (cue) (523.0 ± 28.9 msec) compared with the trials without a cue (723.0 ± 34.2 msec). After chronic MPTP exposure, reaction times increased for both conditions (advance information: 999.0 ± 55.2 msec; no advance information: 1030.0 ± 51.5 msec), and the effect of advance information on reaction time was no longer significant. It is possible that in the uncued four-choice condition of this task, the attentional demands of monitoring the four possible locations of the target were sufficiently high so as to slow reaction times and impair performance. Similarly slowed reaction times in the cued condition may reflect a deficit in the MPTP-exposed monkeys to prepare their response in anticipation of a predetermined movement.

Motor readiness (impulse control) task: Normal monkeys showed an inverse relation between reaction time and hold time and a direct relationship between hold time and number of errors committed (i.e., the longer the hold time, the more errors committed). Reaction times associated with all hold times were significantly increased following MPTP exposure. The effect of the hold time on reaction time was no longer significant after MPTP exposure. Chronic-low-dose MPTP exposure also significantly increased the number of errors made. In particular, monkeys made significantly more mistakes on trials with the shortest hold times. These results suggest a deficit in motor readiness and planning of action. Deficits in attention have been described in PD patients, especially when they have to rely on internal cues [7], and this may be reflected in the results described above. Additionally, even mild PD patients have deficits in a time-estimation task such that they underestimate the duration of a time interval [39]. Such a deficit in chronic-low-dose MPTP-treated monkeys may have also contributed to our findings.

RELEVANCE OF THE CHRONIC-LOW-DOSE MPTP MONKEY MODEL TO OTHER DISORDERS

While the deficits in chronic-low-dose MPTP-treated monkeys described earlier in this chapter are strikingly similar in nature to those described in early Parkinson's disease patients, the frontal/attentional/executive function deficits observed in these animals have significant overlap with, and clinical relevance for, other disorders such as attention deficit hyperactivity disorder (ADHD) and schizophrenia in which frontostriatal dysfunction has been implicated.

A number of similarities between the cognitive deficits of chronic-low-dose (CLD) MPTP-treated monkeys and children with attention deficit disorder have been noted and attributed to frontostriatal dysfunction secondary to deficits in dopaminergic and noradrenergic neurotransmission [40]. Recent studies of children and adolescents with attention deficit hyperactivity disorder (ADHD) indicate that, in addition to inattention and distractibility, a number of executive functioning deficits have also been documented [41–43]. Additionally, the sense of time is reportedly impaired in children with ADHD and is not improved with administration of stimulant medication [44]. Thus, we suggest that the CLD MPTP model in nonhuman primates may also be a valid large-animal model for studying cognitive deficits associated with attention deficit disorder. The CLD MPTP-treated monkey appears to encompass a number of the criteria suggested by Solanto [45] that should be present in a valid model of ADHD, including deficits in measures of attention and impulsivity and amelioration of deficits by clinically effective treatments [46–48].

Schizophrenia is characterized by dysregulation of attention [49]. Four important domains of cognitive functioning have been described as impaired in schizophrenia: attention, verbal fluency, working memory, and executive functioning [50]. Sustained attention appears to be a particular problem in schizophrenia that contributes to problems in other aspects of higher cognitive functioning [51]. Because of the apparent specificity of sustained attention to schizophrenia and its relationship to other important cognitive abilities, this aspect of cognition (which can be modeled in chronic-low-dose MPTP-treated monkeys) may be an important future target of intervention.

Executive functions encompass a number of abilities, including but not limited to the ability to initiate, plan, and sequence behaviors, the ability to abstract a principle or problem-solving strategy, and the ability to be cognitively flexible (i.e., switch cognitive sets). Whereas executive functions are largely subserved by the frontal cortex, they are also related to other parts of the brain that have strong connections to the frontal cortex, such as the striatum and temporal-limbic complex [51]. A recent meta-analysis from 71 studies [52] demonstrated an overall effect size of -1.45 for schizophrenia patients relative to controls on measures of executive functioning. This is a large effect size, suggesting that patients with schizophrenia have significant difficulty on these measures compared with other psychiatric patients (effect size of -0.40). In addition, functional imaging studies have provided support for dysfunction in frontal-basal ganglia circuitry [53] as a neural substrate for attention/executive cognitive dysfunction in schizophrenia. Considering the cognitive deficits known to affect schizophrenics and the attention/executive deficits described in chronic-low-dose MPTP-treated monkeys, these animals may serve as a useful model for the cognitive deficits of schizophrenia and as a good model system for evaluating new treatment modalities.

CONCLUSIONS

Chronic administration of low doses of the neurotoxin MPTP to macaque monkeys results in primarily attentional and executive function-type cognitive deficits in animals devoid of parkinsonian motor deficits. The fronto-executive deficits in these animals may serve as a model for the cognitive deficits of "early" parkinsonism as well as for the cognitive deficits characteristic of other disorders such as ADHD and chronic schizophrenia. Continued administration of low doses of MPTP to these animals can result in the evolution of a parkinsonian motor disorder, making this a useful animal system for studying the progression of symptoms in parkinsonism.

ACKNOWLEDGMENTS

Work described in this chapter was funded in part by NIH grant DA 013452. The author acknowledges the contributions of Emmanuel Decamp to the continued development of the low-dose MPTP model.

REFERENCES

1. Levin, B.E., Llabre, M.M., and Weiner, W.J., Cognitive impairments associated with early Parkinson's disease, *Neurology*, 39, 557–561, 1989.
2. Heitanen, M. and Teravainen, H., Cognitive performance in early Parkinson's disease, *Acta Neurol. Scand.*, 73, 151–159, 1986.
3. Cooper, J.A., Sagar, H.J., Jordan, N., Harvey, N.S., and Sullivan, E., Cognitive impairment in early, untreated Parkinson's disease and its relationship to motor disability, *Brain*, 114, 2095–2122, 1991.
4. Owen, A.M., James, M., Leigh, P.N., Summers, B.A., Marsden, C.D., Quinn, N.P., Lange, K.W., and Robbins, T.W., Fronto-striatal cognitive deficits at different stages of Parkinson's disease, *Brain*, 115, 1727–1751, 1992.
5. Lees, A.J. and Smith, E., Cognitive deficits in the early stages of Parkinson's disease, *Brain*, 106, 257–270, 1983.
6. Taylor, A.E., Saint-Cyr, J.A., and Lang, A.E., Frontal lobe dysfunction in Parkinson's disease, *Brain*, 109, 845–883, 1986.
7. Brown, R.G. and Marsden, C.D., "Subcortical dementia": the neuropsychological evidence, *Neuroscience*, 25, 363–387, 1988.
8. Boller, F., Passafiume, D., Keefe, N.C. et al., Visuospatial impairments in Parkinson's disease: role of perceptual and motor factors, *Arch. Neurol.*, 41, 485–490, 1984.
9. Owen, A.M., Iddon, J.L., Hodges, J.R., Summers, B.A., and Robbins, T.W., Spatial and non-spatial working memory at different stages of Parkinson's disease, *Neuropsychologia*, 35, 519–532, 1997.
10. Flowers, K.A. and Robertson, C., The effect of Parkinson's disease on the ability to maintain a mental set, *J. Neurol. Neurosurg. Psychiatr.*, 48, 517–529, 1985.

11. Downes, J.J., Roberts, A.C., Sahakian, B.J., Evenden, J.L., Morris, R.G., and Robbins, T.W., Impaired extra-dimensional shift performance in medicated and unmedicated Parkinson's disease: evidence for a specific attentional dysfunction, *Neuropsychologia*, 27, 1329–1343, 1989.

12. Freedman, M. and Oscar-Berman, M., Selective delayed response deficits in Parkinson's and Alzheimer's disease, *Arch. Neurol.*, 43, 886–890, 1986.

13. Sharpe, M.H., Distractibility in early Parkinson's disease, *Cortex*, 26, 239–246, 1990.

14. Sharpe, M.H., Auditory attention in early Parkinson's disease: an impairment in focused attention, *Neuropsychologia*, 30, 101–106, 1992.

15. Owen, A.M., Beksinska, M., James, M., Leigh, P., Summers, B.A. et al., Visuo-spatial memory deficits at different stages of Parkinson's disease, *Neuropsychologia*, 31, 627–644, 1993.

16. Owen, A.M., Sahakian, B.J., Semple, J. et al., Visuo-spatial short term recognition memory and learning after temporal lobe excisions, frontal lobe excisions or amygdalo-hippocampectomy in man, *Neuropsychologia*, 33, 1–24, 1995.

17. Owen, A.M., James, M., Leigh, P.N., Summers, B.A., Marsden, C.D., Quinn, N.P., Lange, K.W., and Robbins, T.W., Fronto-striatal cognitive deficits at different stages of Parkinson's disease, *Brain*, 115, 1727–1751, 1992.

18. Bradley, V.A., Welch, J.L., and Dick, V.J., Visuospatial working memory in Parkinson's disease, *J. Neurol. Neurosurg. Psychiatr.*, 52, 1228–1235, 1989.

19. Postle, B.R., Corkin, S., and Growdon, J.H., Dissociation between two kinds of visual working memory in Parkinson's disease, *Soc. Neurosci. Abstr.*, 409, 1002, 1993.

20. Partiot, A., Verin, M., Pillon, B. et al., Delayed response tasks in basal ganglia lesions in man: further evidence for a striato-frontal cooperation in behavioural adaptation, *Neuropsychologia*, 34, 709–721, 1996.

21. Owen, A.M., Cognitive dysfunction in Parkinson's disease: the role of frontostriatal circuitry, *Neuroscientist*, 10, 525–537, 2004.

22. Pillon, B., Dubois, B., Bonnet, A.M., Esteguy, M., Guimaraes, J., Vigouret, J.M., Lhermitte, F., and Agid, Y., Cognitive slowing in Parkinson's disease fails to respond to levodopa treatment: the 15-objects test, *Neurology*, 39, 762–768, 1989.

23. Cools, R., Barker, R.A., Sahakian, B.J., and Robbins, T.W., L-dopa medication remediates cognitive inflexibility but increases impulsivity in patients with Parkinson's disease, *Neuropsychologia*, 41, 1431–1441, 2003.

24. Schneider, J.S., Ferraro, S., Gollomp, S.M., Mancall, E.L., Cambi, F., and Mozley, P.D., Effects of dopamine replacement on cognitive and motor functions in Parkinson's disease: correlation with [Tc-99m]TRODAT-1 labeling of striatal dopamine transporter sites, *Soc. Neurosci. Abstr.*, 27, 2001.

25. Roeltgen, D.P. and Schneider, J.S., Task persistence and learning ability in normal and chronic low dose MPTP-treated monkeys, *Behavioural Brain Res.*, 60, 115–124, 1994.

26. Schneider, J.S. and Kovelowski, C.J., Chronic exposure to low doses of MPTP, I: cognitive deficits in motor asymptomatic monkeys, *Brain Res.*, 519, 122–128, 1990.

27. Schneider, J.S. and Roeltgen, D.P., Delayed matching-to-sample, object retrieval, and discrimination reversal deficits in chronic low dose MPTP-treated monkeys, *Brain Res.*, 615, 351–354, 1993.

28. Decamp, E. and Schneider, J.S., Attention and executive function deficits in chronic low dose MPTP-treated non-human primates, *Eur. J. Neurosci.*, 20, 1371–1378, 2004.

29. Pillom, B., Deweer, B., Vidhailet, M. et al., Is impaired memory for spatial location in Parkinson's disease domain specific or dependent on "strategic" processes? *Neuropsychologia*, 36, 1–9, 1998.

30. Schneider, J.S. et al., Effects of the nicotinic acetylcholine receptor agonist SIB-1508Y on object retrieval performance in MPTP-treated monkeys: comparison with levodopa treatment, *Ann. Neurol.*, 43, 311–317, 1998.

31. Schneider, J.S., Tinker, J.P., Van Velson, M., Menzaghi, F., and Lloyd, G.K., Nicotinic acetylcholine receptor agonist SIB-1508Y improves cognitive functioning in chronic low dose MPTP-treated monkeys, *J. Pharmacol. Exp. Ther.*, 290, 731–739, 1999.

32. Decamp, E. Tinker, J.P., and Schneider, J.S., Attentional cueing reverses deficits in spatial working memory task performance in chronic low dose MPTP-treated monkey, *Behavioural Brain Res.*, 152, 259–262, 2004.

33. Downes, J.J., Roberts, A.C., Sahakian, B.J., Evenden, J.L., Morris, R.G., and Robbins, T.W., Impaired extra-dimensional shift performance in medicated and unmedicated Parkinson's disease: evidence for a specific attentional dysfunction, *Neuropsychologia*, 27, 1329–1343, 1989.

34. Voytko, M.L., Cognitive functions of the basal forebrain cholinergic system in monkeys: memory or attention? *Behavioural Brain Res.*, 75, 13–25, 1996.

35. Witte, E.A. and Marrocco, R.T., Alteration of brain noradrenergic activity in rhesus monkeys affects the alerting component of covert orienting, *Psychopharmacology*, 132, 315–323, 1997.

36. Amalric, M., Moukhles, H., Nieoullon, A., and Daszuta, A., Complex deficits on reaction time performance following bilateral intrastriatal 6-OHDA infusion in the rat, *Eur. J. Neurosci.*, 7, 972–980, 1995.

37. Leplow, B., Holl, D., Zeng, L., Herzog, A., Behrens, K., and Mehdorn, M., Spatial behavior is driven by proximal cues even in mildly impaired Parkinson's disease, *Neuropsychologia*, 40, 1443–1455, 2002.

38. Yamaguchi, S. and Kobayashi, S., Contributions of the dopaminergic system to voluntary and automatic orienting of visuospatial attention, *J. Neurosci.*, 18, 1869–1878, 1998.

39. Pastor, M.A., Artieda, J., Jahanshani, M., and Obeso, J.A., Time estimation and reproduction is abnormal in Parkinson's disease, *Brain*, 115, 211–225, 1992.

40. Roeltgen, D.P. and Schneider, J.S., Chronic low-dose MPTP in nonhuman primates: a possible model for attention deficit disorder, *J. Child. Neurol.*, 6 (Suppl.), S80–S87, 1991.

41. Barkley, R.A., Edwards, G., Laneri, M., Fletcher, K., and Metevia, L., Executive functioning, temporal discounting, and sense of time in adolescents with attention deficit hyperactivity disorder (ADHD) and oppositional defiant disorder (ODD*)*, *J. Abnormal Child Psychol.*, 29, 541–556, 2001.

42. Barnett, R., Maruff, P., Vance, A., Luk, E.S., Costin, J., Wood, C., and Pantelis, C., Abnormal executive function in attention deficit hyperactivity disorder: the effect of stimulant medication and age on spatial working memory, *Psych. Med.*, 31, 1107–1115, 2001.

43. Shallice, T., Marzocchi, G.M., Coser, S., Del Savio, M., Meuter, R.F., and Rumiati, R.I., Executive function profile of children with attention deficit hyperactivity disorder, *Dev. Neuropsychol.*, 21, 43–71, 2002.

44. Barkley, R.A., Koplowitz, S., Anderson, T., and McMurray, M.B., Sense of time in children with ADHD: effects of duration, distraction, and stimulant medication, *J. Int. Neuropsychol. Soc.*, 3, 359–369, 1997.

45. Solanto, M.V., Clinical psychopharmacology of AD/HD: implications for animal models, *Neurosci. Biobehavioral Rev.*, 24, 27–30, 2000.

46. Schneider, J.S., Sun, Z.Q., and Roeltgen, D.P., Effects of dopamine agonists on delayed response performance in chronic low-dose MPTP-treated monkeys, *Pharmacol. Biochem. Behav.*, 48, 235–240, 1994.

47. Schneider, J.S., Giardiniere, M., and Morain, P., Effects of the prolyl endopeptidase inhibitor S 17092 on cognitive deficits in chronic low dose MPTP-treated monkeys, *Neuropsychopharmacology*, 26, 176–182, 2002.

48. Schneider, J.S., Tinker, J.P., Menzaghi, F., and Lloyd, G.K., The subtype-selective nicotinic acetylcholine receptor agonist SIB-1553A improves both attention and memory components of a spatial working memory task in chronic low dose 1-methyl-4-phenyl-1,2,3,6-tetrahydropyridine-treated monkeys, *J. Pharmacol. Experimental Ther.*, 306, 401–406, 2003.

49. Nestor, P.G. et al., Neuromodulation of attention in schizophrenia, *Schizophrenia Bulletin*, 16, 424–432, 1990.

50. Stip, E. and Lussier, I., The heterogeneity of memory dysfunction in schizophrenia, *Can. J. Psychiatry*, 41 (Suppl. 1), 14S–20S, 1996.

51. Hoff, A.L., Neuropsychology in schizophrenia: an update, *Curr. Opin. Psychiatry*, 16, 149–155, 2003.

52. Johnson-Selfridge, M. and Zalewski, C., Moderator variables of executive functioning in schizophrenia: meta-analytic findings, *Schizophrenia Bull.*, 2, 305–316, 2001.

53. Barch, D.M., Carter, C.S., Braver, T.S. et al., Selective deficits in prefrontal cortex function in medication-naive patients with schizophrenia, *Arch. Gen. Psychiatry*, 58, 280–288, 2001.

Section III

Mouse Genetic Models

10 Cognitive Impairment in Transgenic Mouse Models of Amyloid Deposition

Dave Morgan
University of South Florida

CONTENTS

INTRODUCTION

The identification of the Aβ peptide as a major component of amyloid deposited in brain vessels and subsequently of parenchymal plaques in the brains of Alzheimer's victims [1] led to a focus on this molecule as a key element in the pathophysiology of Alzheimer's disease (AD). Subsequent work found that some mutations causing the disease occurred in the amyloid precursor protein (APP) that is processed, in some circumstances, into the Aβ peptide [2]. Ultimately, all mutations causing AD have been demonstrated to result in overproduction of the long variant of the Aβ peptide [3]. Thus the pathology, genetics, and the *in vitro* neurotoxicity of the Aβ peptide has led to a focus on the aggregation and accumulation of this material as a prime target for therapies designed to treat AD.

As for other disorders, development of animal models to understand amyloid pathology became an important research goal. Our lab and many others attempted to mimic the AD condition in rodent brain by direct injections of the peptide [4, 5]. These efforts were largely unsatisfactory. Similarly, a number of other research groups attempted to create transgenic models overexpressing various forms of the APP gene. After many failures to create models that deposit amyloid in a manner similar to AD, and after a couple of retracted claims, the first APP transgenic mouse that deposited amyloid in a manner similar to AD was developed [6]. Because of the platelet-derived growth-factor promoter used, this mouse is referred to as the PDAPP mouse. Shortly thereafter, other APP transgenic mice were successfully developed with amyloid pathology similar to AD: the Tg2576 mouse of Hsiao [7] and the APP23 mouse of Novartis [8]. In a variety of informative ways, these mice have been crossed with other genetically modified mice. One of these was the presenilin-1 (PS1) mouse of Duff [9]. The presenilin gene codes for a critical component of the gamma secretase complex, which determines the length of the $A\beta$ peptide during processing of the APP. Mutations of this gene can also cause early-onset AD in humans. Our group and several others found that crosses between the PS1 mice and the APP mice resulted in accelerated amyloid pathology as the mice age [10]. This chapter focuses on the behavioral abnormalities in the APP transgenic models. It is important to recognize that these are not models of AD, as they lack the tau pathology and the neuron loss found consistently in the human disorder. However, the mouse models mimic in considerable detail the amyloid deposition found in the human disease, and they are all largely similar, irrespective of the mutations used or the promoters employed.

MEMORY DEFICITS CORRELATE WITH Aβ LOAD

The first to demonstrate that APP transgenic mice had both amyloid deposits as well as memory deficits was Hsiao et al. [7]. They demonstrated deficits in a reference-memory version of the open-pool water maze [11], with the deficits first appearing at an age when the amyloid plaques started to appear (10 to 11 months). A number of studies have now found within-age-group correlations between spatial-navigation performance in various forms of the water maze and one or more measures of amyloid load in several different APP lines [12–20]. In some sense, given the perception that rodent memory-assessment methods have some imprecision and that the number of mice included in such studies is typically small (less than 20), the correlation was surprising and argued for an intimate relationship between amyloid load measures and memory dysfunction. This led to the somewhat comforting conclusion that the amyloid plaques and attendant disruptions of neuronal architecture with dystrophic neurites and activation of proinflammatory states of glial cells were responsible for the memory loss. However, even these studies were not always consistent regarding the pool of $A\beta$ that correlated best with memory loss.

There are several domains in which amyloid loads can be measured. The pathologist examines amyloid histologically and identifies two major forms. One form is fibrillar aggregates, stained by Congo red or Thioflavine S, that form compacted amyloid plaques (associated with dystrophic neurites) in the parenchyma or amyloid angiopathy in the vasculature. A second, more widely spread form of $A\beta$ is stained only immunohistochemically and is referred to as diffuse deposits. The neurochemist views $A\beta$ content from the perspective of solubility in different reagents. There are water-soluble fractions, detergent-soluble fractions, and water-insoluble fractions (typically dissolved in concentrated formic acid or guanidinium solutions). Most often, these are analyzed by sandwich enzyme-linked immunosorbent assays

(ELISA). A third domain is that of the physical chemist, who views the $A\beta$ peptide from the perspective of its secondary and tertiary structure. There are many forms of $A\beta$: monomers, multimers varying from 2mers to 20mers that are collectively called oligomers, protofibrillar forms in which $A\beta$ is aggregated in a beta-pleated sheet structure, and finally mature 6- to 9-nm fibrils, which are the forms thought to form the amyloid plaques. One of the great unanswered questions in Alzheimer research is how these different fractions relate to each other. As a further level of complexity, there are two major C-terminal length variants of $A\beta$ ending at amino acid 40 or 42. These appear to be distributed differently, with vascular amyloid deposits being primarily of the $A\beta$ 40 form and diffuse $A\beta$ ending primarily at amino acid 42 [21, 22]. The final common path of mutations causing early-onset AD in humans is to increase the amount of the long form of $A\beta$ [3]. In general, the more-hydrophobic $A\beta42$ variant is more prone to forming aggregates.

The studies correlating memory loss with amyloid increases used a variety of indices for amyloid load. Often, only one of several measurements of amyloid load correlated with behavior, and this was not always the same measure or even in the same brain regions (hippocampus versus cortex). Undermining the association between congophilic amyloid plaques and memory disruption was the increasing number of observations of memory dysfunction in mice that either never developed amyloid pathology [23–28] or exhibited some form of memory disruption prior to the appearance of amyloid deposits [15, 26, 29–32]. In many instances, the severity of the memory deficits worsened as mice grew older [30, 33, 34]. However, the worsening of memory function as mice aged was not linearly related to the increasing amount of $A\beta$ in the brain [15, 17]. Intriguingly, not all reports find memory dysfunction in transgenic rodents that overexpress APP. For example, Savonenko et al. [35] were unable to detect changes in two APP mouse lines that deposit amyloid, and Ruiz-Opazo [36] found that transgenic rats expressing APP actually had protection from age-associated memory deficits.

Although the transgenic mice are very similar genetically, there are large variations in the extent to which they overproduce $A\beta$. In our APP+PS1 model, we find roughly a two-fold range of amyloid load in mice of the same age and gender. In the PDAPP mouse line, some transgenic mice never deposit amyloid. These large variations permit identification of correlations with behavior within an age group. If the amyloid loads were uniform, correlations within age groups would be nearly impossible. However, it is likely that whatever factors are accounting for this variation in amyloid load, they are affecting all of the pools of $A\beta$ similarly. Thus mice with a high level of soluble $A\beta$ when young are likely destined to have high levels of deposited $A\beta$ as they age. So too, all pools of $A\beta$ are likely to covary within individual members of an age cohort; those mice with the highest levels of deposited $A\beta$ are likely to have higher levels of soluble $A\beta$, insoluble $A\beta$, oligomeric $A\beta$, etc. In this sense, a correlation with one pool need not indicate that it is the one causing the memory disruption; indeed, it may be acting as a surrogate for other, more difficult to measure pools. At the moment, there is some evidence that an oligomeric pool of $A\beta$ may be most closely related to the degeneration in AD [37]. However, this pool has been notoriously ephemeral; reliable assays have not been widely adopted, and the oligomers appear to rapidly convert to other forms. It is not certain that different investigators are studying the same entity when each refers to oligomeric pools. At the recent Alzheimer Conference (2004), Karen Ashe presented impressive data supporting an oligomeric form of $A\beta$ (referred to as $A\beta^*$) as being most closely related to memory loss in younger Tg2576 transgenic mice. The reader is advised to monitor progress in this area critically and with some caution, but the data collected thus far is consistent with this hypothesis. For example, some forms of fibrillar $A\beta$ may not be apparent by traditional methods. Richardson et al. [31] found extracellular $A\beta$ fibrils in memory-deficient APP mice well before the appearance of histologically identifiable deposits.

In summary, memory dysfunction is a very consistent part of the APP mouse phenotype. Correlations with amyloid load suggest that this plays a causal role in the memory disruption. However, when examined across age groups, the correlation breaks down, with old mice having much more amyloid than predicted by their memory deficits. Still, given the potential for ceiling effects in the behavioral tests and the likelihood that amyloid effects on memory may be nonlinear and exhibit saturation, these data alone are not sufficient to rule out observable amyloid pools as the major factor in the memory impairments developed by the transgenic mice.

INTERVENTIONS IMPROVING MEMORY IN APP TRANSGENIC MICE

A number of approaches have been proposed to reduce $A\beta$ loads in AD patients and, in many instances, these are first being tested in the APP transgenic mouse models of amyloid deposition (Table 10.1). If the hypothesis that amyloid causes memory loss is to hold merit, then amyloid-reducing treatments should also protect the mice from memory disruptions. To a large extent, this outcome has been demonstrated, although the $A\beta$ pool responsible for the benefits is not clear.

TABLE 10.1
Therapeutic Interventions Improving Memory Performance of APP Transgenic Mice

Treatment	Reference
Anti-$A\beta$ vaccine	Janus et al. [40]
Anti-$A\beta$ vaccine	Morgan et al. [39]
Anti-$A\beta$ monoclonal antibody	Dodart et al. [41]
Anti-$A\beta$ monoclonal antibody	Kotilinek et al. [42]
hApo E3, but not hApo E4	Raber et al. [43]
BACE1 null condition	Ohno et al. [44]
Dominant negative RAGE	Arancio et al. [45]
Ginkgo biloba	Stackman et al. [48]
Memantine	Minkeviciene et al. [49]
Anti-CD40L monoclonal antibody	Todd et al. [51]
Melatonin	Feng et al. [54]
Rolipram	Gong et al. [55]

In 1999, Schenk et al. [38] found that vaccines against the $A\beta$ peptide dramatically reduced the amyloid loads in PDPP transgenic mice. Our research group, believing that inflammation was a contributing factor in the disease, attempted to show that the microglial activation associated with such immunization would worsen the memory, not protect it. Within weeks of the publication, we were vaccinating our APP+PS1 mice against $A\beta$ at an age prior to the onset of spatial-navigation-memory dysfunction. We tested mice at an age before the onset of memory loss in the working-memory version of the radial arm water-maze task. We found all mice capable of robust learning of platform location, even those vaccinated against $A\beta$. However, when tested several months later, we found that the mice given control inoculations had lost the ability to remember platform location, but the anti-$A\beta$-vaccinated mice could ultimately learn as well as the nontransgenic littermate controls [39]. Similar data were collected in parallel by the Toronto group using a different APP mouse model (CRND8 [40]). Thus this surprisingly effective treatment not only reduced amyloid loads, but also protected mice from memory deficits.

Subsequent studies transitioned from the active immunization protocols used by Schenk et al. [38] to passive immunization with anti-$A\beta$ monoclonal antibodies. Dodart et al. [41] and Kotilinek et al. [42] both observed reversal of memory deficits in APP transgenic mice using different antibodies and different models. However, the striking feature of both studies was the rapidity of the reversal; as little as a single dose of antibody reversed the hole-board deficits in PDAPP mice, and three injections reversed water-maze deficits in Tg2576 mice. In neither circumstance was the measurable form of $A\beta$ load reduced by these treatments. These data are some of the strongest evidence for considering small, difficult-to-measure soluble pools — rapidly affected by antibodies — as being the form of $A\beta$ most proximally linked to memory dysfunction in the APP transgenic mice.

Some genetic manipulations have been found to modify the mnemonic dysfunction of APP transgenic mice. One of the first observations was that by Raber et al. [43]. Using the J9 APP mouse, they crossed these mice with transgenic mice expressing hApoE3 or hApoE4 on an mApoE-null background. They found that the ApoE3 protein was capable of reversing the memory deficits in the J9 mice but that the ApoE4 gene could not. These results suggest that ApoE3 can counteract the memory-disrupting action of $A\beta$, while the E4 variant, linked to increased risk for AD, cannot. Another genetic manipulation found to completely eliminate both amyloid production and memory loss was a knockout of the BACE1 gene [44]. BACE1 is a protease mediating the APP cleavage at the N-terminal of $A\beta$. Alternative APP processing by another enzyme, alpha secretase, precludes $A\beta$ formation. One concern in the transgenic field was that APP overexpression, not just $A\beta$ overproduction, was responsible for the memory problems in the mice. The Tg2576 mice raised on the BACE1-null background failed to develop memory loss. Moreover, electrophysiological abnormalities in these mice were avoided. Thus, APP overexpression alone is not responsible for these aspects of the APP mouse phenotype.

Another genetic manipulation impacting cognitive performance in transgenic mice involves the RAGE (Receptor for Advanced Glycation Endproducts) protein, a receptor binding a number of modified proteins that also binds $A\beta$. APP mice overexpressing RAGE in neurons develop memory deficits at a younger age than unmodified APP mice. APP mice expressing a dominant-negative form of RAGE are protected from memory loss [45]. It is unclear to what extent these manipulations also modified $A\beta$ accumulation in these mice. RAGE has been proposed to transport $A\beta$ from blood into brain [46], possibly modifying brain $A\beta$ content.

Drugs have also been found to reverse or prevent the memory dysfunctions in APP transgenic mouse models. *Ginkgo biloba*, an herbal agent often promoted as a remedy for age-associated memory loss and AD, but with minimal support in the human literature [47], was found to prevent memory loss in Tg2576 mice when administered from 8 to 14 months of age [48]. Surprisingly, this was associated with increased protein carbonyl formation but no changes in amyloid loads measured by ELISA. A drug with demonstrated benefits for AD patients (memantine) also alleviates memory deficits after 3 weeks of treatment in APP+PS1 mice [49]. Treatment with antiCD40L antibodies, previously demonstrated to reduce $A\beta$ deposition [50], also protects APP+PS1 transgenic mice from memory impairments [51]. Another treatment argued to reduce $A\beta$ (melatonin [52, 53]) also prevents memory disruptions in the APP695 mouse when administered from 5 to 9 months of age [54]. Most recently, a remarkable delayed effect was observed with the phosphodiesterase-inhibitor rolipram [55]. In this study, APP+PS1 mice were treated for 3 weeks with the agent when the mice were 3 months of age and the first amyloid deposits were appearing. Even though the drug was discontinued for at least 2 months, when tested at 6 to 7 months of age, mice receiving the drug scored significantly better than mice given vehicle treatments on contextual fear conditioning, radial arm water maze, and open-pool water-maze tasks. The treatment also reversed long-term potentiation (LTP) deficits and increased cyclic AMP response-element binding protein (CREB) phosphorylation at 7 to 8 months. No changes in ELISA-measured $A\beta$ were found at 7 to 8 months. Taken at face value, these data argue for a permanent rearrangement of the brain's response to $A\beta$ caused by this agent. Estrogen replacement to ovariectomized mice also improved memory in APP+PS1 mice, but the treatment produced similar effects in nontransgenic mice as well as transgenic animals [56], demonstrating no selectivity of estrogen for the amyloid-depositing mice.

In summary, a number of manipulations impact the memory phenotype of APP transgenic mice. Most importantly, the BACE-null APP mouse is protected from memory loss, verifying that it is the $A\beta$ overproduction associated with the transgenes that is responsible for the impairments. Although some of these manipulations are directed at reducing amyloid, many of the successful treatments have no apparent action on amyloid accumulation (at least in the pools measured). This implies that (a) there is a chain of events associated with amyloid-induced memory deficits in APP mice and (b) interventions may be directed either at the amyloid deposits themselves or at downstream events in the process, leading to consolidations that are otherwise disrupted by the presence of $A\beta$. Thus in AD, as in other chronic degenerative diseases (e.g., heart disease), there will be multiple therapeutic targets to improve memory functions. In fact, the combined efficacy of donepezil and memantine in AD cases is the first step in a graded series of improvements in managing the disease [57].

TABLE 10.2
Electrophysiological Plasticity in APP Transgenic Mice

Basal Synaptic Activity	Long-Term Potentiation	Reference
Normal	Reduced	Chapman et al. [63]
Reduced	Present	Larson et al. [65]
Reduced	Normal	Hsia et al. [26]
Reduced	Normal	Fitzjohn et al. [64]
Reduced	Increased	Jolas et al. [68]
—	Reduced	Dewachter et al. [69]
Reduced	Normal	Roder et al. [66]
—	Normal	Harris-Cerruti et al. [67]
Normal	Faster decay	Gureviciene et al. [70]
—	Reduced	Trinchese et al. [17]

MECHANISMS OF Aβ-ASSOCIATED MEMORY IMPAIRMENT IN TRANSGENIC MICE

Although the APP transgenic mouse is a reasonable model for amyloid deposition, the absence of significant neuron loss implies that it is not a good model for AD [58–62]. Thus, one major contribution to the dramatic cognitive declines observed in AD cases is not present in these murine models. Still, the memory impairments are a consistent feature of the APP mouse phenotype and, to a large extent, have similar characteristics, applying primarily to hippocampus-mediated tasks. Assuming that the mechanism(s) causing memory impairment in APP mice contribute to at least early mnemonic changes found in AD cases, an understanding of these mechanisms is likely to suggest additional therapeutic targets.

An obvious correlate of memory function to examine in APP transgenic mice is long-term potentiation (LTP), a form of synaptic plasticity often argued to underlie learning. Unfortunately, the phenotypic changes in synaptic transmission in APP transgenic mice have not been as uniform as the changes in memory function (Table 10.2). The first study examining LTP in the Tg2576 mice found normal synaptic transmission in 14- to 17-month-old mice but reduced hippocampal LTP [63]. Noting that many *ex vivo* hippocampal slices from APP mice died unless kynurenic acid was included during slice preparation, the researchers also measured LTP *in vivo* and found a similar reduction in APP transgenic mice. When APP mice were combined with nontransgenic mice, there was a correlation between T-maze performance and the extent of LTP. However, when evaluating the same mice, Fitzjohn et al. [64] found normal paired pulse and long-term potentiation but impaired synaptic transmission at 18 months. In 27- to 28-month-old PDAPP mice, Larson et al. [65] found results similar to those described by Fitzjohn et al. [64], with impaired synaptic activity but with LTP maintained. Similar observations were made by Hsia et al. [26] and Roder et al. [66] in 12- and 18-month-old APP23 mice and by Harris-Cerruti et al. [67]. Perhaps most atypical was the report by Jolas et al. [68]. They found that LTP was increased in slices from 5-month-old CRND8 APP mice relative to nontransgenic animals but that field potential slopes declined. Dewachter et al. [69] found an LTP deficit in an APP mouse that was rescued when the mice were bred onto a targeted PS1-null condition. However, the memory deficits in this mouse were not rescued by the PS1-null condition, arguing that something more than LTP changes were involved in the memory dysfunction. Trinchese et al. [17] found LTP deficits in APP+PS1 mice at 3 months when plaques were first appearing. The extent of the LTP deficit progressed as the mice aged, in parallel with reductions in working memory on the radial arm water maze. In a slightly different APP+PS1 mouse, Gureviciene et al. [70] found normal LTP induction and maintenance for 60 min *in vitro*. Yet, when potentiating the perforant path *in vivo*, they found more-rapid decay of the potentiation when measured 24 h later. Although a simple summary of these disparate results is not straightforward, it seems likely that APP mice have some disruption of normal synaptic physiology. It may be hard to measure LTP in the same manner if baseline parameters are not consistent in transgenic and nontransgenic mice. The two *in vivo* studies appear to consistently identify reductions in LTP, and a more rapid decay of a potentiated synapse would appear consistent with the behavioral literature. Still, this is an area requiring more effort before reaching consensus.

As mentioned above, neuron loss is not a significant feature for the APP transgenic mouse phenotype. However, another possibility is that there is synapse loss, argued to be the best pathological correlate of cognitive decline in AD [71]. Here the data are mixed. Initial results suggested loss of synaptophysin fluorescence in the APP mice [6, 72], often in the absence of detectable amyloid deposits [26, 27]. In some instances, it is possible that these changes were associated with reduced volumes of the corresponding structures [12, 73, 74]. Other studies failed to find reductions in synaptophysin staining [58–61] or even increases in synaptophysin staining [12, 75]. Some of the differences may be attributable to different mouse lines or the brain regions studied. Thus far, most studies finding deficits used fluorescence detection, while those not finding differences used peroxidase reaction product for detection. Furthermore, given the enrichment of synaptophysin in the dystrophic neurites surrounding plaques [21], determining how each study in plaque-bearing mice dealt with these sources of synaptophysin reactivity is an important consideration. It is likely that there is some structural loss of presynaptic markers in select regions of the transgenic mice, but this loss is likely modest.

Perhaps more important than structural changes are functional changes in synaptic markers. Several years back, our group compared the gene-expression profiles of memory-deficient 16- to 18-month-old APP+PS1 mice to nontransgenic littermates in both plaque-bearing regions (hippocampus, cortex) and plaque-free regions (cerebellum, brain stem) using both microarrays and real-time PCR (polymerase chain reaction) [76]. We used several criteria (size of difference, statistical significance, selectivity for plaque-bearing regions) to identify a small number of genes (fewer than 50) that were modified only in the plaque-bearing regions of transgenic animals. Many of these were associated with inflammation in the vicinity of the plaques, and some were associated with the transgene itself, but several were unexpected findings and had been previously linked to learning and memory. One category included immediate early genes such as arc and zif 264. Down-regulation of these genes causes consolidation failures in rodent models of memory or synaptic plasticity (referenced in Dickey et al. [77]). Others were postsynaptic proteins, such as the NMDA (N-methyl-D-aspartate) receptor subunit NR2B. PSD-95 (Post Synaptic Density Marker-95), or calmodulin kinase II-, is also linked to neural plasticity. In general, the RNA content for a number of presynaptic markers, including synaptophysin, remained stable. A similar pattern of changes, with deficiencies of postsynaptic markers (debrin, fractin) and stability of presynaptic markers (synaptophysin), was observed by Calon et al. [78] in 16- to 18-month-old Tg2576 mice. For the immediate early genes, our group subsequently identified that the basal level of expression was unaffected in the transgenic animals but that the induction caused by exposure to a novel environment was suppressed [77]. Palop et al. [79] observed a similar reduction in the immediate early gene c-fos in the J20 APP transgenic mouse line. This marker and reductions in the calcium-regulated protein calbindin were significantly correlated with $A\beta 1$-42 levels and memory impairment in these mice.

We also found that sodium potassium ATPase, the enzyme using 40% of the brain's ATP, was decreased at both the message level and by enzymatic assay in plaque-bearing regions [80]. When evaluated immunohistochemically, there was a paucity of immunoreactivity for the enzyme in a penumbral region surrounding each Congo red-stained deposit. Given the overlap of this zone with the location of the swollen neurites, we speculated ("wildly" so, to some reviewers) that a local loss of ionic homeostasis might lead to osmotic imbalance in the vicinity of the plaques, leading to swelling of the neural processes. This might also alter the electrotonic properties of the dendrites associated with these processes, leading to impaired transmission of postsynaptic potentials. Clearly, further efforts will be needed to test this hypothesis, but 30% reductions in the activity of this critical enzyme are likely to have significant impact on neural function.

CONCLUSIONS

The APP transgenic mice are a very good model of the amyloid deposition found in Alzheimer dementias. The patterns of deposition, regional distribution, and even the anatomical localization of the short and long variants mimic the human disease. The APP mouse phenotype also consistently includes progressive memory impairment. This phenotype appears to be due to $A\beta$ accumulation and not overexpression of APP, as the BACE1-null background, which overproduces APP but not $A\beta$, rescues the memory phenotype. Still, none of the readily measurable pools of $A\beta$ seem to correlate linearly with this memory loss, suggesting that an occult pool, possibly oligomeric, is more directly linked to the memory deficits.

A number of manipulations, most notably immunotherapies, have been found to regulate the memory phenotype. Not all successful manipulations modify $A\beta$ levels (at least detectable forms of $A\beta$). This argues that there are steps in memory processing downstream from the site of $A\beta$ action where interventions can be targeted. It is plausible that even treatments targeted at $A\beta$ might have their greatest effect at downstream sites, e.g., reduced inflammation, independent of $A\beta$ reductions.

It is likely that AD, like many other degenerative disorders, will be managed through multiple treatment modalities. An overnight cure seems unlikely.

The mechanisms mediating these memory deficiencies are not clear. Given that we cannot identify a pool of $A\beta$ that is intimately linked to the memory impairments, it is difficult to identify targets for this $A\beta$ pool and their impact on downstream effectors. A number of candidates have been identified, but at the moment these can only be viewed as associated with the memory disruptions. Causal linkages will be difficult to prove.

With respect to AD, the value of the APP transgenic models is to screen drug candidates proposed to act on the human disease by reducing amyloid. It is less certain whether agents protecting from memory loss in the APP mice, independent of their influence on $A\beta$ effects, will translate into the human condition. This will depend upon the extent to which failing memories in AD are due to amyloid-associated changes in neural processing versus structural loss of neurons and synapses.

ACKNOWLEDGMENTS

The author has received support from AG 15490, AG 18478, and NS 48335 from the National Institutes of Health (NIH). The author also wishes to thank Karen Hsiao Ashe and Karen Duff for early access to their transgenic mice.

REFERENCES

1. Glenner, G.G. and Wong, C.W., Alzheimer's disease: initial report of the purification and characterization of a novel cerebrovascular amyloid protein, *Biochem. Biophys. Res. Commn.*, 120, 885, 1984.
2. Goate, A., Chartier-Harlin, M.C., Mullan, M., Brown, J., Crawford, F., Fidani, L., Giuffra, L., Haynes, A., Irving, N., James, L., and Hardy, J., Segregation of a missense mutation in the amyloid precursor protein gene with familial Alzheimer's disease, *Nature*, 349, 704, 1991.
3. Hardy, J. and Selkoe, D.J., The amyloid hypothesis of Alzheimer's disease: progress and problems on the road to therapeutics, *Science*, 297, 353, 2002.
4. Snow, A.D., Sekiguchi, R., Nochlin, D., Fraser, P., Kimata, K., Mizutani, A., Arai, M., Schreier, W.A., and Morgan, D.G., An important role of heparan sulfate proteoglycan (perlecan) in a model system for the deposition and persistence of fibrillar A-Beta amyloid in rat brain, *Neuron*, 12, 219, 1994.
5. Holcomb, L.A., Gordon, M.N., Benkovic, S.A., and Morgan, D.G., $A\beta$ and perlecan in rat brain: glial activation, gradual clearance and limited neurotoxicity, *Mechanisms of Ageing Dev.*, 112, 135, 1999.
6. Games, D., Adams, D., Alessandrini, R., Barbour, R., Berthelette, P., Blackwell, C., Carr, T., Clemens, J., Donaldson, T., and Gillespie, F., Alzheimer-type neuropathology in transgenic mice overexpressing V717F beta-amyloid precursor protein, *Nature*, 373, 523, 1995.
7. Hsiao, K., Chapman, P., Nilsen, S., Eckman, C., Harigaya, Y., Younkin, S., Yang, F., and Cole, G., Correlative memory deficits, Abeta elevation, and amyloid plaques in transgenic mice, *Science*, 274, 99, 1996.

8. Sturchler-Pierrat, C., Abramowski, D., Duke, M., Wiederhold, K.H., Mistl, C., Rothacher, S., Ledermann, B., Burki, K., Frey, P., Paganetti, P.A., Waridel, C., Calhoun, M.E., Jucker, M., Probst, A., Staufenbiel, M., and Sommer, B., Two amyloid precursor protein transgenic mouse models with Alzheimer disease-like pathology, *Proc. Natl. Acad. Sci. U.S.A.*, 94, 13287, 1997.

9. Duff, K., Eckman, C., Zehr, C., Yu, X., Prada, C.M., Perez-tur, J., Hutton, M., Buee, L., Harigaya, Y., Yager, D., Morgan, D., Gordon, M.N., Holcomb, L., Refolo, L., Zenk, B., Hardy, J., and Younkin, S., Increased amyloid-beta42(43) in brains of mice expressing mutant presenilin 1, *Nature*, 383, 710, 1996.

10. Holcomb, L.A., Gordon, M.N., McGowan, E., Yu, X., Benkovic, S., Jantzen, P., Wright, K., Saad, I., Mueller, R., Morgan, D., Sanders, S., Zehr, C., O'Campo, K., Hardy, J., Prada, C.M., Eckman, C., Younkin, S., Hsiao, K., and Duff, K., Accelerated Alzheimer-type phenotype in transgenic mice carrying both mutant amyloid precursor protein and presenilin 1 transgenes, *Nat. Med.*, 4, 97, 1998.

11. Morris, R.G., Garrud, P., Rawlins, J.N., and O'Keefe, J., Place navigation impaired in rats with hippocampal lesions, *Nature*, 297, 681, 1982.

12. Dodart, J.C., Mathis, C., Saura, J., Bales, K.R., Paul, S.M., and Ungerer, A., Neuroanatomical abnormalities in behaviorally characterized APP(V717F) transgenic mice, *Neurobiol. Dis.*, 7, 71, 2000.

13. Chen, G., Chen, K.S., Knox, J., Inglis, J., Bernard, A., Martin, S.J., Justice, A., McConlogue, L., Games, D., Freedman, S.B., and Morris, R.G., A learning deficit related to age and beta-amyloid plaques in a mouse model of Alzheimer's disease, *Nature*, 408, 975, 2000.

14. Gordon, M.N., King, D.L., Diamond, D.M., Jantzen, P.T., Boyett, K.L., Hope, C.E., Hatcher, J.M., DiCarlo, G., Gottschal, P., Morgan, D., and Arendash, G.W., Correlation between cognitive deficits and Aβ deposits in transgenic APP+PS1 mice, *Neurobiol. Aging*, 22, 377, 2001.

15. Westerman, M.A., Cooper-Blacketer, D., Mariash, A., Kotilinek, L., Kawarabayashi, T., Younkin, L.H., Carlson, G.A., Younkin, S.G., and Ashe, K.H., The relationship between Abeta and memory in the Tg2576 mouse model of Alzheimer's disease, *J. Neurosci.*, 22, 1858, 2002.

16. Puolivali, J., Wang, J., Heikkinen, T., Heikkila, M., Tapiola, T., van Groen, T., and Tanila, H., Hippocampal A beta 42 levels correlate with spatial memory deficit in APP and PS1 double transgenic mice, *Neurobiol. Dis.*, 9, 339, 2002.

17. Trinchese, F., Liu, S., Battaglia, F., Walter, S., Mathews, P.M., and Arancio, O., Progressive age-related development of Alzheimer-like pathology in APP/PS1 mice, *Ann. Neurol.*, 55, 801, 2004.

18. Liu, L., Tapiola, T., Herukka, S.K., Heikkila, M., and Tanila, H., Abeta levels in serum, CSF and brain, and cognitive deficits in APP+PS1 transgenic mice, *NeuroReport*, 14, 163, 2003.

19. Ikarashi, Y., Harigaya, Y., Tomidokoro, Y., Kanai, M., Ikeda, M., Matsubara, E., Kawarabayashi, T., Kuribara, H., Younkin, S.G., Maruyama, Y., and Shoji, M., Decreased level of brain acetylcholine and memory disturbance in APPsw mice, *Neurobiol. Aging*, 25, 483, 2004.

20. Morgan, D., Learning and memory deficits in APP transgenic mouse models of amyloid deposition, *Neurochem. Res.*, 28, 1029, 2003.

21. Gordon, M.N., Holcomb, L.A., Jantzen, P.T., DiCarlo, G., Wilcock, D., Boyett, K.L., Connor, K., Melachrino, J.O., O'Callaghan, J.P., and Morgan, D., Time course of the development of Alzheimer-like pathology in the doubly transgenic PS1+APP mouse, *Experimental Neurol.*, 173, 183, 2002.

22. Savage, M., Kawooya, J., Pinsker, L., Emmons, T., Mistretta, S., Siman, R., and Greenberg, B., Elevated $A\beta$ levels in Alzheimer's disease brain are associated with a selective accumulation of $A\beta42$ in parenchymal amyloid deposits, and both $A\beta40$ and $A\beta42$ in cerebrovascular deposits, *Amyloid Int. J. Exp. Clin. Invest.*, 2, 234, 1995.

23. Yamaguchi, F., Richards, S.J., Beyreuther, K., Salbaum, M., Carlson, G.A., and Dunnett, S.B., Transgenic mice for the amyloid precursor protein 695 isoform have impaired spatial memory, *NeuroReport*, 2, 781, 1991.

24. Moran, P.M., Higgins, L.S., Cordell, B., and Moser, P.C., Age-related learning deficits in transgenic mice expressing the 751-amino acid isoform of human beta-amyloid precursor protein, *Proc. Natl. Acad. Sci. U.S.A.*, 92, 5341, 1995.

25. Koistinaho, M., Ort, M., Cimadevilla, J.M., Vondrous, R., Cordell, B., Koistinaho, J., Bures, J., and Higgins, L.S., Specific spatial learning deficits become severe with age in beta-amyloid precursor protein transgenic mice that harbor diffuse beta-amyloid deposits but do not form plaques, *Proc. Natl. Acad. Sci. U.S.A.*, 98, 14675, 2001.

26. Hsia, A.Y., Masliah, E., McConlogue, L., Yu, G. Q., Tatsuno, G., Hu, K., Kholodenko, D., Malenka, R.C., Nicoll, R.A., and Mucke, L., Plaque-independent disruption of neural circuits in Alzheimer's disease mouse models, *Proc. Natl. Acad. Sci. U.S.A.*, 96, 3228, 1999.

27. Mucke, L., Masliah, E., Yu, G.Q., Mallory, M., Rockenstein, E.M., Tatsuno, G., Hu, K., Kholodenko, D., Johnson-Wood, K., and McConlogue, L., High-level neuronal expression of abeta 1-42 in wild-type human amyloid protein precursor transgenic mice: synaptotoxicity without plaque formation, *J. Neurosci.*, 20, 4050, 2000.

28. Kawasumi, M., Chiba, T., Yamada, M., Miyamae-Kaneko, M., Matsuoka, M., Nakahara, J., Tomita, T., Iwatsubo, T., Kato, S., Aiso, S., Nishimoto, I., and Kouyama, K., Targeted introduction of V642I mutation in amyloid precursor protein gene causes functional abnormality resembling early stage of Alzheimer's disease in aged mice, *Eur. J. Neurosci.*, 19, 2826, 2004.

29. Holcomb, L.A., Gordon, M. N., Jantzen, P., Hsiao, K., Duff, K., and Morgan, D., Behavioral changes in transgenic mice expressing both amyloid precursor protein and presenilin-1 mutations: Lack of association with amyloid deposits, *Behav. Gen.*, 29, 177, 1999.

30. Lee, K.W., Lee, S.H., Kim, H., Song, J.S., Yang, S.D., Paik, S.G., and Han, P.L., Progressive cognitive impairment and anxiety induction in the absence of plaque deposition in C57BL/6 inbred mice expressing transgenic amyloid precursor protein, *J. Neurosci. Res.*, 76, 572, 2004.

31. Richardson, J.C., Kendal, C.E., Anderson, R., Priest, F., Gower, E., Soden, P., Gray, R., Topps, S., Howlett, D.R., Lavender, D., Clarke, N.J., Barnes, J.C., Haworth, R., Stewart, M.G., and Rupniak, H.T., Ultrastructural and behavioural changes precede amyloid deposition in a transgenic model of Alzheimer's disease, *Neuroscience*, 122, 213, 2003.

32. Van Dam, D., D'Hooge, R., Staufenbiel, M., Van Ginneken, C., Van Meir, F., and De Deyn, P.P., Age-dependent cognitive decline in the APP23 model precedes amyloid deposition, *Eur. J. Neurosci.*, 17, 388, 2003.

33. Arendash, G.W., King, D.L., Gordon, M.N., Morgan, D., Hatcher, J.M., Hope, C.E., and Diamond, D.M., Progressive behavioral impairments in transgenic mice carrying both mutant APP and PS1 transgenes, *Brain Res.*, 891, 45, 2001.

34. Kelly, P.H., Bondolfi, L., Hunziker, D., Schlecht, H.P., Carver, K., Maguire, E., Abramowski, D., Wiederhold, K.H., Sturchler-Pierrat, C., Jucker, M., Bergmann, R., Staufenbiel, M., and Sommer, B., Progressive age-related impairment of cognitive behavior in APP23 transgenic mice, *Neurobiol. Aging*, 24, 365, 2003.

35. Savonenko, A.V., Xu, G.M., Price, D.L., Borchelt, D.R., and Markowska, A.L., Normal cognitive behavior in two distinct congenic lines of transgenic mice hyperexpressing mutant APP SWE, *Neurobiol. Dis.*, 12, 194, 2003.

36. Ruiz-Opazo, N., Kosik, K.S., Lopez, L.V., Bagamasbad, P., Ponce, L.R., and Herrera, V.L., Attenuated hippocampus-dependent learning and memory decline in transgenic TgAPPswe Fischer-344 rats, *Mol. Med.*, 10, 36, 2004.

37. Gong, Y., Chang, L., Viola, K.L., Lacor, P.N., Lambert, M.P., Finch, C.E., Krafft, G.A., and Klein, W.L., Alzheimer's disease-affected brain: presence of oligomeric A beta ligands (ADDLs) suggests a molecular basis for reversible memory loss, *Proc. Natl. Acad. Sci. U.S.A.*, 100, 10417, 2003.

38. Schenk, D., Barbour, R., Dunn, W., Gordon, G., Grajeda, H., Guido, T., Hu, K., Huang, J., Johnson-Wood, K., Khan, K., Kholodenko, D., Lee, M., Liao, Z., Lieberburg, I., Motter, R., Mutter, L., Soriano, F., Shopp, G., Vasquez, N., Vandevert, C., Walker, S., Wogulis, M., Yednock, T., Games, D., and Seubert, P., Immunization with amyloid-beta attenuates Alzheimer-disease-like pathology in the PDAPP mouse, *Nature*, 400, 173, 1999.

39. Morgan, D., Diamond, D.M., Gottschall, P.E., Ugen, K.E., Dickey, C., Hardy, J., Duff, K., Jantzen, P., DiCarlo, G., Wilcock, D., Connor, K., Hatcher, J., Hope, C., Gordon, M., and Arendash, G.W., A beta peptide vaccination prevents memory loss in an animal model of Alzheimer's disease, *Nature*, 408, 982, 2000.

40. Janus, C., Pearson, J., McLaurin, J., Mathews, P.M., Jiang, Y., Schmidt, S.D., Chishti, M.A., Horne, P., Heslin, D., French, J., Mount, H.T., Nixon, R.A., Mercken, M., Bergeron, C., Fraser, P.E., George-Hyslop, P., and Westaway, D., A beta peptide immunization reduces behavioural impairment and plaques in a model of Alzheimer's disease, *Nature*, 408, 979, 2000.

41. Dodart, J.C., Bales, K.R., Gannon, K.S., Greene, S.J., DeMattos, R.B., Mathis, C., DeLong, C.A., Wu, S., Wu, X., Holtzman, D.M., and Paul, S.M., Immunization reverses memory deficits without reducing brain Abeta burden in Alzheimer's disease model, *Nat. Neurosci.*, 5, 452, 2002.

42. Kotilinek, L.A., Bacskai, B., Westerman, M., Kawarabayashi, T., Younkin, L., Hyman, B.T., Younkin, S., and Ashe, K.H., Reversible memory loss in a mouse transgenic model of Alzheimer's disease, *J. Neurosci.*, 22, 6331, 2002.

43. Raber, J., Wong, D., Yu, G.Q., Buttini, M., Mahley, R.W., Pitas, R.E., and Mucke, L., Apolipoprotein E and cognitive performance, *Nature*, 404, 352, 2000.

44. Ohno, M., Sametsky, E.A., Younkin, L.H., Oakley, H., Younkin, S.G., Citron, M., Vassar, R., and Disterhoft, J.F., BACE1 deficiency rescues memory deficits and cholinergic dysfunction in a mouse model of Alzheimer's disease, *Neuron*, 41, 27, 2004.

45. Arancio, O., Zhang, H.P., Chen, X., Lin, C., Trinchese, F., Puzzo, D., Liu, S., Hegde, A., Yan, S.F., Stern, A., Luddy, J.S., Lue, L.F., Walker, D.G., Roher, A., Buttini, M., Mucke, L., Li, W., Schmidt, A.M., Kindy, M., Hyslop, P.A., Stern, D.M., and Du Yan, S.S., RAGE potentiates Abeta-induced perturbation of neuronal function in transgenic mice, *EMBO J.*, 23, 4096, 2004.

46. Zlokovic, B.V., Clearing amyloid through the blood-brain barrier, *J. Neurochem.*, 89, 807, 2004.

47. Gold, P.E., Cahill, L., and Wenk, G.L., The lowdown on *Ginkgo biloba*, *Sci. Am.*, 288, 86, 2003.

48. Stackman, R.W., Eckenstein, F., Frei, B., Kulhanek, D., Nowlin, J., and Quinn, J.F., Prevention of age-related spatial memory deficits in a transgenic mouse model of Alzheimer's disease by chronic *Ginkgo biloba* treatment, *Exp. Neurol.*, 184, 510, 2003.

49. Minkeviciene, R., Banerjee, P., and Tanila, H., Memantine improves spatial learning in a transgenic mouse model of Alzheimer's disease, *J. Pharmacol. Exp. Ther.*, 311, 677, 2004.

50. Tan, J., Town, T., Crawford, F., Mori, T., DelleDonne, A., Crescentini, R., Obregon, D., Flavell, R.A., and Mullan, M.J., Role of CD40 ligand in amyloidosis in transgenic Alzheimer's mice, *Nat. Neurosci.*, 5, 1288, 2002.

51. Todd, R.J., Volmar, C.H., Dwivedi, S., Town, T., Crescentini, R., Crawford, F., Tan, J., and Mullan, M., Behavioral effects of CD40-CD40L pathway disruption in aged PSAPP mice, *Brain Res.*, 1015, 161, 2004.

52. Matsubara, E., Bryant-Thomas, T., Pacheco, Q.J., Henry, T.L., Poeggeler, B., Herbert, D., Cruz-Sanchez, F., Chyan, Y.J., Smith, M.A., Perry, G., Shoji, M., Abe, K., Leone, A., Grundke-Ikbal, I., Wilson, G.L., Ghiso, J., Williams, C., Refolo, L.M., Pappolla, M.A., Chain, D.G., and Neria, E., Melatonin increases survival and inhibits oxidative and amyloid pathology in a transgenic model of Alzheimer's disease, *J. Neurochem.*, 85, 1101, 2003.

53. Lahiri, D.K., Chen, D., Ge, Y.W., Bondy, S.C., and Sharman, E.H., Dietary supplementation with melatonin reduces levels of amyloid beta-peptides in the murine cerebral cortex, *J. Pineal Res.*, 36, 224, 2004.

54. Feng, Z., Chang, Y., Cheng, Y., Zhang, B.L., Qu, Z.W., Qin, C., and Zhang, J.T., Melatonin alleviates behavioral deficits associated with apoptosis and cholinergic system dysfunction in the APP 695 transgenic mouse model of Alzheimer's disease, *J. Pineal Res.*, 37, 129, 2004.

55. Gong, B., Vitolo, O.V., Trinchese, F., Liu, S., Shelanski, M., and Arancio, O., Persistent improvement in synaptic and cognitive functions in an Alzheimer mouse model after rolipram treatment, *J. Clin. Invest.*, 114, 1624, 2004.

56. Heikkinen, T., Kalesnykas, G., Rissanen, A., Tapiola, T., Iivonen, S., Wang, J., Chaudhuri, J., Tanila, H., Miettinen, R., and Puolivali, J., Estrogen treatment improves spatial learning in APP+PS1 mice but does not affect beta amyloid accumulation and plaque formation, *Exp. Neurol.*, 187, 105, 2004.

57. Tariot, P.N., Farlow, M.R., Grossberg, G.T., Graham, S.M., McDonald, S., and Gergel, I., Memantine treatment in patients with moderate to severe Alzheimer disease already receiving donepezil: a randomized controlled trial, *JAMA*, 291, 317, 2004.

58. Irizarry, M.C., McNamara, M., Fedorchak, K., Hsiao, K., and Hyman, B.T., APPSw transgenic mice develop age-related A beta deposits and neuropil abnormalities, but no neuronal loss in CA1, *J. Neuropathol. Exp. Neurol.*, 56, 965, 1997.

59. Irizarry, M.C., Soriano, F., McNamara, M., Page, K.J., Schenk, D., Games, D., and Hyman, B.T., Abeta deposition is associated with neuropil changes, but not with overt neuronal loss in the human amyloid precursor protein V717F (PDAPP) transgenic mouse, *J. Neurosci.*, 17, 7053, 1997.

60. Calhoun, M.E., Wiederhold, K.H., Abramowski, D., Phinney, A.L., Probst, A., Sturchler-Pierrat, C., Staufenbiel, M., Sommer, B., and Jucker, M., Neuron loss in APP transgenic mice, *Nature*, 395, 755, 1998.

61. Takeuchi, A., Irizarry, M.C., Duff, K., Saido, T.C., Hsiao, A.K., Hasegawa, M., Mann, D.M., Hyman, B.T., and Iwatsubo, T., Age-related amyloid beta deposition in transgenic mice overexpressing both Alzheimer mutant presenilin 1 and amyloid beta precursor protein Swedish mutant is not associated with global neuronal loss, *Am. J. Pathol.*, 157, 331, 2000.

62. Bondolfi, L., Calhoun, M., Ermini, F., Kuhn, H.G., Wiederhold, K.H., Walker, L., Staufenbiel, M., and Jucker, M., Amyloid-associated neuron loss and gliogenesis in the neocortex of amyloid precursor protein transgenic mice, *J. Neurosci.*, 22, 515, 2002.

63. Chapman, P.F., White, G.L., Jones, M.W., Cooper-Blacketer, D., Marshall, V.J., Irizarry, M., Younkin, L., Good, M.A., Bliss, T.V., Hyman, B.T., Younkin, S.G., and Hsiao, K.K., Impaired synaptic plasticity and learning in aged amyloid precursor protein transgenic mice, *Nat. Neurosci.*, 2, 271, 1999.

64. Fitzjohn, S.M., Morton, R.A., Kuenzi, F., Rosahl, T.W., Shearman, M., Lewis, H., Smith, D., Reynolds, D.S., Davies, C.H., Collingridge, G.L., and Seabrook, G.R., Age-related impairment of synaptic transmission but normal long-term potentiation in transgenic mice that overexpress the human APP695SWE mutant form of amyloid precursor protein, *J. Neurosci.*, 21, 4691, 2001.

65. Larson, J., Lynch, G., Games, D., and Seubert, P., Alterations in synaptic transmission and long-term potentiation in hippocampal slices from young and aged PDAPP mice, *Brain Res.*, 840, 23, 1999.

66. Roder, S., Danober, L., Pozza, M.F., Lingenhoehl, K., Wiederhold, K.H., and Olpe, H.R., Electrophysiological studies on the hippocampus and prefrontal cortex assessing the effects of amyloidosis in amyloid precursor protein 23 transgenic mice, *Neuroscience*, 120, 705, 2003.

67. Harris-Cerruti, C., Kamsler, A., Kaplan, B., Lamb, B., Segal, M., and Groner, Y., Functional and morphological alterations in compound transgenic mice overexpressing Cu/Zn superoxide dismutase and amyloid precursor protein (Correction), *Eur. J. Neurosci.*, 19, 1174, 2004.

68. Jolas, T., Zhang, X.S., Zhang, Q., Wong, G., Del Vecchio, R., Gold, L., and Priestley, T., Long-term potentiation is increased in the CA1 area of the hippocampus of APP(swe/ind) CRND8 mice, *Neurobiol. Dis.*, 11, 394, 2002.

69. Dewachter, I., Reverse, D., Caluwaerts, N., Ris, L., Kuiperi, C., Van den Honte, C., H.C., Spittaels, K., Umans, L., Serneels, L., Thiry, E., Moechars, D., Mercken, M., Godaux, E., and Van Leuven, F., Neuronal deficiency of presenilin 1 inhibits amyloid plaque formation and corrects hippocampal long-term potentiation but not a cognitive defect of amyloid precursor protein [V717I] transgenic mice, *J. Neurosci.*, 22, 3445, 2002.

70. Gureviciene, I., Ikonen, S., Gurevicius, K., Sarkaki, A., van Groen, T., Pussinen, R., Ylinen, A., and Tanila, H., Normal induction but accelerated decay of LTP in APP+PS1 transgenic mice, *Neurobiol. Dis.*, 15, 188, 2004.

71. Terry, R.D., Masliah, E., Salmon, D.P., Butter, N., DeTeresa, R., Hill, R., Hansen, L.A., and Katzman, R., Physical basis of cognitive alterations in Alzheimer's disease: synapse loss is the major correlate of cognitive impairment, *Ann. Neurol.*, 30, 572, 1991.

72. Fonseca, M.I., Zhou, J., Botto, M., and Tenner, A.J., Absence of C1q leads to less neuropathology in transgenic mouse models of Alzheimer's disease, *J. Neurosci.*, 24, 6457, 2004.

73. Gonzalez-Lima, F., Berndt, J.D., Valla, J.E., Games, D., and Reiman, E.M., Reduced corpus callosum, fornix and hippocampus in PDAPP transgenic mouse model of Alzheimer's disease, *NeuroReport*, 12, 2375, 2001.

74. Weiss, C., Venkatasubramanian, P.N., Aguado, A.S., Power, J.M., Tom, B.C., Li, L., Chen, K.S., Disterhoft, J.F., and Wyrwicz, A.M., Impaired eyeblink conditioning and decreased hippocampal volume in PDAPP V717F mice, *Neurobiol. Dis.*, 11, 425, 2002.

75. King, D.L. and Arendash, G.W., Maintained synaptophysin immunoreactivity in Tg2576 transgenic mice during aging: correlations with cognitive impairment, *Brain Res.*, 926, 58, 2002.

76. Dickey, C.A., Loring, J.F., Montgomery, J., Gordon, M.N., Eastman, P.S., and Morgan, D., Selectively reduced expression of synaptic plasticity-related genes in amyloid precursor protein + presenilin-1 transgenic mice, *J. Neurosci.*, 23, 5219, 2003.
77. Dickey, C.A., Gordon, M.N., Mason, J.E., Wilson, N.J., Diamond, D.M., Guzowski, J.F., and Morgan, D., Amyloid suppresses induction of genes critical for memory consolidation in APP+PS1 transgenic mice, *J. Neurochem.*, 88, 434, 2004.
78. Calon, F., Lim, G.P., Yang, F., Morihara, T., Teter, B., Ubeda, O., Rostaing, P., Triller, A., Salem, N., Jr., Ashe, K.H., Frautschy, S.A., and Cole, G.M., Docosahexaenoic acid protects from dendritic pathology in an Alzheimer's disease mouse model, *Neuron*, 43, 633, 2004.
79. Palop, J.J., Jones, B., Kekonius, L., Chin, J., Yu, G.Q., Raber, J., Masliah, E., and Mucke, L., Neuronal depletion of calcium-dependent proteins in the dentate gyrus is tightly linked to Alzheimer's disease-related cognitive deficits, *Proc. Natl. Acad. Sci. U.S.A.*, 100, 9572, 2003.
80. Dickey, C.A., Gordon, M.N., Wilcock, D.M., Herber, D.L., Freeman, M.J., Barcenas, M., and Morgan, D., Dysregulation of Na+/K+ ATPase by amyloid in APP+PS1 transgenic mice, *BMC Neurosci.*, 6, 7, 2005.

11 Cholinergic Receptor Knockout Mice

Lu Zhang
Pfizer, Inc.

CONTENTS

The cholinergic system's essential involvement in both preclinical and clinical aspects of cognition processes has been proposed and reviewed extensively elsewhere. Terry and Buccafusco[1] provide the latest review on this subject. According to this review, all currently approved FDA drugs for the treatment of Alzheimer's disease are cholinesterase inhibitors, which exert their efficacy apparently through stimulation of both muscarinic acetylcholine receptors (mAChRs) and nicotinic acetylcholine receptors (nAChRs).

Gene targeting (knockout) technologies allow us to replace the gene of interest with one that is inactive, altered, or irrelevant.[2] In the case of a completed deletion, a gene is knocked out *in vivo*, and a mutant organism with a deficit in the gene product is generated. The lack of suitable embryonic stem (ES) cell lines prevents the application of these technologies into rats.[3] However, mice and humans are both mammals, and both species contain a similar number of genes that show a high degree of similarity.[4,5] Therefore, in rodents, mouse is a better species for employing the knockout approach compared with rat. In theory, knockout (KO) mice contain a targeted gene that is deleted; therefore, no product of the mutated gene is synthesized in the null mutants.[6] These mutant mouse lines, which have inactivating mutations of the individual genes, can be studied in a battery of physiological, pharmacological, behavioral, biochemical, and neurochemical tests to confirm or invalidate a hypothesis on the effects of specific proteins in the function of the brain.

Müller[7] provided an excellent review article regarding targeted mouse mutants from vector design to phenotype analysis. The cited article provides detailed technical information on how to generate KO mice, including selection markers and screening strategies, potential problems and pitfalls, as well as construct design. Moreover, Bolivar et al.[8] have summarized behavioral profiles in all available knockout mice. They also provide updated information on available KO mice through their Web site, which is identified in their article. Additional information about Internet resources regarding transgenic rodent production is provided by Wells and Carter.[3]

Because gene-targeting techniques enable us to analyze diverse aspects of gene function in whole animals, measurable phenotypes relevant to human pathology could be obtained in a good mouse model of human disease. However, for most diseases in the central nervous system (CNS), the coexistence of malfunctions in multiple subtypes of the same receptor or in multiple neurotransmitters typically contribute to a complex phenotype such as cognition. For example, Buccafusco and Terry[9] demonstrate that multiple CNS targets are needed to elicit beneficial effects on memory and cognition. Therefore, generating multiple mouse mutants that mimic all facets of a multifactorial disease would help us to address individual subsets of symptoms of the disease.

In short, knockout technologies provide a powerful tool to reveal and refine treatment strategies for human disorders by building bridges between genetics and the pathogenesis of disease. Behavioral phenotypes discovered in the mutant can be used to evaluate (screen) the efficacy of potential new pharmacological therapies. For example, by evaluating the results of specific behavioral tests — including learning and memory tests in mAChR, nAChR, and acetylcholinesterase (AChE)

KO mice — our knowledge of cognitional impairment would be broadened. This, in turn, will improve our ability to identify potential new drug therapies for the treatment of cognitional aspects of such diseases as Alzheimer's disease, attention deficit hyperactivity disorder (ADHD), and schizophrenia.

This chapter focuses on:

Available cholinergic receptor KO mouse models, including muscarinic acetylcholine receptor (mAChR) KO mice, nicotinic acetylcholine receptor (nAChR) KO mice, and acetylcholinesterase (AChE) KO mice

Cognitional-related data for mAChR KO mice

Limitations on KO mouse models and future directions

MUSCARINIC ACETYLCHOLINE RECEPTORS (MACHRS)

Detailed information on mAChRs is presented in Chapter 2 of this book. For the purpose of the discussion presented here, it is enough to state that five subtypes of mAChR (M_1, M_2, M_3, M_4, and M_5) have been identified so far, with M_1, M_3, and M_5 receptors coupled to the $G_{q/11}$ protein, which activates phospholipase C, and with the M_2 and M_4 receptors coupled to the $G_{i/0}$ protein, which inhibits adenylate cyclase activity. Most tissues express multiple mAChRs, and those mAChRs are often coupled to the same subset of G proteins; this has led to the generation of double knockout mice, such as $M_2-/-$ $M_4-/-$ and $M_1-/-$ $M_3-/-$.

GENERATION OF MACHR KO MICE

Knockout mAChR mice are genetically altered mice with no functional single-subtype mAChR or double-subtype mAChRs. The lack of highly selective ligands (agonist and antagonists) for mAChRs limits us from realizing the fullest potential of the classical pharmacological approaches that are used to investigate the functional roles of individual mAChR subtypes. The fact that most brain regions express several different mAChRs does not help in overcoming this obstacle either. Alternatively, one way to make a correlation between specific pharmacological activities and subtype-specific mAChRs is to evaluate parameters of interest in a mutant mouse lacking in subtype-specific mAChRs with no dramatic changes in its vital signs.

In 1997, the first mAChR KO mouse (129SvJxC57BL/6) was generated by Nathanson's group at the University of Washington.[10] In brief, the homologous recombination technology is employed in embryonic stem (ES) cells to generate mice with a selective deficiency in the M_1 mAChR gene. The lack of an M_1 receptor is verified by immunoprecipitation analysis, and there must be no alterations in the levels of the M_2, M_3, and M_4 receptors. In addition, there must be no significant changes in brain morphology or in the pattern or levels of expression of the M_2, M_3, and M_4 receptors via immunocytochemical analysis.

Within the next five years after introduction of the first mAChR KO mouse, Wess led groups within the National Institutes of Health (NIH) to create all five subtypes of mAChR KO mice:

M_1 mAChR KO mice were produced by Miyakawa et al.[11] with genetic background 129SvEvxC57BL/6J and by Fisahn et al.[12]

M_2 mAChR KO mice were produced by Gomeza et al.[13] with genetic background 129J1xCF-1.

M_3 mAChR KO mice were produced by Yamada et al.[14] with genetic background 129SvEvxC57BL/6 and 129SvEv.

M_4 mAChR KO mice were produced by Gomeza et al.[15] with genetic background 129SvEvxCF-1.

M_5 mAChR KO mice were produced by Yamada et al.[16] with genetic background 129SvEvxCF-1.

Others laboratories also produced different types of single mAChR KO mice. For example,

M_1 mAChR KO mice were established by Gerber et al.[17] with genetic background C57BL/6.

M_2 mAChR KO mice were produced by Matshui et al.[18]

M_3 mAChR KO mice were created by Matshui et al.[19] with genetic background 129SvJxC57BL/6).

M_4 mAChR KO mice were created by Karasawa et al.[20]

M_5 mAChR KO mice were generated by Yeomans et al.[21] with genetic background 129SvJxCD1 and by Takeuchi et al.[22]

Subsequently, double subtypes of mAChR KO mice were generated:

M_1/M_3 mAChR double mutants were created by Ohno-Shosaku et al.[23]

M_2/M_3 mAChR double KOs were produced by Matshui et al.[24] via first crossing M_2–/– (N_3 generation) and M_3–/– (N_2 generation) to generate M_2+/– M_3+/– mice; then, intercrossing between these mice yielded pups of various genotypes for M_2 and M_3 alleles, including M_2–/– M_3–/– mutants.

M_2/M_4 mAChR mutants were generated by Zhang et al.[25] with genetic background 129J1x129SvEvxCF1 via intermating homozygous M_2–/– and M_4–/–.

GENERAL OBSERVATIONS IN mAChR KO MICE

Although multiple abnormalities were observed in all single or double mAChR KO mice, their general appearance was normal compared with their wild-type counterparts, and they did not show any obvious morphological or behavioral deficits. A summary of the major phenotypes displayed by M_1 to M_5 mAChR-deficient mice is provided by Wess.[26] A brief review of the functional analysis of mAChRs in mAChR KO mice is provided by Matsui et al.[27]

COGNITIONAL DATA WITH mAChR KO MICE

Muscarinic agonists have been proposed as potential pharmacotherapeutic targets for the treatment of cognitive impairment (for a review, see Felder et al.[28]).

M_1 mAChR KO Mice (M_1 Receptor-Deficient Mice)

The M_1 mAChRs medicate neurotransmitter signaling in the cortex and hippocampus.[29] In addition, M_1 mAChRs actively participate in higher cognitive processes, such as learning and memory processes (Bymaster et al.,[29] Sarter and Bruno,[30] Iversen,[31]). Therefore, M_1 mAChR agonists have been proposed to have a potential clinical utility in ameliorating cognitive deficits associated with Alzheimer's disease.[31]

To determine the relationship between the lack of M_1 mAChRs and cognitive deficits, M_1 KO mice were evaluated in three different hippocampus-dependent learning tasks, as described by Miyakawa et al.[11] Three major findings were reported from those tests:

1. M_1 mAChR KO mice have no performance deficit in the Morris water test compared with their wild-type (Wt) counterparts.
2. M_1 mAChR KO mice exhibit performance deficits in the eight-arm radial-maze test.
3. M_1 mAChR KO mice display reduced freezing in a fear-conditioning test.

Highlights from each of these tests are summarized as follows.

Morris water maze with a hidden platform is a test that is used to evaluate spatial reference memory in rodents. Overall, M_1 mAChR KO mice performed at the same level as their Wt controls in this test paradigm. For example, there were no significant differences detected between M_1 mAChR KO mice and Wt controls in escape latencies (time required to reach the platform) in both the original training and the reversal training, in swimming speed, or in time spent in the perimeter of the pool. Moreover, when the platform was removed, there were no significant differences in the performance of the genotypes for three additional parameters: searching the correct area where the platform was originally located, time spent in the training quadrant compared with the other quadrants, and frequencies of crossing the training site compared with equivalent sites in the other three quadrants.

For assessment of spatial working memory, an eight-arm radial maze was used. Two primary end points were measured in this test. The first end point was the number of revising errors, defined as test subjects returning to the arms that had been visited previously to retrieve a food pellet. M_1 mAChR KO mice had significantly high numbers during trials without delay (1st to 14th trials) compared with their Wt controls. However, there were no significant differences identified in numbers between genotypes during the trials with delay (30-sec delay in the 15th trial and 2-min delay in the 16th to 18th trials. The second end point was the number of different arms chosen during the first eight choices, which represents working memory. In this measurement, M_1 mAChR KO mice exhibited similar levels of performance compared with their Wt controls in trials without delay and with delay.

In a contextual and cued fear-conditioning test, M_1 mAChR KO mice displayed lower levels of freezing after foot shocks compared with their Wt littermates during the conditioning period. However, similar levels of freezing during context testing conducted 24 h after conditioning were observed between M_1 mAChR KO mice and their Wt controls. In contrast, a significant reduction in freezing levels was detected in M_1 mAChR KO mice compared with their Wt controls when context testing was conducted 4 weeks after conditioning. Moreover, M_1 mAChR KO mice exhibited significantly reduced levels of freezing when the conditioned stimulus (tone) was challenged in an altered context 48 h after conditioning (cue testing).

M_1 mAChR KO mice exhibit hyperactivity under normal, social, or stressful conditions.[11] This pronounced increase in locomotor activity is apparently a unique phenotype to M_1 mAChR KO mice, since hyperactivity was not detected during an open field test in either M_2 mAChR KO or M_3 mAChR KO mice, although M_4 mAChR KO mice exhibited a relatively moderate elevation in locomotor activity compared with the hyperactivity of the M_1 mAChR KO mice. Therefore, the authors propose that this phenotype might be the primary reason for the performance deficits of the M_1 mAChR KO mice in both the eight-arm radial maze and the contextual and cued fear-conditioning test. They further suggest that M_1 mAChRs may play a less critical role in cognitive function and a more prominent role in the regulation of locomotor activity. Given the proposed link between cognition and hyperactivity in M_1 mAChR mice, the authors then speculate that M_1 mAChR KO mice may be an appropriate animal model to study attention deficit hyperactivity disorder (ADHD), with key symptoms including hyperactivity and impermanent cognition.[32]

Anagnosaras et al.[33] also conducted several learning and memory tests in M_1 mAChR KO mice. In their studies, compared with Wt controls, M_1 mAChR KO mice displayed better performance in a context conditioning test; normal performance in the Morris water-maze test (hidden platform); and severe deficits in winshift and social discrimination learning, in which M_1 mAChR KO mice failed to show significant discrimination ability, while the Wt controls did. The authors credit their results as a notable finding, since all of those memories are assumed to be dependent on similar processes in the hippocampus and thus are distinguishable by the M_1 mAChR null mutation. Therefore, the authors suggest that the effect of M_1 mAChRs on memory function should not be categorized as either the acquisition or maintenance of information. Brief analyses on data from these studies are summarized below.

In contrast to the impaired performance of M_1 mAChR KO mice observed in the win-shift radial arm maze or social discrimination tests, there were no genotype differences detected in the Morris water maze (hidden platform) for the following parameters: (1) latencies throughout acquisition and training, (2) search time in each quadrant during the first probe trial at the end of day 3 and the second probe trial taken at the end of day 6, (3) preference for the target quadrant, and (4) good retention in the additional probe trial 10 days after probe 2.

According to Anagnosaras et al.,[33] assessment of spatial reference memory could be conducted in either the Morris water maze (hidden-platform version) or the win-shift radial arm maze. In addition, both contextual fear conditioning and the Morris water maze (hidden-platform version) are considered as reference-memory tasks and could be viewed as a matching-to-sample problem, leading to the concept of generalization. On the other hand, win-shift working memory, which is dependent on the hippocampus and prefrontal cortex, could be considered as a nonmatching-to-sample problem, which potentially requires the buildup of inhibition against matching to previously visited arms; therefore, a nonmatching-to-sample task is a more complicated process compared with a matching-to-sample task. M_1 mAChR KO mice display deficits in both win-shift and social discrimination tests (nonmatching-to-sample tasks), which require the prefrontal cortex; thus the authors suggest that M_1 mAChR KO mice may have a prefrontal or other cortical deficiency. Furthermore, M_1 mAChR KO mice initially showed enhanced performance in contextual fear conditioning, then followed by impaired performance in consolidation over the time period when the task becomes independent of hippocampus. Therefore, the authors propose that M_1 mAChR KO mice have a deficiency in either cortical memory or hippocampal-cortical interaction, which is specific to working memory and to remote reference memory. In turn, the authors conclude that M_1 mAChR receptors are important in cortical memory function and in the interaction between the cortex and the hippocampus, but that they are not essential for memory acquisition by the hippocampus *per se.*

A nonselective mAChR antagonist, scopolamine, was used by the authors to evaluate scopolamine-induced hyperactivity. Exploratory activity was measured on a large, dark, open field. Both M_1 mAChR KO and Wt mice showed hyperactivities after both saline and scopolamine administration. Because scopolamine had no different effect on hyperactivity in M_1 mAChR KO mice compared with Wt littermates, scopolamine-induced hyperactivity may mediate through other subtypes of mAChRs in addition to M_1 mAChRs. Furthermore, in conditioning chambers, scopolamine did induce hyperactivity in M_1 mAChR KO mice. However, after transferring those same M_1 mAChR KO mice to the fear-conditioning chambers for 4 min without scopolamine, hyperactivity disappeared. These data suggest that M_1 mAChR KO mice can be hyperactive under certain conditions, such as during exploration under low anxiety, and that the magnitude of this hyperactivity is milder than scopolamine-induced hyperactivity. Therefore, Anagnosaras et al.[33] conclude that the hyperactivity phenotype of M_1 mAChR KO mice should not prevent them from being used as an animal model to assess memory, since M_1 mAChR KO mice had normal learning and memory in tasks sensitive to hyperactivity.

Cognition impairment was also examined in M_1 mAChR KO mice and Wt controls with scopolamine administration.[33] Scopolamine induced impaired watermaze performance and had an equivalent effect on both M_1 mAChR mice and Wt controls. These data are in line with the work of Miyakawa et al.,[11] which provides further evidence that scopolamine does not produce memory deficits by acting through M_1 mAChRs alone.

Data from both cell biology and biochemistry fields suggest that there is a link between hippocampal long-term potentiation (LTP) and learning and memory.[4] A modest role of M_1 mAChRs in synaptic plasticity was demonstrated via evaluating Schaffer collateral LTP in the hippocampal slice, a cellular model of learning. Anagnosaras et al.[33] reported that, under physiologically relevant conditions, there was a genotype difference, with pronounced reduced LTP induced by two theta burst stimulations (TBS) in M_1 mAChR KO mice compared with those in Wt controls.

In summary, the data presented here indicate that the lack of M_1 mAChR receptors may not produce severe cognitive deficits, since only relatively mild cognitive deficits were observed in M_1 mAChR KO mice. Moreover, other subtypes of mAChRs besides M_1 mAChRs may play a role in learning and memory as well, a supposition supported by (a) the data showing that scopolamine, a nonselective mAChR antagonist, induced comparable cognitive deficits in both M_1 mAChR KO mice and Wt controls in the Morris water-maze test and (b) the observations that some learning abilities remained intact following scopolamine administration. However, several research groups report that deletion of one specific mAChR gene does not have major effects on the expression levels of the remaining four mAChRs at the limited tissues examined.[26] Furthermore, M_1 mAChRs may have a more complicated role than is generally assumed, a role that is not limited to the hippocampus. For example, based on Anagnosaras et al.,[33] the interaction between the hippocampus and cortex for processing information requires the involvement of M_1 mAChRs. In addition, it is possible that compensative processes may have taken place during the development of M_1 mAChR KO mice. For example, multiple neuronal systems are believed to be involved in cognition processes[1]; thus, other non-mACh neurotransmitters may play a critical cognition role in M_1 mAChR KO mice. Moreover, the contribution of M_1 mAChRs to cognitive functions may only be recognized when other compensatory receptor systems are disrupted simultaneously.[26] In conclusion, the exact role of M_1 mAChRs in learning and memory has yet to be elucidated.

M_2 mAChR KO Mice (M_2 Receptor-Deficient Mice)

A large body of evidence suggests that cholinergic hippocampal pathways are involved in cognitive processes (for review, see Terry and Buccafusco[1]). In addition, M_2 mAChR is one of the major autoreceptors acting in hippocampal circuits (for review, see Felder et al.[28]). Therefore, Tzavara et al.[34] conducted some experiments to determine the role of M_2 mAChRs in learning and memory processes in relationship to its regulation of ACh release in hippocampus. They reported that basal ACh levels in the hippocampus were similar in both M_2 mAChR KO mice and Wt controls. However, scopolamine induced less hippocampal ACh release in M_2 mAChR KO mice compared with that in Wt controls. When placing test subjects into a novel environment, M_2 mAChR KO mice had significant elevation of hippocampal ACh levels in both amplitude and duration compared with their counterparts. In a passive-avoidance paradigm, M_2 mAChR KO mice displayed inferior performance, evidenced by shorter latencies to enter the darkened chamber on day 2, compared with their Wt littermates. Therefore, the authors suggest that M_2 mAChRs are involved in regulation of ACh efflux in the hippocampus, which translates into their participation in cognitive processes. This finding has broadened the functional role of M_2 mAChRs beyond their generally recognized contribution in mediating cardiovascular function.

Furthermore, Zhang et al.[25] suggest that M_2 mAChRs are the predominant inhibitory mACh autoreceptor in both the hippocampus and cerebral cortex. In their experiments, they found that agonist-dependent inhibition of stimulated [^{3}H]ACh release was completely abolished in hippocampal and cortical preparations from M_2

mAChR KO mice, but not from M_4 mAChR KO mice. The authors thus propose that centrally activated M_2 mAChR antagonists may be considered for the treatment of Alzheimer's disease, since Alzheimer's disease patients normally have reduced levels of ACh in both the hippocampus and cerebral cortex.

M_2 mAChRs may also have a role in learning and memory during the aging process, as seen in data based on rat,[35] where cognitive deficits were observed in aged and memory-impaired (AI) rats in a Morris water maze. BIBN-99, an M_2 antagonist, reversed the impaired ACh release in a dose-dependent manner, as evidenced via *in vivo* dialysis. In addition, BIBN-99 countered scopolamine-induced amnesia in young rats. Therefore, the authors suggest that the role of M_2 mAChRs in autoregulation of ACh levels in synaptic cleft may contribute to the learning deficits observed in the age-impaired group.

M_3 mAChR KO Mice

M_3 mAChRs are richly expressed in the brain. However, little CNS work has been conducted to assess the functional roles of M_3 mAChRs.[29] Possible links between M_3 mAChRs and cognition could be a fruitful area of exploration.

M_4 mAChR KO Mice

In addition to M_2 mAChRs, M_4 mAChRs are the other autoreceptors in hippocampal circuits.[28] To determine the role of M_4 mAChRs in cognitive processes, Tzavara et al.[34] conducted several experiments in the hippocampus using M_4 mAChR KO mice. They reported that basal ACh levels were elevated significantly in M_4 mAChR KO mice compared with their Wt controls. However, scopolamine induced elevation in ACh release in both M_4 mAChR KO mice and Wt, with no significant difference in magnitude. Moreover, in passive-avoidance tests, M_4 mAChR KO mice did not show impaired memory retention compared with Wt littermates. After mice were placed in a novel environment, hippocampal ACh levels were increased in M_4 mAChR KO mice. Based on these data, the authors suggest that M_4 mAChRs play a predominate role in a tonic autoregulatory fashion to mediate ACh release, which means that M_4 mAChR has a potential to be involved in learning and memory processes. However, its exact role in cognitive processes is still unclear at this point.

Other evidence also suggests a potential role of M_4 mAChRs in cognitive processes. For example, the distribution and density of M_4 mAChRs in the brain are similar to those for M_1 mAChRs.[29] Moreover, almost all antagonists that display high affinity for M_1 receptors also show high affinity for M_4 receptors.[15] Therefore, M_4 mAChR KO mice could be utilized to further explore the role of M_4 mAChRs in learning and memory processes.

M_5 mAChR KO Mice

Deficits in cholinergic dilation of cerebral blood vessels have been suggested to play a role in the pathophysiology of Alzheimer's disease. Yamada et al.[16] have proposed that M_5 mAChRs mediate the diameter of cerebral arterioles and arteries. They base this proposal on their data showing the absence of cholinergic dilation of cerebral blood vessels in M_5 mAChR KO mice, while Wt controls have this function. As a result, the authors suggest that selective M_5 mAChR agonists may have a potential clinical utility for increasing cerebral blood follow in certain diseases, including Alzheimer's disease and cerebral ischemia.

Double M_1 and M_3 mAChR KO Mice

Although double M_1/M_3 mAChR KO mice have been generated,[23] few cognitive tests have been conducted in this animal model. Iversen[31] has reported that mAChR agonists with functional selectivity for M_1/M_3 receptors reversed scopolamine-induced effects in rats on several test paradigms, including passive avoidance, conditioned suppression of drinking, and delayed matching to position. Therefore, double M_1/M_3 mAChR KO mice may be useful in further exploring the physiological function roles of M_1 or M_3 mAChRs in cognition.

Double M_2 and M_4 mAChR KO Mice

In vivo microdialysis data[34] showed significant increases in basal ACh levels in double M_2/M_4 mAChR KO mice compared with Wt controls. In addition, the magnitude of the elevation of basal ACh levels in double M_2/M_4 mAChR KO mice was greater than that in M_4 mAChR KO mice, indicating the involvement of M_2 mAChRs in inhibition of basal ACh release. Moreover, a scopolamine-induced increase in ACh levels in hippocampus was completely abolished in M_2/M_4 mAChR KO mice, while there was a diminished effect in M_2 mAChR KO mice, suggesting a contribution by M_4 mAChRs as well. In the passive-avoidance paradigm, Tzavara et al.[34] discovered that double M_2/M_4 mAChR KO mice as well as M_2 mAChR KO mice had deficits in memory retention, but not M_4 mAChR KO mice. Therefore, they conclude that M_2 mAChRs are essential for the full development of learning and memory in the passive-avoidance test.

NICOTINIC ACETYLCHOLINE RECEPTORS (NACHRS)

Extensive information on nAChRs, including characterization of nAChR subunits at the molecular level, is available in the literature as well as in Chapter 3 of this book. So far, nine ($\alpha2$ through $\alpha7$ as well as $\beta2$ through $\beta4$) genes have been identified as having expression in mammals' CNS. However, their *in vivo* functional roles are not yet fully understood. Furthermore, the large number of subunits of nAChR significantly adds to the complexities of this task. Moreover, as is the situation with mAChRs, highly selective ligands are not readily available at the present time. Therefore, genetically altered mice with functionally deficit nAChRs provide an opportunity to investigate both the function as well as composition of individual nAChRs.

GENERATION OF NACHR KO MICE

Details of the generation of nAChR KO mice are available in the literature.

$\alpha3$ nAChR KO mice[36]
$\alpha4$ nAChR KO mice[37,38]
$\alpha5$ nAChR KO mice[39]
$\alpha6$ nAChR KO mice[40]
$\alpha7$ nAChR KO mice[41]
$\alpha9$ nAChR KO mice[42]
$\beta2$ nAChR KO mice[43,44]
$\beta3$ nAChR KO mice[45]
$\beta4$ nAChR KO mice[44]
Double $\beta2$ and $\beta4$ nAChR KO mice ($\beta2-/- \beta4-/-$)[44]
Double $\alpha5$ and $\beta4$ nAChR KO mice (double-null mutants with $\alpha5$ and $\beta4$ nAChR subunits deficiencies)[46]

GENERAL OBSERVATIONS IN nAChR KO MICE

Characterization of available nAChR KO mice and their potential functional roles in the CNS have been summarized by Champtiaux and Changeux.[47] In addition, Picciotto's group[48] has provided a review on the physiological and behavioral phenotypes and possible clinical implications in nAChR KO mice.

In general, nAChR KO mice, with the exception of $\alpha3$ nAChR KO mice, have no obvious abnormalities based on external observation. The $\alpha3$ subunit is essential for survival, as 40% of $\alpha3$ nAChR KO mice died from unknown causes in the first three days of life. In addition, almost all of the surviving $\alpha3$ nAChR KO mice died over an interval of six to eight weeks after weaning[36]; therefore, $\alpha3$ nAChR KO mice may not be useful for meaningful behavioral studies. However, if the genetic targeting could be limited to the CNS only (a conditional mutation), then such modified $\alpha3$ nAChR KO mice might be useful for evaluating the function of the $\alpha3$ subunit of nAChRs in the central tissue.

The double $\alpha5-/- \beta4-/-$ mutants had no visible abnormality or neurological deficits.[46] In contrast, the double $\beta2/\beta4$ nAChR KO mice die during the first three weeks of life, although $\beta2$ nAChR KO mice and $\beta4$ nAChR KO mice are viable and develop normally. In addition, this double $\beta2/\beta4$ nAChR KO mouse has a similar phenotype as that in $\alpha3$ nAChR KO mice.[44] Therefore, Xu et al.[44] proposed that there might be a partial redundancy between $\beta2$ and $\beta4$ subunits of nAChRs.

COGNITIONAL DATA WITH nAChR KO MICE

Data from several groups suggest that nicotinic agonists improve cognition in animal models.[9,49-51] Furthermore, based on data in adult female rats, $\alpha4\beta2$ and $\alpha7$ receptors in the ventral hippocampus appear to have critical roles in working memory.[52] However, Levin[51] pointed out that due to availability of relatively selective ligands that exist only for $\alpha4\beta2$ and $\alpha7$, research has been principally limited to those two receptors. Furthermore, other brain regions such as the frontal cortex and the amygdala may also be of interest in nicotinic actions on memory.[53] A large body of evidence presented elsewhere also demonstrates that nicotine has a clinical utility for cognitive enhancement. Buccafusco and Terry[9] provided summary data on memory enhancement in human and nonhuman primates as well as a comparison of efficacy data from potentially cognitive-enhancing agents. Moreover, the number of $\alpha4$-containing receptors seems to be reduced in Alzheimer's disease. Furthermore, in schizophrenia, there is a decrease in both $\alpha4\beta2$ and $\alpha7$ receptors.[51] Levin and Rezvani[53,54] have reviewed nicotinic treatments for Alzheimer's disease, schizophrenia, and attention deficit hyperactivity disorder along with relevant preclinical data. Finally, after analyzing many epidemiological studies, Fratiglioni and Wang[55] have found that cigarette smokers are about 50% less likely to have Alzheimer's disease than age- and gender-matched nonsmokers, indicating nicotine (cigarette) may have an effect against the development of Alzheimer's disease.

$\beta2$ nAChR KO Mice

$\beta2$ nAChRs play a role in passive-avoidance learning. In a passive-avoidance paradigm, $\beta2$ nAChR KO mice exhibited enhanced latency of entry into a dark compartment, where they previously had been punished during the training, compared with Wt controls, indicating that at least some $\beta2$ nAChRs are endogenously active. However, 24 h postadministration of low doses of nicotine, $\beta2$ nAChR KO mice did not further increase the previously observed latency; in other words, $\beta2$ nAChR KO mice failed to show nicotine-induced enhancement of performance, suggesting that $\beta2$ nAChRs contribute to this phenotype.[43]

A fear-conditioning task was conducted in $\beta2$ nACh KO mice by Caldarone et al.[56] Based on their data, the authors conclude that $\beta2$ nAChRs are not essential for normal performance in this test paradigm in the absence of other abnormalities. However, $\beta2$ nAChRs have a critical role during the aging process to maintain neuronal function for performance in a fear-conditioning task. Some highlights from their work are summarized below.

In either contextual or tone-conditioned fear tests, performance between young (2 to 4 months) $\beta2$ nAChR KO mice and Wt controls were indistinguishable. However, aged (9 to 20 months) $\beta2$ nAChR KO males displayed deficits in freezing in both context and tone tests compared with age-matched Wt males. In contrast, no differences in fear conditioning were detected between aged KO and Wt females. The authors point out that data from their test regarding unimpaired fear conditioning in young $\beta2$ nAChR KO mice are in line with intact spatial memory data generated by Picciotto's group[43] from both Morris water-maze and passive-avoidance tests. Furthermore, in the latent-inhibition study, $\beta2$ nAChR KO mice showed a similar level of latent inhibition as that in Wt controls when both were preexposed to the tone. Moreover, both tone-preexposed and -nonpreexposed $\beta2$ nAChR KO mice displayed less freezing to the context as well as the tone compared with their Wt controls. In contrast, in the contextual and cued conditioning tests, there were no differences in freezing observed between $\beta2$ nAChR KO mice and Wt littermates.

According to Caldarone et al.,[56] $\beta2$ nAChRs have a protective role in aging. Their data demonstrate impaired fear conditioning in aged male $\beta2$ nAChR KO mice, but not in young $\beta2$ nAChR KO mice. Therefore, they conclude that the $\beta2$ nAChR KO mouse is a good animal model for evaluating age-related cognitive disorders such as Alzheimer's disease.

$\alpha4$ nAChR KO Mice

There is little cognition-related data on $\alpha4$ nAChR KO mice. Ross et al.[38] evaluated motor learning with a rotarod in $\alpha4$ nAChR KO mice and found that $\alpha4$ nAChR KO mice performed equally well on rotarod compared with Wt controls.

$\alpha7$ nAChR KO Mice

Nicotine improves sustained attention in mice, as evidenced in a five-choice serial reaction time (5-CSR) task in $\alpha7$ KO mice.[57] In this study, $\alpha7$ nAChR KO mice not only acquired the task much more slowly than their Wt counterparts, but also displayed a higher level of omission on attaining asymptotic performance. Because $\alpha7$ nAChR KO mice take more time to acquire the task compared with their Wt controls and display a higher level of omission in a less-demanding version of the task, the authors suggest that the $\alpha7$ nAChR mediates nicotine-induced improvements in sustained attention. Moreover, this was the first study to demonstrate nicotine-induced improvements in sustained attention in normal mice. This result further validates the hypothesis that nicotine-induced improvement in attention in normal humans is characterized by a reduction in omission levels and by an increase in the proportion of correct responses.

Furthermore, it appears that the $\alpha7$ subunit of nAChR does not have an essential role in aging-related cognition processes, since young adult $\alpha7$ nAChR KO mice perform equally as well as their Wt controls in spatial learning (Morris water maze), in contextual and auditory fear conditioning, and in anxiety tests.[58]

According to Orr-Urtreger et al.,[41] $\alpha7$ nAChRs are richly expressed in the hippocampus. The authors concluded that $\alpha7$ nAChR KO mice lack binding sites for α-bungarotoxin (α-BGT), a selective nAChR antagonist, and thus do not have hippocampal fast nicotinic currents. Furthermore, they suggest a possible action of nicotine on the hippocampus through activation, desensitization, or modification of α-BCT binding sites containing the $\alpha7$ subunit.

CHOLINESTERASES

Two kinds of cholinesterases, namely acetylcholinesterase (AChE) and butyrylcholinesterase (BChE), have been identified in the human CNS.[59] In contrast to rich information on AChE's roles in CNS, there is very limited knowledge on BChE's roles.

GENERATION OF AChE KO MICE
(ACETYLCHOLINESTERASE −/− MICE OR NULLIZYGOUS)

Details of the generation of AChE KO mice are available in the literature.[60,61]

GENERAL OBSERVATIONS IN AChE KO MICE

AChE KO mice live to adulthood, although they also die early (21 days) from seizures.[60,61] Detailed information on characterization of AChE KO is provided by

Duysen et al.[61] BChE activity was reported to be normal in AChE KO mice by Xie et al.[60] In addition, AChE KO mice were highly sensitive to the BChE-specific inhibitor bamabuterol. Moreover, Mesulam et al.[59] reported that mice lacking AChE inhibitor could use BChE to hydrolyze acetylcholine. These findings indicate that BChE and possibly other enzymes may be able to compensate for some functions of AChE.

COGNITIONAL DATA WITH ACHE KO MICE

Chronic administration of cholinesterase inhibitors is the primary therapy strategy for the treatment of Alzheimer's disease. In fact, all FDA-approved drugs for the treatment of Alzheimer's disease are AChE inhibitors (for a review, see Buccafusco and Terry[9]).

Volpicelli-Daley et al.[62] discovered that the absence of AChE changes mAChR expression, cell-surface availability, and function in brain regions associated with learning and memory, such as the hippocampus and cortex. For example, in AChE KO mice, the M_1, M_2, and M_4 mAChRs exhibited 50 to 80% decreased expression in the hippocampus and cortex. Moreover, M_1 and M_2 mAChRs decreased localization to dendrites, the cell surface, and presynaptic terminals. These data suggest that chronic use of AChE inhibitors may lead to mAChR down regulation. These results may also explain the observations of a modest symptomatic improvement after chronic administration of AChE inhibitors for the treatment of Alzheimer's disease. Furthermore, AChE KO mice exhibited increased sensitivity to mAChR antagonist-induced elevation in locomotor activity, indicating functional mAChR down regulation and validating the possibility of adaptation.[62] Thus, the authors suggest that future studies should focus on understanding the factors regulating expression and localization of molecules involved in cholinergic transmission to improve efficacy of long-term use of AChE inhibitors.

Li et al.[63] have proposed that other adaptation mechanisms may exist that have not yet been tested. For example, nicotine receptors could be down regulated in AChE KO mice, and rates of ACh synthesis and release may change in AChE KO mice as well.

Similar levels of BChE activity were observed in both AChE KO mice and their Wt controls.[64] In addition, AChE KO mice have high levels of BChE activity in many tissues, and AChE KO mice live normally. Therefore, Li et al.[64] have suggested that BChE may have an essential function in AChE KO mice and probably in Wt mice as well, implicating a potential new therapy of selective BChE inhibitors for the treatment of Alzheimer's disease.

LIMITATIONS WITH GENE-TARGETING (KNOCKOUT) APPROACH AND FUTURE DIRECTIONS

Although a mutant mouse with an identical molecular lesion generated by the gene-targeting technology enables us to analyze diverse aspects of gene function *in vivo*, this technology has some limitations as well. A summary of key points, including limitations and potential solutions identified by different research groups, is presented below.

GENETIC BACKGROUND

Conventional homologous recombination technologies are commonly used to introduce null mutations of interesting genes into embryonic stem (ES) cells. Homozygous mutant (KO) mice are generated by applying standard transgenic and mouse-breeding techniques. For technical reasons, two mouse strains are often used in gene targeting to produce null-mutation mice. Therefore, hybrid knockout mice are genetically different from their Wt controls. Consequently, observed alterations in phenotypes may not be considered solely as the result of null mutation, since genetic background may contribute to alterations as well.[65] Moreover, the same mutation with different mouse strains may produce different phenotypes.[66] To address the issue of genetic background, Gerlai[65] has proposed two solutions. One strategy is to create a congenic strain that carries the mutation on the desired genetic background via repeatedly backcrossing the mutant hybrid animals (e.g., heterozygous mice) to the strain of choice. However, the drawback of this approach is that it is not an economically sound approach from both timeline and budget perspectives. The other option is not to utilize the hybrid mouse, considering the fact that it might not be an appropriate animal model for some behavioral tests.

Wolfer et al.[67] have proposed another solution, a so-called reverse F_2 strategy, to address the problem of genetic background. According to the authors, employing their breeding schemes would not add specific technology, significant resources, and timeline concerns to the existing best practice for conventional gene targeting.

To minimize the influence of background alleles, at least F_2 animals should be used when conducting studies in KO mice with mixed genetic background. Moreover, Müller[7] suggests that the same mutation with different genetic backgrounds can enable us to discover gene functions that are not noticed on a single background. Consequently, mapping and cloning of these strain-specific modifier genes will help create better animal models for evaluating human diseases. Some examples of genetic backgrounds of mAChR KO mice are listed below.

M_1 mAChR KO mice:
   129SvJxC57BL/6[10]
   C57BL/6[17]
   129SvEvxC57BL/6[11]
M_3 mAChR KO mice:
   129SvJxC57BL/6[19]
   129SvEvxC57BL/6 (mixed genetic background) and 129SvEv (pure genetic background; so-called isogenic mice)[14]
M_5 mAChR KO mice:
   129SvEvxCF-1[16]
   129SvJxCD1[21]

Most of the nAChR KO mice have been generated via backcrossing to the C57B16 strain ($\alpha3$, $\alpha4$, $\alpha6$, $\alpha7$, and $\beta2$ and $\beta4$).[47]

DEVELOPMENTAL AND FUNCTIONAL COMPENSATION

Compensatory processes could be trigged by alteration of the targeted gene, by molecular pathways associated with the targeted gene, as well as by the genetic background.[65] However, conducting studies in KO mice could reveal functional contributions of individual subtypes of mAChRs or nAChRs; two inherited issues, namely developmental adaptation and functional compensation, need to be addressed.[47] Furthermore, in general, many compensatory changes could be considered as "normalizing," since situations of "overshooting the mark" do occur.[68]

In theory, developmental compensation could occur at multiple levels, involving either the single-gene level, the genetic pathway level, or systemic adaptive mechanisms. Moreover, it is most likely that related proteins functionally compensate any deficit associated with the loss of one family member.[7] When mutation occurs at the earliest stages of development, other genes may alter their own development courses in response to the absence of the target gene.[6] In such cases, developmental compensation may mask adult physiological or behavioral phenotypes due to the lack of an endogenous nAChR subunit in nAChR KO mice.[4,7] To address this issue, better control over gene ablation needs to be considered. For example, the approach of confining gene disruption into a specific timing, such as postnatal stage, should in principle help minimize the impact of potential developmental compensation and enable us to assess acute phenotypes in a true deficit state.[2] In addition, tissue-restricted deletion could prevent death in some cases, such as in $\alpha 3$ nAChR KO mice; if the deletion could be limited to the brain tissue, then the mutants would live long enough to conduct studies of the deletion's potential CNS roles.

The literature provides several examples of functional compensation in mAChR KO mice. Karasawa et al.[20] reported that there were no genotype differences on the cataleptic response between M_4 mAChR KO mice and their Wt controls. Thus, they concluded that a lack of signaling mediated by M_4 mAChRs does not have a major effect on haloperidol-induced catalepsy. They further speculated that a chronic adaptation of the nervous system in the M_4 mAChR mice may contribute to this observation.

Gerber et al.[17] found increased striatal dopaminergic transmission in M_1 mAChR KO mice. They suggest that inhibition of the dopaminergic cells dampened due to the lack of M_1 mAChRs in the KO mice. Furthermore, compensation may occur in M_1 mAChR KO mice, since only mild learning and memory deficits were observed in this animal model. The potential compensation could occur in several places, such as other mAChR or nAChR subtypes or even in neuronal systems other than the cholinergic system.

Although the possibility of compensation does exist, data from several laboratories in mAChR KO mice demonstrate that deletion of one subtype mAChR gene does not lead to significantly altered expression levels of the remaining four mAChRs.[26] Moreover, Shapiro et al.[69] did not find compensation in other mAChR subtypes when they studied mAChR (M_1, M_2, and M_4) KO mice in relationship with ion channels, although they did detect loss of signaling. Thus, they propose that loss of signaling from one subtype of mAChR may not have enough strength to trigger a compensatory process because sympathetic neurons modulate their ion channels through a wide spectrum of G-protein-coupled receptors.

REDUNDANT MECHANISM

Gene redundancy is another problem with the genetic knockout method. One piece of evidence on gene redundancy is that all five mAChR KO mutant mouse lines are viable with no major visible abnormalities, indicating a functional redundancy among the subtypes.[26] Another example is that the $\beta2+/+$ $\beta4-/-$ nAChR KO mice do not show an obvious phenotype in terms of survival and growth. In contrast, $\beta2-/-$ $\beta4-/-$ nAChR KO mice display a very severe phenotype. Thus Xu et al.[44] suggest that either partial redundancy may exist or that the $\beta4$ gene is not essential. Furthermore, Champtiaux and Changeux[47] speculate that redundancy may exist among different nAChR subtypes at the level of neurotransmitter systems. Thus, the lack of a β subunit of nAChR in β nAChR KO mice does not guarantee detection of dramatic behavioral performance, since other neurotransmitter systems maintain alterations of behavioral performance.

More than one gene usually regulates most behaviors and physiologic process,[68] and this might be one of the rationales for creating double- or triple-knockout mice. Moreover, a single deletion may not produce a specific phenotypic change when a redundant gene exists. For example, an intact response in $\alpha5$ nAChR KO mice should not be interpreted as proof that $\alpha5$ subunit nAChR is not normally involved in this function.[70] To overcome this obstacle, Müller[7] suggests comparing phenotypes produced by single and double or triple gene deletions. Furthermore, a behavioral phenotype opposite to that of the predicted behavior could be the result of overcompensation by one or more redundant genes. In addition, an observed phenotype may reflect the role of the compensation gene, not the targeted mutation.[6] On the other hand, a single gene is often involved in multiple behaviors and physiologic processes. Thus, abolition of this gene could lead to dramatic changes, including possible death.[68]

REFINED GENE-TARGETING TECHNOLOGY

Phenotype modification occurs when the expression of one gene alters the expression of another gene at multiple traits, from transcription to molecular or cellular levels, up to organ or system levels.[66] Furthermore, inactivation of a gene via the conventional gene targeting occurs in all tissues of the body from the onset of development and throughout the entire lifespan. Therefore, to obtain better control over timing and location of mutation in a time- and tissue-dependent fashion, several new approaches have been developed.[7]

Inducible knockout technology enables us to turn the gene of interest "on" or "off" at particular times. Therefore, the developmental compensation issue, as well as a lethal or otherwise adverse phenotype that prevents a more detailed analysis inherent to conventional knockouts, could be addressed. Moreover, separating chronic versus acute phenotypes and identifying functions along the developmental course are now possible through the use of this technology.[7] However, interpretation of the phenotype from this approach still requires caution, since redundancy may still exist, even when a gene is turned "on" or "off" at a later stage of life.[7,68] Furthermore, when inaction of a gene occurs at the time point of adulthood, the function of the learning gene is going to be intact throughout ontogeny. As a result, in principle, adaptive responses cannot be initiated, and therefore phenotypes are expected to be more severe in conditional compared with conventional knockout mice.[7]

In contrast to conventional knockout's widespread expression, region-specific gene knockout technology allows evaluation a given gene in a predefined region by linking the gene of interest to a tissue-specific promoter.[7,68] For example, a gene might be expressed in both brain and some peripheral organs. Thus, it is impossible to assign a phenotype in the knockout mice to a specific brain structure or pathway, or even to the nervous system.[6] Tissue-specific "conditional" knockouts have been invented to solve this problem. By applying this technology, we can study the role of an individual mAChR or nAChR in a defined brain region without compromising other functions in the organism. For example, the $\alpha 3$ subunit of nAChR is essential for survival, since 40% of $\alpha 3$ nAChR KO mice die within the first three days of life, and almost all of the surviving $\alpha 3$ nAChR KO mice die over an interval of six to eight weeks after weaning.[36] Because this type of KO mouse is too weak for behavioral studies, a conditional mutation limited to the CNS might be helpful in examining the role(s) of $\alpha 3$ nAChR in the brain.

Conventional gene-targeting technology does not have either spatial or temporal restrictions. For example, Mayford et al.[71] have pointed out that it is impossible to correlate a phenotype, such as aggressiveness, to a single anatomical location or circuit in the brain of $5HT_{1B}$ KO mice. However, further efforts are required to improve temporally regulated gene targeting, controlled by the administration of inducers. There have been some limited successes for some organs, but the technique is not yet fully ready for the brain.[7]

Rescue strategies have been proposed by Phillips et al.[72] for the ultimate proof that a specific phenotype is the result of a mutated gene. This is accomplished by introducing the functional gene and reversing the observed effect.

CLASSICAL PHARMACOLOGICAL APPROACHES

Although conventional gene-targeting technology has several issues, the unique features of the knockout approach are relatively precise and free of the side effects that are encountered with ligand (agonists and antagonists) probes.[68] Moreover, application of the knockout approach helps identify and verify various parallel and distributed pathways of cognition processes by eliminating key molecules and key functional pathways.[71] On the other hand, studies in whole-animal models, a critical tool for evaluation of learning and memory, provide insights on localization and interactions with other neural systems. Many results obtained from knockout mice complement data from pharmacology studies,[51] and each test has its own advantages as well as limitations. Thus, employing several tests for each behavioral domain of interest is the optimal approach to prevent false conclusions.[6] For example, the Morris water maze has been used to evaluate spatial learning in knockout mice. However, it is widely known that mice with motor and visual deficits do not perform well in this test.[68] Thus careful selection of appropriate knockout mouse models is key to obtaining valid conclusions from this behavior paradigm. In short, we can advance our knowledge of how individual receptor subtypes contribute to cognition processes by combining knockout technologies and traditional pharmacology tests.

REFERENCES

1. Terry, A.V., Jr. and Buccafusco, J.J., The cholinergic hypothesis of age and Alzheimer's disease-related cognitive deficits: recent challenges and their implications for novel drug development, *J. Pharmacol. Exp. Ther.*, 306, 821, 2003.
2. Majzoub, J.A. and Muglia, L.J., Molecular medicine, knockout mice, *New England J. Med.*, 334, 904, 1996.
3. Wells, T. and Carter, D.A., Genetic engineering of neural function in transgenic rodents: towards a comprehensive strategy? *J. Neurosci. Meth.*, 108, 111, 2001.
4. Picciotto, M.R. and Wickman, K., Using knockout and transgenic mice to study neurophysiology and behavior, *Physiol. Rev.*, 78, 1131, 1998.
5. Zambrowicz, B.P. and Sands, A.T., Knockouts model the 100 best-selling drugs — will they model the next 100? *Nature*, 2, 38, 2003.
6. Crawley, J.N., Behavioral phenotyping of transgenic and knockout mice: experimental design and evaluation of general health, sensory functions, motor abilities, and specific behavioral tests, *Brain Res.*, 835, 18, 1999.
7. Müller, U., Ten years of gene targeting: targeted mouse mutants, from vector design to phenotype analysis, *Mechanisms Dev.*, 82, 3, 1999.
8. Bolivar, V., Cook, M., and Flaherty, L., List of transgenic and knockout mice: behavioral profiles, *Mammalian Genome*, 11, 260, 2000.
9. Buccafusco, J.J. and Terry, A.V., Multiple central nervous system targets for eliciting beneficial effects on memory and cognition, *J. Pharmacol. Exp. Ther.*, 295, 438, 2000.
10. Hamilton, S.E., Loose, M.D., Levey, A.I., Hille, B., McKnight, G.S., Idzerda, R.I., and Nathanson, N.M., Disruption of the M_1 receptor gene ablates muscarinic receptor-dependent M current regulation and seizure activity in mice, *Proc. Natl. Acad. Sci. U.S.A.*, 94, 13311, 1997.
11. Miyakawa, T., Yamada, M., Duttaroy, A., and Wess, J., Hyperactivity and intact hippocampus-dependent learning in mice lacking the M1 muscarinic acetylcholine receptor, *J. Neurosci.*, 21, 5239, 2001.
12. Fisahn, A., Yamada, M., Duttaroy, A., Gan, J.W., Deng, C.X. et al., Muscarinic induction of hippocampal gamma oscillations requires coupling the M_1 receptor to two mixed cation channels, *Neuron*, 33, 615, 2002.
13. Gomeza, J., Shannon, H., Kostenis, E., Felder, C., Zhang, L., Brodkin, J., Grinberg, A., Sheng, H., and Wess, J., Pronounced pharmacologic deficits in M_2 muscarinic acetylcholine receptor knockout mice, *Proc. Natl. Acad. Sci. U.S.A.*, 96, 1692. 1999.
14. Yamada, M., Miyakawa, T., Duttaroy, A., Yamadaka, A., Moriguchi, T., Makita, R., Ogawa, M., Chou, C.J., Xia, B., Crawley, J.N., Felder, C.C., Deng, C.X., and Wess, J., Mice lacking the M_3 muscarinic acetylcholine receptor are hypophagic and lean, *Nature*, 410, 207, 2001.
15. Gomeza, J., Zhang, L., Kostenis, E., Felder, C., Bymaster, F., Brodkin, J., Shannon, H., Xia, B., Deng, C., and Wess, J., Enhancement of D_1 dopamine receptor-mediated locomotor stimulation in M_4 muscarinic acetylcholine receptor knockout mice, *Proc. Natl. Acad. Sci. U.S.A.*, 96, 10483, 1999.
16. Yamada, M., Lamping, K.G., Duttaroy, A., Zhang, W., Cui, Y., Bymaster, F.P., McKinzie, D.L., Felder, C.C., Deng, C.X., Faraci, F.M., and Wess, J., Cholinergic dilation of cerebral blood vessels is abolished in M_5 muscarinic acetylcholine receptor knockout mice, *Proc. Natl. Acad. Sci. U.S.A.*, 98, 14096, 2001.
17. Gerber, D.J., Sotnikova, T.D., Gainetdinov, R.R., Huang, S.Y., Caron, M.G., and Tonegawa, S., Hyperactivity elevated dopaminergic transmission, and response to amphetamine in M_1 muscarinic acetylcholine receptor-deficient mice, *Proc. Natl. Acad. Sci. U.S.A.*, 98, 15312, 2001.

18. Matshui, M., Motomura, D., Fujikawa, T., Jiang, J., Takahashi, S., Manabe, T., and Taketo, M.M., Mice lacking M_2 and M_3 muscarinic acetylcholine receptors are devoid of cholinergic smooth muscle contractions but still viable, *J. Neurosci.*, 22, 10627, 2002.

19. Matshui, M., Motomura, D., Karasawa, H., Fujikawa, T., Jiang, J., Komiya, Y., Takahashi, S., and Taketo, M.M., Multiple functional defects in peripheral autonomic organs in mice lacking muscarinic acetylcholine gene for the M_3 subtype, *Proc. Natl. Acad. Sci. U.S.A.*, 97, 9579, 2000.

20. Karasawa, H., Taketo, M.M., and Matshi, M., Loss of anti-cataleptic effect of scopolamine in mice lacking muscarinic acetylcholine receptor subtype 4, *Eur. J. Pharmacol.*, 468, 15, 2003.

21. Yeomans, J., Forster, G., and Blaha, C., M_5 muscarinic receptors are needed for slow activation of dopamine neurons and for rewarding brain stimulation, *Life Sci.*, 68, 2449, 2001.

22. Takeuchi, J., Fulton, J., Jia, Z.P., Abramov-Newerly, W., Jamot, L., Sud, M., Coward, D., Ralph, M., Roder, J., and Yeomans, J., Increased drinking in mutant mice with truncated M_5 muscarinic receptor genes, *Pharmacol. Biochem. Behav.*, 72, 117, 2002.

23. Ohno-Shosaku, T., Matsui, M., Fukudome, Y., Shosaku, J., Tsubokawa, H., Taketo, M.M., Manabe, T., and Kano, M., Postsynaptic M_1 and M_3 receptors are responsible for the muscarinic enhancement of retrograde endocannabinoid signaling in the hippocampus, *Eur. J. Neurosci.*, 18, 109, 2003.

24. Matshu, M., Motomura, D., Fujikawa, T., Jian, J., Takahashi, S., Manabe, T., and Taketo, M.M., Mice lacking M_2 and M_3 muscarinic acetylcholine receptors are devoid of cholinergic smooth muscle contraction but still viable, *J. Neurosci.*, 22, 10627, 2002.

25. Zhang, W., Basile, A.S., Gomeza, J., Volpicelli, L.A., Levey, A.I., and Wess, J., Characterization of central inhibitory muscarinic autoreceptors by the use of muscarinic acetylcholine receptor knock-out mice, *J. Neurosci.*, 22, 1709, 2002.

26. Wess, J., Muscarinic acetylcholine receptor knockout mice: novel phenotypes and clinical implications, *Ann. Rev. Pharmacol. Toxicol.*, 44, 423, 2004.

27. Matsui, M., Yamada, S., Oki, T., Manabe, T., Taketo, M.M., and Ehlert, F.J., Functional analysis of muscarinic acetylcholine receptor using knockout mice, *Life Sci.*, 75, 2971, 2004.

28. Felder, C.C., Bymaster, F.P., Ward J., and Delapp, N., Therapeutic opportunities for muscarinic receptors in the central nervous system, *J. Med. Chem.*, 43, 4333, 2000.

29. Bymaster, F.P., McKinzie, D.L., Felder, C.C., and Wess, J., Use of M_1–M_5 muscarinic receptor knockout mice as novel tools to delineate the physiological roles of the muscarinic cholinergic system, *Neurochem. Res.*, 28, 437, 2003.

30. Sarter, M. and Bruno, J.P., Cognitive functions of cortical acetylcholine: toward a unifying hypothesis, *Brain Res. Brain Res. Rev.*, 23, 28, 1997.

31. Iversen, S.D., Behavioral evaluation of cholinergic drugs, *Life Sci.*, 60, 1145, 1997.

32. Paule, M.G., Rowland, A.S., Ferguson, S.A., Chelonis, J.J., Tannock, R., Swanson, J.M., and Castellanos, F.X., Attention deficit/hyperactivity disorder: characteristics, interventions and models, *Neurotoxicol. Teratol.*, 22, 631, 2000.

33. Anagnosaras, S.G., Murphy, G.G., Hamilton, S.E., Mitchell, S.L., Rahnama, N.P., Nathanson, N.M., and Silva, A.J., Selective cognitive dysfunction in acetylcholine M_1 muscarinic receptor mutant mice, *Nat. Neurosci.*, 6, 51, 2003.

34. Tzavara, E.T., Bymaster, F.P., Felder, C.C., Wade, M., Gomeza, J., Wess, J., McKinzie, D.L., and Nomikos, G.G., Dysregulated hippocampal acetylcholine neurotransmission and impaired cognition in M_2, M_4 and M_2/M_4 muscarinic receptor knockout mice, *Molecular Psychiatry*, 8, 673, 2003.

35. Quirion, R., Wilson, A., Rowe, W., Aubert, I., Richard, J., Doods, H., Parent, A., White, N., and Meaney, M.J., Facilitation of acetylcholine release and cognitive performance by an M_2-muscarinic receptor antagonist in aged memory-impaired rats, *J. Neurosci.*, 15, 1455, 1995.

36. Xu, W., Gelber, S., Orr-Urtreger, A., Armstrong, D., Lewis, R.A., Ou, C.N., Patrick, J., Role, L., De Biasi, M., and Beaudet, A.L., Megacystis, mydriasis, and ion channel defect in mice lacking the $\alpha 3$ neuronal nicotinic acetylcholine receptor, *Proc. Natl. Acad. Sci. U.S.A.*, 96, 5746, 1999.

37. Marubio, L.M., del mar Arroyo-Jimenez, M., Cordero-Erausquin, M., Lena, C., Le Novere, N., de Kerchove d'Exaerde, A., Huchet, M., Damaj, M.J., and Changeux, J.P., Reduced antinociception in mice lacking neuronal nicotinic receptor subunits, *Nature*, 398, 805, 1999.

38. Ross, S.A., Wong, J.Y., Clifford, J.J., Kineslla, A., Massalas, J.S., Home, M.K., Scheffer, I.E., Kola, I., Waddington, J.L., Berkovic, S.F., and Drago, J., Phenotypic characterization of an $\alpha 4$ neuronal nicotinic acetylcholine receptor subunit knockout mouse, *J. Neurosci.*, 20, 6431, 2000.

39. Salas, R., Orr-Urtreger, A., Broide, R.S., Beaudet, A., Paylor, R., and De Biasi, M., The nicotinic acetylcholine receptor subunit $\alpha 5$ mediates short-term effects of nicotine *in vivo*, *Mol. Pharmacol.*, 63, 1059, 2003.

40. Champtiaux, N., Han, Z.Y., Bessis, A., Rossi, F.M., Zoli, M., Marubio, L., McIntosh, J.M., and Changeux, J.P., Distribution and pharmacology of $\alpha 6$-containing nicotinic acetylcholine receptors analyzed with mutant mice, *J. Neurosci.*, 2, 1208, 2002.

41. Orr-Urtreger, A., Goldner, F.M., Saeki, M., Lorenzo, I., Goldberg, L., De Biasi, M., Dani, J.A., Patrick, J.W., and Beaudet, A.L., Mice deficient in the $\alpha 7$ neuronal nicotinic acetylcholine receptor lack α-bungarotoxin binding sites and hippocampal fast nicotinic currents, *J. Neurosci.*, 17, 9165, 1997.

42. Vetter, D.E., Liberman, M.C., Mann, J., Barhanin, J., Boulter, J., Brown, M.C., Saffiote-Kolman, J., Heinemann, S.F., and Elgoyhen, A.B., Role of $\alpha 9$ nicotinic ACh receptor subunits in the development and function of cochlear efferent innervation, *Neuron*, 23 (1), 93, 1999.

43. Picciotto, M.R., Zoli, M., Lena, C., Bessis, A., Lallemand, Y., leNovere, N., Vincent, P., Pich, E.M., Brulet, P., and Changeux, J.P., Abnormal avoidance learning in mice lacking functional high-affinity nicotine receptor in the brain, *Nature*, 374, 65, 1995.

44. Xu, W., Orr-Urtreger, A., Nigro, F., Gelber, S., Sutcliffe, C.B., Armstrong, D., Patrick, J.W., Role, L.W., Beaudet, A.L., and De Biasi, M., Multiorgan autonomic dysfunction in mice lacking the $\beta 2$ and the $\beta 4$ subunits of neuronal nicotinic acetylcholine receptors, *J. Neurosci.*, 19, 9298, 1999.

45. Cui, C., Booker, T.K., Allen, R.S., Graady, S.R., Whiteaker, P., Marks, M., Salminen, O., Tritto, T., Tutt, C.T., Allen, W.R., Stitzel, J.A., McIntosh, J.M., Boulter, J., Collins, A.C., and Heinemann, S.F., The $\beta 3$ nicotinic receptor subunit: a component of α-conotoxin MII-binding nicotinic acetylcholine receptors that modulate dopamine release and related behaviors, *J. Neurosci.*, 23 (35), 11045, 2003.

46. Kedmi, M., Beaudet, A.L., and Orr-Urtreger, A., Mice lacking neuronal nicotinic acetylcholine receptor beta4-subunit and mice lacking both $\alpha 5$ and $\beta 4$ subunits are highly resistant to nicotine-induced seizures, *Physiol. Genomics*, 17, 221, 2004.

47. Champtiaux, N. and Changeux, J.P., Knockout and knockin mice to investigate the role of nicotine receptors in the central nervous system, *Prog. Brain Res.*, 145, 235, 2004.

48. Picciotto, M.R., Caldarone, B.J., Brunzell, D.H., Zachariou, V., Stevens, T.R., and King, S.L., Neuronal nicotine acetylcholine receptor subunit knockout mice: physiological and behavioral phenotypes and possible clinical implications, *Pharmacol. Ther.*, 92, 89, 2001.

49. Arendash, G.W., Sanberg, P.R., and Sengstock, G.J., Nicotine enhances the learning and memory of aged rats, *Pharmacol. Biochem. Behav.*, 52, 517, 1995.
50. Levin, E.D. and Simon, B.B., Nicotinic acetylcholine involvement in cognitive function in animals, *Psychopharmacology*, 138, 217, 1998.
51. Levin, E.D., Nicotinic receptor subtypes and cognitive function, *J. Neurobiol.*, 53, 633, 2002.
52. Levin, E.D., Dradley, A., Addy, N., and Sigurani, N., Hippocampal alpha7 and alpha4beta2 nicotinic receptors and working memory, *Neuroscience*, 109, 757, 2002.
53. Levin, E.D. and Rezvani, A.H., Nicotinic treatment for cognitive dysfunction, *Curr. Drug Targets — CNS Neurological Disorders*, 1, 423. 2002.
54. Levin, E.D. and Rezvani, A.H., Development of nicotinic drug therapy for cognitive disorders, *E. J. Pharmacol.*, 393, 141, 2000.
55. Fratiglioni, L. and Wang, H.X., Smoking and Parkinson's and Alzheimer's disease: review of the epidemiological studies, *Behav. Brain Res.*, 113, 117, 2000.
56. Caldarone, B.J., Duman, C.H., and Picciotto, M.R., Fear conditioning and latent inhibition in mice lacking the high affinity subclass of nicotinic acetylcholine receptors in the brain, *Neuropharmacology*, 39, 2779, 2000.
57. Young, J.W., Finlayson, K., Spratt, C., Marston, H.M., Crawford, N., Kelly, J.S., and Sharkey, J., Nicotine improves sustained attention in mice: evidence for involvement of the $\alpha7$ nicotinic acetylcholine receptor, *Neuropsychopharmacology*, 29, 891, 2004.
58. Paylor, R., Nguyen, M., Crawley, J.N., Patrick, J., Beaudet, A., and Orr-Urtreger, A., $\alpha7$ nicotinic receptor subunits are not necessary for hippocampal-dependent learning or sensorimotor gating: a behavioral characterization of Acra7-deficient mice, *Learn Mem.*, 5, 302, 1998.
59. Mesulam, M.M., Guilozet, A., Shaw, P., Levey, A., Duysen, E.G., and Lockridge, O., Acetylcholinesterase knockouts establish central cholinergic pathways and can use butylcholinesterase to hydrolyze acetylcholine, *Neuroscience*, 110, 627, 2002.
60. Xie, W., Stribley, J.A., Chatonnet, A., Wilder, P.J., Rizzino, A., McComb, R.D., Taylor, P., Hinrichs, S.H., and Lockridge, O., Postnatal developmental delay and supersensitivity to organophosphate in gene targeted mice lacking acetylcholinesterase, *J. Pharmacol. Exp. Ther.*, 293, 896, 2000.
61. Duysen, E.G., Stibley, J.A., Fry, D.L., Hinrichs, S.H., and Lockridge, O., Rescue of the acetylcholinesterase knockout mouse by feeding a liquid diet; phenotype of the adult acetylcholinesterase deficient mouse, *Brain Res. Dev. Brain Res.*,137, 43, 2002.
62. Volpicelli-Daley, L.A., Duysen, E.G., Lockridge, O., and Levey, A.I., Altered hippocampal muscarinic receptors in acetylcholinesterase-deficient mice, *Ann. Neurol.*, 53, 788, 2003.
63. Li, B., Duysen, E.G., Volpicelli-Daley, L.A., Levey, A.I., and Lockridge, O., Regulatory of muscarinic acetylcholine receptor function in acetylcholinesterase knockout mice, *Pharmacol., Biochem. Behav.*, 74, 977, 2003.
64. Li, B., Stribley, J.A., Ticu, A., Xie, W., Schopfer, L.M., Hammond, P., Brimijoin, S., Hinrichs, S.H., and Lockridge, O., Abundant tissue butyrylcholinesterase and its possible function in the acetylcholinesterase knockout mice, *J. Neurochem.*, 75, 1320, 2000.
65. Gerlai, R., Gene-targeting studies of mammalian behavior: is it the mutation or the background genotype? *Trends Neurosci.*, 19, 177, 1996.
66. Nadeau, J.H., Modifier genes in mice and humans, *Nat. Rev. Gen.*, 2, 165, 2001.
67. Wolfer, D.P., Crusio, W.E., and Lipp, H.-P., Knockout mice: simple solutions to the problems of genetic background and flanking genes, *Trends Neurosci.*, 25, 336, 2002.

68. Holschneider, D.P. and Shih, J.S., Genotype to phenotype: challenges and opportunities, *Int. J. Dev. Neurosci.*, 18, 615, 2000.
69. Shapiro, M.S., Loose, M.D., Hamilton, S.E., Nathanson, N.M., Gomeza, J., Wess, J., and Hille, B., Assignment of muscarinic receptor subtypes mediating G-protein modulation of Ca2+ channels by using knockout mice, *Proc. Natl. Acad. Sci. U.S.A.*, 96, 10899, 1999.
70. Wang, N., Orr-Urtreger, A., Chapman, J., Rabinowitz, R., Nachman, R., and Korczyn, A.D., Autonomic function in mice lacking $\alpha 5$ neuronal nicotinic acetylcholine receptor subunit, *J. Physiol.*, 542, 347, 2002.
71. Mayford, M., Abel, T., and Kandel, E., Transgenic approaches to cognition, *Curr. Opinion Neurobiol.*, 5, 141, 1995.
72. Phillips, T.J., Hen, R., and Crabbe, J.C., Complications associated with genetic background effects in research using knockout mice, *Psychopharmacology*, 47, 5, 1999.

12 Assessments of Cognitive Deficits in Mutant Mice

Ramona Marie Rodriguiz and William C. Wetsel
Duke University Medical Center

CONTENTS

223

Although most behavioral experiments have been conducted in rats, mice are rapidly becoming the preferred rodent of study in many labs because their genetics are well known, their genome has been sequenced, and they can be genetically manipulated. To date, several different approaches have been used to generate a behavioral phenotype for study.

In "forward" genetics, the analysis proceeds from phenotype to genotype. This is a classical approach, and it includes mice where spontaneous mutations have been identified in certain genes [1] or where mice have been subjected to radiation [2] or chemical mutagenesis [3]. This approach also pertains to mouse strains that have been shown to display a certain phenotype or to animals that have been selectively bred for a given behavioral trait [4–6]. Although forward genetics can provide very interesting animal models, the genetic basis of the abnormal behavior is often obscure. This limitation requires the site(s) of the mutation(s) to be mapped and sequenced [7–9], which can be quite laborious and time consuming. Consequently, many investigators have adopted "reverse" genetics, where the analysis proceeds from genotype to phenotype. Here, a specific gene is targeted for disruption or modification, and the mutants are evaluated for behavioral abnormalities. Investigators typically employ transgenesis to produce either gain of function through expression of hybrid genes and duplication of endogenous genes or loss of function by expressing dominant-negative hybrid genes, toxic genes, or disrupting endogenous genes [10–14]. These gene-targeting approaches in embryonic stem cells, or in one-cell embryos, may lead to alterations in expression of other members of the same gene family, with behavioral compensation occurring during development and adulthood [15]. This developmental compensation is a common criticism of transgenic experiments. However, it should be emphasized that such compensation is rarely the basis of study in mutant mice *per se*, and there are many incidences where compensation by nonmutant family members does not appear to contribute to the phenotype [16]. Nevertheless, to obviate this criticism of developmental compensation, some investigators have begun using systems that induce or suppress expression of specific genes at certain ages or within a given brain region [17–19]. More recently, reduction in gene expression *in vivo* has been accomplished through the introduction of RNA interference that targets a specific RNA species [20]. Although this approach does not completely suppress expression of the target gene, it can reduce it (80%) to levels sufficient to produce quantifiable biochemical and behavioral changes.

Together, forward and reverse genetic approaches have provided important insights into the roles that selected genes play in the composition of a given behavioral phenotype. Most of these approaches have some limitations because behaviors in humans are controlled not by a single gene, but by many genes interacting in concert with the environment. Analyses are currently proceeding where (a) qualitative trait loci in mice are identified, (b) mice with known genetic mutations are outcrossed to other mutants, or (c) mice with known genetic backgrounds are exposed to differing environmental conditions [21–23]. This multitude of approaches with mice has and will continue to yield novel insights into the genetic and molecular antecedents that affect behavior.

PRELIMINARY PHENOTYPIC SCREENING

Behavioral assessment in mice has come a long way in the past century with increasing refinement of paradigms [24–27]. The behavioral phenotype of mice is complex; consequently, these animals should first be evaluated on numerous behavioral domains that include the animal's general health and well-being, reflexive and motor capabilities, emotionality, anxiety, affective and social behaviors, consummatory responses, and learning and memory. Partially due to the use of mutant mice, the past decade has witnessed numerous advances in identification of molecular mechanisms that underlie a wide variety of behaviors in animals. Over the years, learning and memory processes have received the greatest emphasis for study. Investigators have produced many strains of mutants to examine the roles of different genes in cognitive behavior. However, the focus on this domain of behavior neglects the other possible responses for these animals, some of which may influence performance on cognitive tasks. For instance, N-methyl-D-aspartate (NMDA) receptors have been shown to play an important role in hippocampal long-term potentiation [28], and administration of a NMDA receptor antagonist into the lateral cerebral ventricles impairs spatial memory [29]. Indeed, disruption of the NR1 subunit gene of the NMDA receptor is lethal [30, 31]. On the other hand, mice with targeted disruption of this gene in the CA1 region of the hippocampus survive but show severe impairment in spatial learning and memory [32]. Knockdown of the NR1 subunit also produces a schizophrenia-like phenotype [12]. Hence, it is important that investigators explore the behavioral phenotypes of their mice more completely in the course of cognitive testing, as additional behavioral deficiencies may contribute to the impairments in learning and memory.

To deal with this concern, we have designed a broad series of tests that are administered to all mice brought for testing into the Mouse Behavioral and Neuroendocrine Analysis Core Facility at the Duke University Medical Center. These assessments include basic tests of sensory and motor function, neurophysiological status, and emotionality. These tests can be conducted in the week before beginning cognitive testing (Table 12.1). In our experience, we have found this test battery to be informative because results from these tests often provide clues as to the proper control experiments that should be run for subsequent investigations [13, 33, 34]. Additionally, results from this initial behavioral screen often allow us to select additional behavioral domains for further study. For the purposes of the present chapter, cognitive behavior will refer to the ability of the mouse to acquire, process, store, retrieve, and act upon information gathered from the environment. The techniques and observational methods necessary to investigate cognitive function in mice will be examined and discussed. Given that the history of cognitive testing in the mouse encompasses eight decades of research, it is far too ambitious to describe all of the cognitive tests presently used for mice. Instead, we will concentrate upon tests that we have found to be useful in the Mouse Behavioral and Neuroendocrine Analysis Core Facility at Duke University Medical Center.

TABLE 12.1
Preliminary Behavioral Screening Prior to Cognitive Assessment

Order[a]	Test	Duration	Purpose	Reference
1st	Zero maze	5-min, single test	Anxiety	[35, 37]
2nd	Open field	30 min to 3 h, single test	Motor performance, anxiety	[37]
3rd	Neurophysio-logical screen	15–25 min, single screen of 36 discrete tests	General health, reflexes, motor coordination and strength, reactivity	[13, 33, 34]

[a] Each test should be separated by 36 to 48 h.

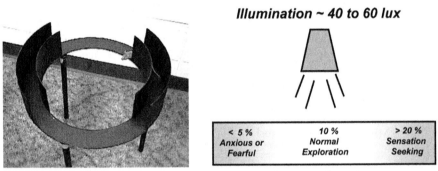

Illumination ~ 40 to 60 lux

| < 5 % Anxious or Fearful | 10 % Normal Exploration | > 20 % Sensation Seeking |

% Time in Open Arms

FIGURE 12.1 Zero maze for testing anxietylike behaviors. The zero maze is elevated approximately 18 in above the floor, and it consists of two open and two closed areas. Mice are permitted free exploration of the maze for 5 min. Illumination of the maze is critical and should be maintained between 40 to 60 lux. Anxious mice remain in the closed areas and do not venture onto the open areas of the maze. The percent time spent in the open areas is taken as an index of anxiety in the mouse.

ANXIETY

The first test that we typically administer to mice first entering the Core Facility is the zero maze. The elevated zero maze was initially described and pharmacologically validated for anxiety-like behaviors in mice by Shepherd and colleagues [35]. The maze is elevated approximately 18 in above the floor and consists of a circular runway divided into two open and two closed quadrants (Figure 12.1). A camera is mounted directly over the apparatus to allow video-taping of behavior for later analyses with programs such as the Noldus Observer or Ethovision (Noldus, The Netherlands). Lighting for the maze should be indirect and even, but between 40 to 60 lux when a light meter is placed level to the surface of the open arm. The apparatus and mouse are shielded from the observer and any other activity in the room by a wall or curtain. In our testing facility, we have found that the best and most reliable results are obtained when the animals are brought into the room several hours before testing and the room remains undisturbed until completion of testing. The mouse is transported to the maze in a solid-bottom container (pipette box) and is placed into the closed arm of the maze facing the wall. The animal is allowed 5 min to freely explore the apparatus. Behavioral measures include the amount of time spent in the open areas, number of entries into the open areas, and the frequency of traversing from the closed area through the open area to the opposite closed area. In addition, the frequency of head dipping over the edge of an open area, rearing, and stretch-attend postures (rear feet in closed arm but body stretches forward such that the front paws are in the open area) are scored, and these ethological behaviors signify measures of "risk-assessment" or exploration by the animal. We also score the time the animal spends freezing and grooming in the apparatus. Animals with anxietylike phenotypes typically exhibit a low propensity to explore the open areas relative to control animals, and may also demonstrate reduced frequencies of the risk assessment or exploratory behaviors with or without enhanced freezing or grooming [36, 37].

In the event a given mouse genotype presents an anxietylike phenotype, responses to anxiolytics are examined, and the animal is evaluated in other tests of emotional responses that include assessments of "depressivelike" behaviors, social interaction, drug abuse, and fear conditioning.

OPEN FIELD

Several days following zero-maze testing, spontaneous activities of the mice are examined in the open field for 1 hour. Many types of open fields are commercially available for mice, and they typically involve a large arena that can be divided into smaller areas. The lighting across the floor of the arena should be even, and we typically test our mice at 350 to 600 lux [37]. Both horizontal (locomotion) and vertical activities (rearing) are monitored by infrared beams, and this information is relayed to a computer with software that automatically records the location and activity of the animal. Behavioral output includes the numbers of horizontal beam-breaks or the distance traveled and the number of vertical beam-breaks or rears. In addition, activities in the center and peripheral zones of the open field as well as activity maps can be generated. Open-field testing typically begins by placing the animal into one of the four corners of the arena and allowing the mouse to freely explore the arena. Overall reduced activity in the open field may be indicative of motor impairment or weakness, particularly if little or no rearing behavior is observed. Alternatively, low activity can also indicate increased anxiety or neopho-bia. These latter behaviors may be associated with reduced time spent in the center of the open field and increased time spent in the corners or along the perimeter of the arena. Analyses of the behaviors in the open field can be quite informative in planning subsequent testing, particularly when motor disturbances or ataxia may be present [16, 38, 39] or when monoaminergic function may be perturbed [13, 37].

NEUROPHYSIOLOGICAL SCREEN

The neurophysiological screen for mice is a series of very short tests used to assess several dimensions of neurological functioning and behavior that include sensory and motor function, autonomic reflexes, emotional responses, and rudimentary cognition [13, 16, 27, 34, 40]. Although procedures may differ among laboratories, most tests are based upon the original behavioral screen described by Irwin [41]. The tests are conducted in five phases that consist of 36 discrete measures (Table 12.2) scored on a multiple point scale (Table 12.3). In addition to the tests from the neurophysiological screen, our laboratory also evaluates grip strength by automated meter. Although our neurophysiological screen is a comprehensive test battery, it is not exhaustive, and additional tests can be incorporated into the screen (depending upon the mutation) or be administered later. For example, Abi2 homozygous mutants (knockout [KO] mice) have eye defects [33]. Since many tests for cognitive performance in mice rely upon vision, it was important to conduct a detailed neurophysiological examination of vision to determine whether the KO mice respond to light, have depth perception, and can discriminate patterned stimuli. Pupillary responses

TABLE 12.2
Phases and Tests of Neurophysiological Screen

Phase	Test	Scoring Criteria
Initial evaluation	Skin color	Vascularity of footpads: ruddy to pale
	Body tone	Soft, normal, firm
	Lacrimation	Discharge from the eyes
	Tremor	Presence of shaking or shuddering
	Convulsions	If present: mild, moderate, severe
	Heart rate	Palpated: slow, normal, fast
	Respiration rate	Palpated: slow, normal, fast
General assessment	Body posture	Four-footed posture, slight curve to back, pelvis and tail at normal elevation (2–3 mm from observation surface)
	Piloerection	Fur standing on end
	Palperbral closure	Eye opening well formed without drooping appearance
	Exopthalamus	Slight "eye-popping" appearance
	Barbering	Bald spots from excessive grooming by cagemates; whiskers may be absent
	Tail elevation	Tail maintained in horizontal position 2–3 mm above observation surface
	Pelvic elevation	Pelvis elevated when in four-footed posture but also not elevated more than 3 mm above observation surface
	Stereotypic or unusual behaviors	Includes excessive jumping, self grooming to point of hair loss or mutilation, spontaneous vocalizations when touched, tonic immobility, twirling behavior
	Temperature	Core body temperature
	Weight	Weight (g)
	Length	Head-to-rump length, leg length
Orienting and reflexive behavior	Visual orientation	Orients toward and tracks a small mirror passed through mouse's visual field
	Visual placement	Mouse is lowered to grid, and height at which the mouse reaches for grid is recorded

TABLE 12.2
Phases and Tests of Neurophysiological Screen (continued)

Phase	Test	Scoring Criteria
	Whisker stop	Whiskers touched with swab orients animal, whiskers stop moving
	Whisker reflex	Whiskers brushed with swab produce twitch
	Eye reflex	Blinking response to touch with swab
	Ear (pinna) reflex	Ear-flick response to touch with swab
	Postural adjustments	Maintains upright posture while observation cage is moved in either a horizontal or vertical plane
	Righting reflex	Self-rights in a Plexiglas tube rapidly rotated
	Startle reflex	Visibly startles following air-puff to the face
Grip strength and coordination	Forepaw reflex	Automatically grips grid when pulled away
	Rear-paw reflex	Automatically grips a 3-mm wire when touched on rear paws
	Rear-paw coordination	In rear-paw reflex, grip is simultaneous for both paws
	Grip strength	Difficultly in removing mouse from grid or 3-mm wire in above tests
Motor coordination and balance	Pole climb down	Latency, duration, difficulty in climbing down an elevated vertical pole
	Pole climb up	Latency, duration, difficulty in climbing up an elevated vertical pole
	Pole walk	Latency, duration, difficulty in walking across an elevated horizontal pole
	Wire hang	Duration in clinging to a suspended 3-mm wire; coordination of paws is also noted
	Stride test	Mouse is footprinted and stride measurements are determined through a series of footprint indices previously described [16]

TABLE 12.3
Scoring for Neurophysiological Screen

Score	Response
−6	No response or feature is absent
−4	Moderate deficiency or only present on one side of the body
−2	Mild deficiency or response is weak/delayed
0	Normal response
+2	Slightly exaggerated response, normal but quick, or rapidly repeated to a single stimulus
+4	Moderately exaggerated response or exaggerated response only present on one side
+6	Appearance is severely exaggerated

to light were first evaluated. The mouse was held by hand and a 1.13-W pen light was shined into the eye for 3 sec, followed by a 5-sec intertrial interval over three trials. The mice were filmed with a high-resolution video camera, and the film segments were digitized (Dazzle Video Creator, Pinnacle Solutions, Mountain View, CA) and analyzed frame by frame for each animal's response. No differences between wild type (WT) and KO mice were observed. We next tested responses to light and dark by placing the mice in a passive-avoidance apparatus. No genotype differences were noted in the latency for the mice to leave a lighted compartment and enter the darkened chamber, or in the time spent in each of the chambers. These data suggested that both genotypes could discriminate light from dark and that levels of emotionality were similar between the animals. To examine depth perception, we slowly lowered the mice to the bench top. Both WT and KO mice extended their forepaws upon being lowered to the lab bench. Finally, we tested whether the mice could track a moving object within the visual field. As no genotype differences were discerned in any of the tests for vision, we concluded that rudimentary vision was not impaired in these KO mice.

COGNITIVE TESTING

Task selection is perhaps the most crucial decision an investigator can make. However, with the multitude of cognitive paradigms available for mice and the availability of mutant animals, it becomes important for the experimenter to decide *a priori* which tests are the most suitable for study. As the behavioral phenotype of many mutants may be heterogeneous, it is critical to examine multiple aspects of cognition that cover different domains of functioning, including preattention and attention, and various aspects of learning and memory (Figure 12.2). Tests of preattentive functioning have been described for mice [42, 43], and most utilize a simple testing paradigm called prepulse inhibition (PPI). Additional paradigms include simple screens using object discrimination tests [37, 44] or more complex paradigms such as go/no-go testing [45–46], five-choice serial attention tasks [47], or latent inhibition [48, 49]. Finally, tests of learning and memory can be designed to assess more specific areas of functioning, including associative learning, nonspatial or spatial learning, short- and long-term memory, as well as neurologically specific deficits as revealed by fear or eyelid conditioning. As testing across multiple cognitive domains is preferable and because the numbers of mice available for testing may be limited, the investigator may need to adopt two different strategies. First, he/she should be mindful of the order in which the behavioral tests are administered, especially if they are given in series. Under a multiple-test regimen, the least stressful tests are conducted first. Second, once behavioral deficits have been identified, it is important to replicate these results in naïve mice so that prior test experience can be excluded as a confounding variable.

PREATTENTIVE PROCESSES

Cognitive performance is enhanced if animals can focus their attention on the most salient information in the environment [50]. Inability to filter information is thought

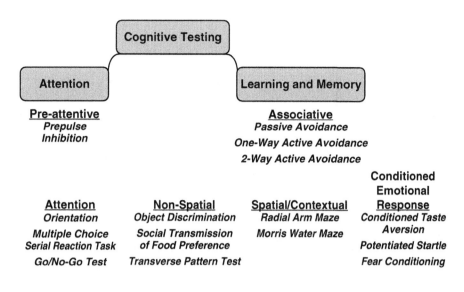

FIGURE 12.2 Cognitive testing for the mouse. There are many cognitive paradigms available for assessment of mouse cognition. As the phenotype of the mutant may be heterogeneous, several tests may be required to identify deficits.

to promote sensory overload and cognitive fragmentation [51], which are thought to contribute to cognitive impairment in several different psychotic disorders [52]. This filtering process is termed sensorimotor gating and is evaluated by PPI. This task refers to the ability of a weak stimulus to reduce the magnitude of response to a subsequently stronger stimulus (Figure 12.3). In the typical auditory PPI paradigm, a prepulse that is 4, 8, or 12 dB above the background noise is presented prior to exposure to a stronger stimulus that reliably elicits a startle response. The auditory prepulse inhibits the magnitude of the startle response and, as the intensity of the prepulse stimulus increases, inhibition of the startle response becomes more enhanced [53]. A distinct advantage in using PPI is that both animals and humans can be evaluated [43]. More importantly, PPI deficiencies are evident in many psychiatric disorders, including schizophrenia [54], thereby rendering it a replicable test for disturbances in preattentive functioning [55].

To date, most PPI studies in mice use a limited range of test parameters. In the standard test, broad-band white-noise stimuli are continually present to provide a stable background. Prepulse acoustic stimuli are presented 4 to 18 dB over this background, and they typically precede the 100- to 120-dB startle stimulus by 60 to 140 msec [54, 56]. Most studies in rats and mice, however, have used the magnitude of inhibition of PPI responses as a single response index. More recently, some investigators have examined the temporal properties and saliency of the prepulse stimuli in rats with disrupted dopaminergic tone [56, 57]. In our laboratory, the PPI paradigm consists of 20-msec prepulse acoustic stimuli that are 4, 8, or 12 dB above a 64-dB white-noise background. The prepulse precedes the 40-msec 120-dB startle stimulus by 100 msec. At the beginning of the experiment, the mouse is

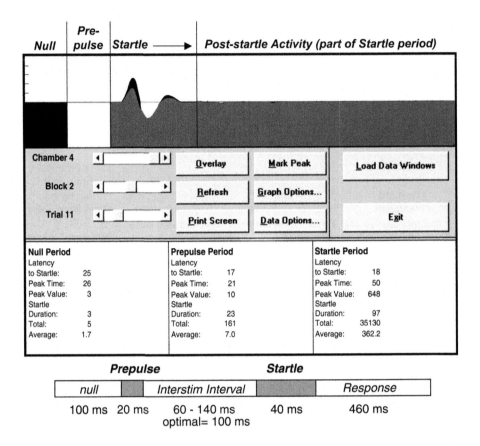

FIGURE 12.3 Prepulse inhibition. Presentation of a weak auditory prepulse inhibits the response to a subsequent stronger stimulus. Testing is typically conducted with an interstimulus interval between prepulse and startle stimuli of 100 msec. By manipulating this interstimulus interval, dynamic shifts in sensorimotor processing can be investigated [57]. The startle response is shown in black in the top portion of the figure, and the reduction in this startle response due to the presentation of the prepulse stimulus is shown in gray. Data are shown for a single test trial as recorded by the Med Associates (St. Albans, VT) PPI software.

placed in a circular Plexiglas tube (the animal can turn around) and is acclimated to the apparatus for 5 min. Thereafter, it is given 64 test trials separated by an intertrial interval of 8 to 20 sec. Testing commences with ten startle-only trials followed by combinations of the three prepulse trials, six startle-only trials, and eight null trials (no startle or prepulse stimuli) in a pseudorandom order; testing is completed with ten startle-only trials. PPI provides a good assessment of preattentive functioning, and PPI performance can alert the investigator to deficits in sensorimotor gating that might impact performance on subsequent cognitive tests.

ATTENTION

Attention is often defined as a heuristic concept, involving many cognitive dimensions, and it can only be evaluated through multiple test procedures [58]. Attempts have been made to classify attention into various categories including, but not limited to, reflexive attention [59, 60]; visual orientation [50]; learned orientation [61]; vigilance [62]; habituation [63]; or selective [64], sustained, and divided attention [47]. In theory, each process can be tested independently of each other; however, in practice most studies examine attention across several dimensions simultaneously [47]. For example, no single behavior executed by an animal can be labeled as "vigilance." Instead, vigilance tests such as the continuous performance task (CPT) require the animal to maintain vigilance, orient toward the visual stimuli, scan the stimuli for change, and make a response to the proper stimulus [62]. The animal must execute all of these behaviors so that "vigilance" can be assessed. Hence, attention subsumes many processes, and these aspects of behavior cannot always be easily dissociated for study.

Orienting Responses

The simplest tests of attention involve orientation. Although tests of visual orientation are the most common [60, 63], tests have also been developed using tactile or auditory stimuli. Visuospatial orientation provides one of the most effective means to study elementary forms of information processing in animals [65]. Currently, a great deal is known about processes of visual perception [66] and the relationship of this behavior to other forms of attentional processing. Another reason for interest in orientation is due to the fact that inefficient saccades, smooth eye movements, and orienting responses have been linked to deficits in information processing for patients diagnosed with schizophrenia and schizotypical disorders [67], anhedonia [68, 69], depression and anxiety [70], obsessive-compulsive disorder (OCD) [71], attention deficit hyperactivity disorder (ADHD) [72], and Parkinson's disease [73]. In addition, neural circuits that regulate this response are fairly well described on an anatomical [63], electrophysiological [74], and neurochemical basis [58, 59]. As orientation across different species is similar, this test can provide important insights into attentional processes.

In species other than the rodent, visual orientation utilizes fovea tracking [50, 75]. This task involves cuing the animal or human to a particular position in space prior to the presentation of a visual stimulus. When the visual stimulus is presented following the cue, information about the stimulus is processed more efficiently. This improved processing is interpreted as being due to directing attention to a specific location [50]. Unfortunately, fovea tracking cannot be easily assessed in rodents with regard to visuospatial orientation, since they have a natural propensity to orient their head or whole body toward a novel stimulus [61, 63]. Several methods have been developed in rats and mice to measure the visual orienting response. The simplest of these techniques involves bringing a single object into the field of vision and observing the ability of the animal to track the object (Figure 12.4). In our laboratory we conduct a simple screen for visual orientation by placing the mouse

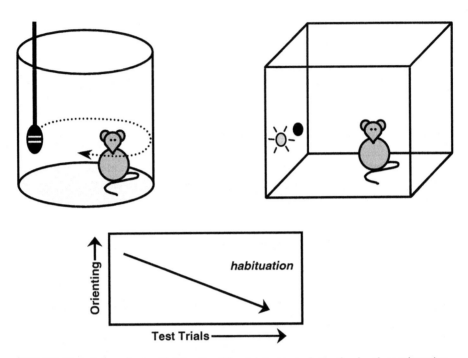

FIGURE 12.4 Orientation and habituation. The simplest test of attention involves orientation, which can be examined by visual tracking (top left) or various operant techniques (top right). In each case, a visual stimulus is presented over 10 to 20 trials, and the physical orientation of the mouse toward the object is recorded. As the number of presentations increases, the number of orienting responses decreases until the animal has habituated to the object. Modifications of the procedures allow for the assessment of anticipation or sustained attention.

in an enclosed circular arena. After 2 min of habituation, a small round object (back of an infant feeding spoon) is moved horizontally at eye level around the perimeter of the arena for 12 sec. Each presentation of the spoon constitutes a test trial, and during the trial, the orienting responses of the mouse are coded at either 1- or 2-sec intervals. Orientations are defined as the mouse orienting its face toward the spoon and tracking the spoon as it moves around the perimeter of the arena, such that the animal approaches, grabs, or manipulates the spoon as it is moved. Twenty presentations of the spoon are usually sufficient to induce habituation. A variation of this rapid screen involves adding an additional ten test trials following habituation where the mouse is randomly presented with either the familiar spoon or a new object of a different color or visual pattern. Introduction of the novel object typically reinstates the orienting response toward the object, but it does not significantly alter the habituation of the mouse to the familiar object.

Orienting responses to visual stimuli can also be assessed by operant techniques (Figure 12.4) [76]. In these paradigms, the rat or mouse orients toward a light cue that is paired to a particular sequence of auditory tones or clicks. The orienting response is simply defined as the animal pointing its face and snout toward the source of the visual stimulus. Habituation represents a loss of this response over testing, and it can be used as an index of novelty recognition and basic learning [76]. Moreover, the orienting response can be reinstated following habituation if a novel stimulus is introduced or paired with the familiar pattern of stimuli, creating a mismatch signal.

Modifications of the orienting response can be used to access different aspects of attention, including anticipation (which increases response efficiency during testing) and sustained attention. Anticipation is measured by delivering an auditory cue before presentation of a light cue. If the mouse learns to orient toward the location of the light cue before its illumination, then subsequent responses to lever press or nose poke as signaled by the light cue will become more efficient and rapid. Sustained attention can also be examined by increasing the time between the predictive auditory cue and illumination of a nose-poke aperture. This procedure allows the investigators to measure how long a mouse will sustain visual orientation toward the aperture in the absence of the light cue. Both anticipatory and sustained attention are important indicators of executive control and self-regulation [77, 78], and these processes may be deficient in patients diagnosed with schizophrenia [79], Tourette's syndrome [80], and ADHD [81]. Thus, assessments of attentional control constitute an additional, but important, component of orienting-response testing for mice.

Studies using lesions in rodents or neuroimaging in primates and cats have identified a limited number of brain regions that regulate the visual orienting response. These areas include the superior colliculus, and the occipital, parietal, and frontal cortex [62–64, 82]. The superior colliculus and parietal-frontal sensorimotor cortex appear to be especially involved in the regulation of visual orientation, as lesions or brain injury of these areas produce a loss in the ability to shift visual attention [83]. The neurotransmitter systems most closely associated with forebrain control of orienting responses are acetylcholine and norepinephrine [61, 62, 84]. Interestingly, response amplitudes of noradrenergic neurons in the locus coeruleus are enhanced when the stimuli are novel rather than familiar [62].

Another brain region implicated in mediating the visual orienting response is the amygdala. While damage to the central nucleus of the amygdala does not affect spontaneous orienting responses or their habituation, this brain area appears involved in reinforced or learned orienting responses [61]. The amygdala can exert indirect influences on orientation through the dorsolateral striatum via input to the midbrain dopamine neurons [61]. Although tests of visual orientation cannot provide definitive answers concerning the mechanisms of attention, these paradigms may be useful as a first level of screening to determine whether orientation deficits are evident and whether further investigations into more specific aspects of attention and information processing are required.

Multiple-Choice Serial-Reaction Test

This test is similar to human continuous performance tasks [85] where the individual scans an array of objects that are briefly presented. Correct selection of the location where the visual target was presented results in a reward. The five-choice serial-reaction time task (5-CSRTT) is a test of attentional performance and vigilance

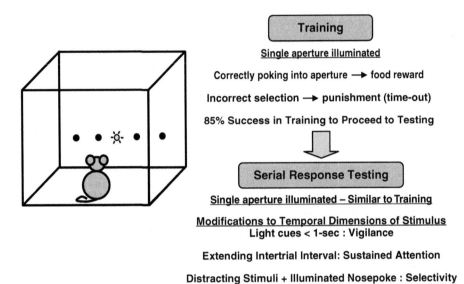

FIGURE 12.5 Multiple-choice serial-reaction test of attention and vigilance. Measuring the accuracy and speed of nose poking into an illuminated aperture during the multiple-choice serial-reaction test has been used to assess attentional capacity. Inhibition of inappropriate responses can be used as a measure of attentional control. Behavioral impulsivity (nose poking before illumination of the aperture) or perseveration (repeated poking into an aperture upon completion of a test trial) can also be assessed.

where the mouse is required to simultaneously monitor three to five locations for the presentation of a brief visual stimulus [47] (Figure 12.5). The testing chamber typically consists of a square box where, on one side, five nose-poke apertures with lights are positioned in a horizontal line equidistant and at eye level to the mouse. On the wall directly opposite the apertures is a small opening for the delivery of food reward. The nose-poke array consists of the five apertures, where one of them is randomly illuminated. At the start of training, an aperture is typically illuminated for 20 to 30 sec, and this duration is reduced to approximately 2 to 4 sec as the mouse becomes more proficient at the task. Correct responses involve nose poking into the lighted aperture; nose poking into nonilluminated apertures results in a time-out, where the nose-poke apertures and house lights are extinguished for 5 to 10 sec beyond the duration of the 5-sec intertrial interval. The accuracy and speed of nose poking, taken as measures of attentional capacity, and the inhibition of inappropriate responses, taken as a measure of attentional control (e.g., impulsivity, perseveration), are used as core measures in this test [86].

Before testing commences, the mouse is placed on food restriction and is maintained at 90% of its free-feeding weight throughout the experiment. Following food magazine training, animals are trained to nose poke for a food reward by randomly illuminating one of the nose-poke apertures. When the mice reach a criterion of 85% success over three consecutive days, serial response testing begins. The first trial of each test day is identical to that used in training, where the mice have 60 sec to make an appropriate response to the single illuminated aperture. Test trials are scored as successes or failures, and responses in each trial can be scored as impulsive/premature or perseverative. With regard to test trials, successful trials are scored when the mouse correctly nose pokes into the illuminated aperture. Failures are recorded if the animal fails to nose poke or responds by nose poking into a non-illuminated aperture. For each trial, responses are recorded as impulsive if the mouse begins nose poking before the light cue is presented or if the animal head-pokes into the food magazine before food delivery. By contrast, perseverative responses refer to behaviors that the mouse emits repeatedly within a single test trial without any reinforcement. The scoring of impulsive or perseverative responses is important during multiple-choice serial-reaction tests, as these behaviors are postulated to reflect a loss of executive control or self-regulation that is thought to contribute to deficits in attention [81].

Modifications in temporal dimensions of the stimulus cues can increase attentional load, thereby providing assessments of vigilance and sustained attention. For example, shortening the light cue to 1 sec or less will create a situation where the mouse must continuously monitor all five nose-poke apertures in order to detect the brief presentation of the cue. In addition, by extending intertrial intervals under this procedure, a maximum time limit will be reached that the mouse can sustain attention toward the five nose-poke apertures. After this time is reached, performance deteriorates. Finally, presentation of distracting stimuli such as bursts of white nose or flashing house lights either concurrently with or prior to the visual cues can be used to test attention selectivity. Animals that can focus and selectively attend to the relevant cues will exhibit higher success rates over testing compared to mice that switch attention to the distracting stimuli and, consequently, fail to notice illumination of the nose-poke aperture.

The ability of the mouse to perform various aspects of the 5-CSRTT depends upon neurological integrity of the prefrontal cortex and upon certain monoaminergic and cholinergic pathways [86]. For instance, lesions of the anterior cingulate and prelimbic cortices impair selective attention by reducing choice accuracy of the mouse. By contrast, lesions of the postgenual anterior cingulate or the infralimbic cortex promote impulsivity without impairment in other measures of attention. Prelimbic cortical lesions produce not only deficits in attention, but they also increase perseveration, particularly when the duration of the target stimulus is reduced. These findings led Dalley and colleagues [86] to suggest that attentional selectivity seems to be controlled primarily by dorsomedial areas of the prefrontal cortex, whereas ventral or lateral areas appear responsible for inhibitory control. Aside from specific brain areas, certain neurotransmitter systems also contribute to various aspects of performance. For example, lesions of ascending cholinergic pathways impair the ability of the mouse to discriminate among or between stimuli, and this debility is especially evident when the load on attention is increased by shortening the duration of the target stimulus or by the presentation of distractors [87]. Destruction of noradrenergic projections to the frontal cortex impedes accuracy of attention [62, 88], whereas depletion of both norepinephrine and dopamine from the medial prefrontal cortex results in selective attention losses that can be corrected by amphetamine [64]. Lesions of the dorsal raphe nucleus resulting in reduced serotonergic

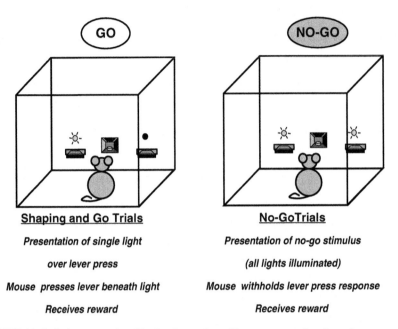

Shaping and Go Trials	No-Go Trials
Presentation of single light over lever press	Presentation of no-go stimulus (all lights illuminated)
Mouse presses lever beneath light	Mouse withholds lever press response
Receives reward	Receives reward

FIGURE 12.6 Go/no-go testing. Testing is conducted in an operant chamber where a mouse is presented with either a go stimulus, where lever pressing is rewarded, or a no-go stimulus, where the animal must learn to suppress the lever-pressing response to earn a reward. The test is used to assess attention, impulsivity, and perseveration.

tone in the neocortex and striatum lead to an enhancement in impulsive responses [45]. Together, these results show that certain aspects of attention are regulated through neural pathways that project to the forebrain [86].

Go/No-Go (GNG) Testing

This test is a variant of the stop-signal reaction time test that utilizes a recognition task where reaction time is analyzed (Figure 12.6). Briefly, the test can include only "go" trials or can include combinations of "go" and "no-go" trials. On a "go" trial, the animal is presented with a stimulus (e.g., light) and it must respond to this stimulus with a lever press or a nose poke. This response is rewarded. For "no-go" trials, a signal such as a tone or flashing light is presented prior to the "go" cue, and the animal must learn to withhold its response to receive reinforcement. Presentation of the "no-go" and "go" stimuli in succession within the same test trial creates a "familiarity-based conflict" and provides a context where error detection can be examined by comparing correct and incorrect responses made by the animal [89, 90].

GNG testing is usually conducted in operant chambers equipped with stimulus lights positioned above two lever presses or with two illuminated nose-poke apertures. In our laboratory we have obtained the best success with mice when the levers or nose-poke apertures are situated at eye level to the mouse and mounted on either side of the food magazine. Food rewards include a magazine that dispenses sucrose pellets (20 mg) or a liquid dipper system that delivers a small volume of sweetened condensed milk. Mice are maintained on food restriction at 90% their free-feeding weights. Testing is conducted over three phases: shaping, "go training" where the mouse learns to respond to a "go signal," and GNG testing where the animal learns to withhold its response when presented with a "no-go signal." During shaping, mice are trained to press a lever or nose poke for a food reward. Once animals are trained to feed from a food magazine and reliably lever press or nose poke for reward under a continuous reinforcement schedule, they next receive training on the "go test" for at least five consecutive days. This test consists of 40 trials/day. Training begins with the presentation of the "go signal," which is a light illuminated either directly over a single lever or within a nose-poke aperture. Delivery of reward occurs when the mouse responds to the appropriate cue within a certain period of time (e.g., 60 sec). The animal then has a limited amount of time to locate and consume the food reward (20 sec). Failure on either part of the task results in a time-out (10 sec) before the intertrial interval (10 sec) is imposed. Once mice reach a criterion of at least 85% success over three consecutive days, they are introduced to GNG testing. In this phase, mice receive 30 daily trials of "go testing" randomly interspersed with 30 trials of "no-go testing" over ten days. "No-go" trials are usually signified by a tone or flashing light, and the mouse must withhold its response (15 sec) to receive a food reward. If the animal responds with a lever press or nose poke during or following presentation of the "no-go" signal, the trial is terminated and recorded as a failure, and a time-out is imposed without the delivery of reward.

In a variant of the "no-go" trials, the "no-go" signal can be presented following the "go stimulus." Under this paradigm, vigilance and impulsivity can be simultaneously assessed by varying the length of time between the "go" and "no-go" signals, and the maximum "withholding" time achieved by the animal can be assessed [91]. Another variant of the test includes lengthening the amount of time the mouse is required to withhold its response before a food reward is delivered. Animals that demonstrate difficulty in withholding a response for a certain period (e.g., 15 sec) should also be tested at shorter intervals (2 to 10 sec) to determine whether the deficit is due to an inability to withhold its response for an extended period or an inability to acquire the no-go task.

Behavioral performance in both "go" and "no-go" testing is assessed by determining the percent successful trials and the average latency to make a correct behavioral response. The data can also be analyzed with signal-detection methods [92]. Incorrect responses in GNG testing may result from impulsive or perseverative responses. Impulsive errors include repetitive lever pressing or nose poking during the intertrial interval before onset of the next trial or repeated head entry into the food magazine when the lever lights or nose-poke apertures are lit. Perseverative errors consist of the mouse repeatedly pressing a lever or nose poking into an aperture following reward delivery or persistent head entries into the food magazine following retrieval of reward.

Although the GNG test has been classified as a nonspatial recognition memory test [93], it can also be used to assess attention dysfunction and impulsivity [94]. Accordingly, GNG testing has been used to model neurological and behavioral deficits in attentional and inhibitory control found in patients with ADHD [95].

Studies in both rats [91] and humans [96] reveal that lesions of the basal forebrain and other cortical areas impair "no-go" responses while leaving "go" responses intact, suggesting that deficits observed in GNG testing could be attributable to cognitive impulsivity [81] and cognitive control [96]. Both processes appear to be regulated in the anterior cingulate cortex, a brain region attributed to conflict monitoring [97]. The frontal cortex and striatum also contribute to cognitive control [91, 98–101]. Studies in monkeys have revealed that reducing midbrain dopamine activity can impair responding on "no-go" trials and hinder the ability of the animal to switch responding between "go" and "no-go" trials, resulting in a high rate of perseverative responses during "go" trials [102]. Perseverative responses leading to deficits on "no-go" trials are also attributed to central serotonin depletion, even when levels of dopamine and norepinephrine are not appreciably changed [45]. The deficits in responding following depletion are very difficult to overcome, as this deficiency persists even when incorrect responses in "no-go" trials are punished with foot shocks [103]. Interestingly, serotonin depletion of the prefrontal cortex results in increased perseverative responses to a rewarded stimulus and an inability of the animals to withhold its response during reversal learning, even while other executive tasks remain intact [104]. Similar results have been shown in human studies with tryptophan depletion [96]. In this regard, it is important to note that cognitive perseveration and inflexibility are associated with reduced prefrontal activity in patients diagnosed with ADHD [96], schizophrenia [105], and OCD [106].

Latent Inhibition (LI)

LI is a test that examines selective attention while also addressing aspects of learning and memory. LI refers to the retardation of conditioning to a stimulus due to its prior repeated nonreinforcement [49]. LI testing consists of two phases. In the first phase, animals are divided into two groups: one receives multiple presentations of a stimulus in the test apparatus, while the other is merely exposed to the apparatus. In the second phase, the previous stimulus is associated with an aversive event. Typically, LI is evident when animals pre-exposed to the stimulus, as compared with those only exposed to the test apparatus, exhibit a delay in acquiring the conditioned response. Strategies for assessing LI include the use of appetitive behaviors with conditioned suppression [107] or the use of aversive procedures such as foot-shock avoidance [47, 49, 108] or conditioned fear [109].

In our lab, animals are divided into two different groups. One group is placed into a two-chambered shuttle apparatus and allowed to explore it for one hour. A second group is placed in the apparatus for the same period of time and is exposed to an intermittent tone and/or light as the conditioned stimulus (CS). The next day, mice are tested in shuttle avoidance, where the CS predicts foot shock. If the mouse exits the chamber during the 8-sec tone, the trial is scored as an avoidance response. If it leaves the chamber after the initial 8 sec and during the subsequent 8-sec presentation of tone and foot shock, the behavior is scored as an escape. Alternatively, if it does not exit the chamber at either time, the trial is scored as a failure. Using this system, "learning curves" can be constructed across time so that the avoidance, escape, and failure responses of the mice can be followed in detail.

In an early study, rats were administered 6-hydroxy-dopamine into the dorsal noradrenergic bundle to induce norepinephrine depletion in the forebrain [110]. LI was found to be deficient in these animals. Later studies revealed that ablation of hippocampus prevented LI and that this effect was primarily mediated through the nucleus accumbens [111]. Interestingly, N-methyl-D-aspartate (NMDA) lesions extending from the entorhinal cortex to the ventral subiculum prevented LI, and these effects were reversed by systemic administration of haloperidol [112]. Aside from this pathway, more-recent studies have revealed that lesions of the basolateral amygdala produce persistent LI and these effects, including those induced by lesions of the nucleus accumbens, are ameliorated with atypical antipsychotics [113] — possibly through antagonism of 5-HT$_{2A}$ receptors.

LEARNING AND MEMORY

ASSOCIATIVE LEARNING

Investigations of associative learning are usually conducted as paradigms of classical conditioning, where two previously noncontiguous stimuli are paired. In these tests, one stimulus is designated as the unconditioned stimulus (UCS), and presentation of the UCS elicits a reliable and measurable unconditioned response (UCR). A second stimulus is considered "neutral" at the start of testing, but when paired with the UCS, this CS elicits a conditioned response (CR). The UCS can be food or water, in which case it is appetitive conditioning, while foot shock or other noxious stimuli constitute aversive conditioning [114]. Another important paradigm for studying learning and memory is instrumental conditioning, which differs from classical conditioning in that the animal's behavior is instrumental to the production of reward or avoidance of punishment [115]. Instrumental behaviors tend to be voluntary rather than reflexive or autonomic.

Classical and instrumental conditioning procedures are commonly used to investigate mechanisms that underlie learning and memory in rodents. Memory can be classified as either implicit (nondeclarative) or explicit (declarative), and these distinctions are based upon how information is stored and recalled (Figure 12.7) [114, 116].

Implicit or nondeclarative memory refers to nonconscious learning that is evident through performance and does not require access to any conscious memory contents. This type of memory includes procedural habits and skills, priming, simple classical conditioning that involves the skeletal musculature and emotional responses, and nonassociative learning where the striatum, neocortex, cerebellum and amygdala, and reflex pathways mediate these respective responses [117, 118]. By comparison, explicit or declarative memories involve knowledge of discrete events, places, or facts. This information is recalled with conscious effort and is highly plastic, permitting the creation of new associations [114].

Explicit memories involve three distinct processes that include encoding, storage, and retrieval [114, 118]. Encoding pertains to the processing of information to be stored and includes input of details about a stimulus and the environment. The encoding stage has two components: acquisition and consolidation. Acquisition

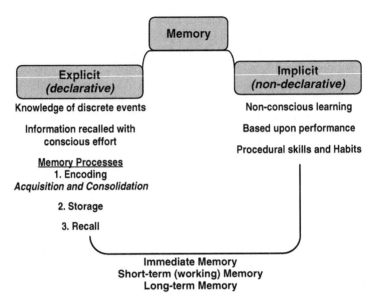

FIGURE 12.7 Learning and memory. Processes of learning and memory in the mouse are analogous to those found in humans and nonhuman primates. Memory is classified as either explicit or implicit, with explicit memory involving the processes of encoding, storage, and recall. Both explicit and implicit memories can be assessed with regard to immediate, short-term (working), and long-term memory.

refers to registration of inputs in sensory buffers and preliminary analysis of this information. Consolidation may involve the reorganization of newly encoded information so that it has stronger representation in memory. Storage is the mechanism by which information is retained over time, whereas retrieval uses stored information to create a conscious representation of the event or to execute a learned motor response.

In addition to these processes, explicit or declarative memory can also categorized as immediate, short term, or long term. Immediate memory, also referred to as a sensory register [116, 119], lasts less than several seconds but has a large capacity that can receive input simultaneously from multiple sensory modalities. Short-term memory is also termed working memory and is defined as a process that recruits knowledge on a short-term basis for rehearsal, elaboration, recoding, and comparison in order to solve a current problem [116, 120]. Long-term memory is recognized as a mechanism for storing information over a prolonged period of time that can last from several hours to a lifetime [116, 120]. Although different classical or instrumental conditioning paradigms have been used with mice to examine various aspects of learning and memory, such as short- and long-term memory [33], working memory [121], consolidation [122], or emotional memory [123], it should be apparent that each paradigm requires the use of several memory components simultaneously. As will be demonstrated in the following section, the ability to examine various aspects of learning and memory within the same paradigm can provide valuable opportunities to investigate the mechanisms underlying these processes.

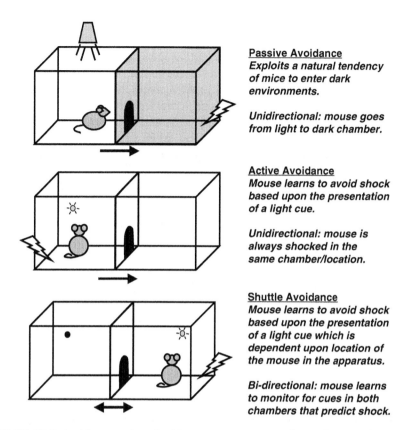

Passive Avoidance
Exploits a natural tendency of mice to enter dark environments.

Unidirectional: mouse goes from light to dark chamber.

Active Avoidance
Mouse learns to avoid shock based upon the presentation of a light cue.

Unidirectional: mouse is always shocked in the same chamber/location.

Shuttle Avoidance
Mouse learns to avoid shock based upon the presentation of a light cue which is dependent upon location of the mouse in the apparatus.

Bi-directional: mouse learns to monitor for cues in both chambers that predict shock.

FIGURE 12.8 Avoidance testing. Avoidance tests are commonly used to examine various memory functions in mice. In passive avoidance, the mouse learns in a single trial to suppress the natural tendency to enter a darkened chamber. In one-way active avoidance, multiple test trials are given over days to examine acquisition, retention, and extinction. In this avoidance test, the mouse is started in the same chamber on each test trial. Two-way (shuttle) avoidance evokes conflict resolution for the mouse, in that the animal must learn to monitor the cues that predict shock independent of the testing chamber (context). This test allows examination of complex cognitive function as associative learning.

AVOIDANCE TESTS

Passive and active avoidance are commonly used to examine various memory functions in mice, including acquisition, short-term or working memory, consolidation, and long-term memory. Although both tests require the mouse to avoid shock, the paradigms differ from one another, in that each paradigm evaluates unique cognitive attributes of the animal (Figure 12.8).

In passive avoidance, the mouse learns to avoid an adjoining chamber where the shock was previously delivered. Hence, the animal has to suppress its natural tendency to enter darkened, confining spaces. Active avoidance requires the mouse to make a proactive response to escape a chamber or area within the chamber where foot shock was previously administered [124]. Although passive avoidance constitutes one-trial learning and assesses the ability of the mouse to retain and recall information about the environment and foot shock, active-avoidance tests require multiple learning trials. For instance, in two-way shuttle avoidance, shock can be administered to either chamber, depending upon the current location of the mouse. This condition creates a conflict for the mouse, as the location where the mouse was previously shocked on one trial may become the "safe" chamber during subsequent trials. To this end, the mouse must learn to suppress the tendency to avoid a compartment where shock was previously administered and to use the CS to predict and avoid shock [125]. Active-avoidance procedures offer the opportunity to examine acquisition and consolidation within a single animal over testing [126, 127], whereas multiple animals are required in passive-avoidance experiments for the same purpose [33].

Passive Avoidance

Memories of single experiences are a rapid form of learning that provides recollections of events or places that can be adaptive to the organism [128, 129]. Passive avoidance is a test of rapid one-trial learning, where an animal is conditioned with a single aversive event and is later tested for recollection of that experience. Typically, recollection is based on the mouse's ability to avoid a test environment where it previously received a noxious stimulus. The conditioning can be examined over very short periods (e.g., 5 min) to much longer time points (e.g., days to months) [33, 130]. By testing different mice at various time points, certain aspects of memory can be examined that include acquisition, consolidation, retention, and recall. Furthermore, extinction can also be evaluated.

Passive avoidance is typically conducted in a two-chambered apparatus in which one chamber is illuminated with a light while the adjacent chamber is in darkness (Figure 12.8). Many apparatuses are commercially available with automated gates and computerized systems that monitor the location of the mouse, making the test relatively easy and automated. Video cameras, some of which can fit directly inside the testing chambers, can be used if ethological analyses are desired. Passive-avoidance testing commences with placing the mouse into the illuminated chamber with the door separating the light and dark chamber closed. After a brief period (5 sec), the door is raised. When the mouse completely enters the darkened chamber, the door separating the chambers is closed. The latency of the mouse to enter into the darkened chamber is recorded from the moment that the door between the two chambers opens until the mouse crosses the threshold into the dark chamber with all four feet. Immediately after the door is closed, the animal receives a 2-sec scrambled foot shock (0.1- to 0.4-mA intensity). Parenthetically, we have found that some mutant strains of mice are differentially responsive to foot shock, so we always test a small subset of mice in a shock-threshold test to determine the optimal parameters for testing (see Appendix to this chapter) [33]. Thirty seconds after the mouse receives foot shock, it is removed from the dark chamber and returned to the home cage until testing, which consists of placing the mouse back into the lighted chamber and raising the door separating the lighted and darkened chambers. During the test, no shock is given and the animal is observed for 5 min. The latency to cross to the darkened chamber is used as an indicator of memory. In addition to the latency to cross, the number of crosses between the two chambers, the total time spent in each chamber, and the frequency of head pokes into the darkened chamber until the mouse actually enters the chamber can be recorded. It is also important to note freezing behavior (especially in the lighted chamber), as this behavior can interfere with typical passive avoidance.

An important consideration in the test design is the selection of the interval between conditioning and testing. The length of this interval depends upon whether the investigator wants to assess acquisition, working memory, or long-term recall. For acquisition, durations as short as 5 min can be used, whereas intervals between 30-min to several hours may be indicative of short-term memory processes [130]. Intact retention at 5 min but deteriorating or disrupted performance at 30 or 60 min can reflect inadequate memory consolidation. Long-term memory typically refers to processes invoked after 8 h following conditioning [131]; however, 24 h is more traditionally used [33, 130, 132]. We usually evaluate learning and memory in this test at 24 h and subsequently investigate other intervals based upon the findings of this test. If mutant animals demonstrate deficiencies at 24 h, then shorter intervals may be considered to exclude deficiencies in acquisition of the task or working memory. If no differences are discerned from the control animals, then longer intervals may be assessed (e.g., 48 or 96 h) or extinction can be examined by repeatedly exposing the animal to the darkened chamber. Regardless of paradigm, it should be apparent that passive avoidance cannot be presumed to evaluate only one "type" of memory.

One-Way Active Avoidance

Active avoidance can be conducted in either a two-chamber apparatus, where the mouse is required to cross to the adjoining chamber to escape shock, or in a single chamber with a single vertical pole that the mouse can jump onto when shock is administered (Figure 12.8) [124]. During training, mice are given 20 trials/day. Each test trial is separated by 20 sec, and it consists of a 10-sec tone that is followed by a 0.1-mA scrambled foot shock. The foot shock is administered for 20 sec or until the mouse escapes either into the adjoining chamber (if a two-chamber apparatus is used) or jumps onto a centrally placed pole (if a single chamber is used). On subsequent trials the latency to respond is recorded for each animal. Avoidance responses signify crossing to the other chamber or jumping onto the pole during the 10-sec tone, and escape responses occur after the tone is extinguished (e.g., 11 to 20 sec after trial initiation). Trials where mice do not avoid or escape the foot shock are failures. Daily training is continued until the mice reach a criterion that reflects learning [126]. Learning curves within and across days reveal acquisition and consolidation of avoidance, with deficient mice requiring more training to reach criterion relative to controls. Once mice reach criterion, retention of the avoidance response can be measured by testing animals at a higher criterion. Extinction is examined by exposing the mice to daily test sessions without foot shock. Controls typically learn shock avoidance within several days and require approximately 20 trials to demonstrate retention [126]. Moreover, controls retain the response in the absence of foot shock for several days during extinction. Although active avoidance allows the investigator to examine acquisition, retention, and extinction within the same animals, this test is protracted and may require several weeks to complete. As with passive avoidance, it is important to study a small subset of mice before testing in order to determine the optimal level of foot shock to use [33, 126].

Two-Way Active (Shuttle) Avoidance

Although two-way shuttle avoidance is considered a more complex cognitive task than one-way active avoidance because it evokes "conflict resolution" for the mouse, the test is generally administered over 1 to 2 days. Hence, it is more time efficient than one-way active avoidance. Conflict resolution occurs in two-way avoidance because the mouse must reenter the test chamber where foot shock was previously administered to successfully avoid foot shock on the present test trial (Figure 12.8).

Two-way shuttle avoidance begins when a mouse is placed into one side of a two-chambered shuttle apparatus. A door dividing the two chambers is opened and the mouse is allowed 5 min to explore the apparatus. Following acclimatization, testing commences over 70 trials. In the Duke Core Facility, trials are given over 70 min with a variable intertrial interval of 30 to 90 sec. Each trial begins with presentation of the CS (0.5 sec of a 72-dB 2900-Hz tone with a 5-sec 1-mA house-light) in the chamber where the mouse is located. If the mouse does not cross to the adjoining chamber after 8 to 10 sec, a scrambled foot shock is administered. Both the light and foot shock are terminated when the animal crosses to the alternate chamber or after 10 sec. Behavioral responses are coded as avoidances, escapes, or failures. Successful avoidances consist of trials where the mouse crosses to the adjoining chamber following the onset of the CS, but before the foot shock. Escapes include trials when the mouse crosses to the adjacent chamber during the foot shock. If the mouse does not cross to the other chamber within 10 sec after foot shock, the trial is considered a failure.

As with any test using foot shock, it is important to determine the level of shock to produce a forward ambulating response, without locomotor reactivity, freezing, or vocalization (see Appendix to this chapter). If the mice demonstrate an inordinate number of escapes or failures during testing, the investigator may wish to observe subsequent animals to determine whether they have locomotor problems or difficulty orienting and detecting the CS [133]. In some cases, mice may move more slowly or cautiously during CS presentation, resulting in the foot shock being administered when the animal reaches the opening to the other chamber. In this case, the mouse may associate punishment or foot shock with approaching the adjoining chamber. Increasing the avoidance phase by 1 to 2 sec, before administering the foot shock, may remedy the problem.

Besides evaluating overall levels of avoidances, escapes, and failures, examination of learning curves over the 70 test trials can reveal information regarding acquisition or loss of each response. In most studies, the two-way shuttle avoidance is used to assess learning and short-term memory within a single test (70 trials); however, this test can be easily adapted to examine long-term memory by retesting the mice 24 h after the first test. Additional time points can also be used, with intervals between the two tests as long as 6 months being reported [134].

NONSPATIAL LEARNING

Nonspatial learning typically relies upon the ability of the animal to learn relationships between different stimuli or to demonstrate flexibility during recall, and these attributes are independent of the spatial context in which initial training occurred. A discrimination between familiar and novel stimuli is used as an index of learning and memory [93]. Several methods to test this form of cognition have been developed for mice, including simple object discrimination [37, 44, 135], social transmission of food preference [136–138], or tests of matching and nonmatching stimuli [139–141]. Each of these tests exhibit several key characteristics of declarative memory, in that (a) novel information can be acquired quickly with few exposures and (b) the memory is flexible, as the acquired information can be used in situations that are quite different from the initial conditioning environment.

Object-Discrimination Test

The object-discrimination test permits rapid screening of recognition memory in mice [135]. By varying intervals between initial object exposure and the subsequent test, general assessments of short- and long-term memory can be made [131]. It should be emphasized that more-detailed studies of retention are required if deficits are detected [44, 135]. Although Tang and colleagues [135] originally habituated their animals to the test chamber for three days, we have found that exposure of the mice to the test arena for 30 min prior to object introduction is sufficient to habituate them to the test [37]. When utilizing larger test arenas, multiple days of habituation may be required before testing begins. Regardless of whether large or small arenas are used, the test area should be opaque and covered with a lid to prevent the mouse from using spatial cues outside the cage. All phases of testing are videotaped for subsequent analysis of object exploration. A potential problem with this test is that the mouse may have an innate preference for one object over another. To limit this potential confound, a small group of mice from each genotype is usually exposed to a series of different objects before instituting formal testing. Objects that elicit similar exploration times and have a similar duration of contact are used in the formal test with a new cohort of mice.

In our lab, mice are habituated to the test arena for 30 min. After this time, two identical objects are placed in opposite corners of the arena without disturbing the animal (Figure 12.9). Parenthetically, we have found that multipatterned Legos objects work well in this test. It is important that the objects be similar or smaller in size relative to the mouse, as this facilitates exploration; larger objects can promote neophobia and reduced exploration. The mouse is given 5 to 10 min to explore the objects. All incidences of contacts are coded, as well as the time spent with each object. Object contacts include sniffing and climbing on the object or attempts to manipulate the object. After this exposure, mice are returned to their home cage until retention testing. In between animals or tests, the test cage and objects are cleaned with a disinfectant to remove any odor cues. Retention testing is conducted with two Lego objects minutes to days later, where a familiar object that was used previously is paired with a novel object of approximately the same size and height.

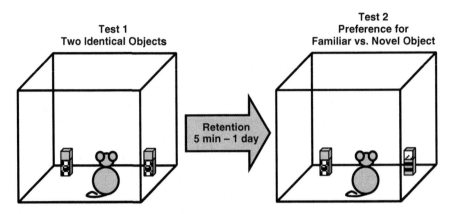

FIGURE 12.9 Object discrimination. Used as a rapid screen, the object-discrimination task allows for the examination of recognition memory in mice. By varying the interval between the initial exposure to the objects (test 1) and the preference test where the mouse must select between familiar and novel objects (test 2), inferences can be made regarding short- and long-term memory.

The mice are returned to the arena for 30 min, and then the familiar and novel objects are introduced to permit investigation for 5 to 10 min. The data are expressed as a preference ratio by subtracting the amount of time spent exploring the novel object from the time spent with the familiar object, divided by the total amount of time spent exploring both objects. Hence, values close to +1 indicate a preference for the familiar object, whereas negative numbers signify preference for the novel object. Typically, mice exhibit a preference for the familiar object at the 5- to 10-min interval. This suggests that acquisition of the Lego's features is intact. Tests conducted 1 to 3 h later permit assessment of short-term or working memory, whereas tests given at 24 h reflect long-term memory [35]. It is noteworthy that preference testing can also be conducted in other sensory modalities, including olfaction or tactile sensation [137].

Social Transmission of Food Preference

Although this task has been available for some time for rats [142], Alcino Silva's lab was one of the first to adapt it for mice [136]. Since that time, the test has been used as a method for studying memory and retention by several investigators [137, 138]. The advantages of the test are several-fold. First, this is an appetitive task that does not use aversive stimuli such as foot shock. Second, the behavioral measures take advantage of naturally occurring propensities of rodents to sample novel food sources and to develop food preferences from interactions with social partners. Finally, the test is devoid of spatial references [143]. Similar to the object-discrimination test, this task allows preferences of the animals to be examined at several time points, permitting assessments for acquisition processes and short- or long-term memory.

Mice are housed in pairs for several days before testing, during which time animals are placed on food restriction to reduce their body weights to 95% of their

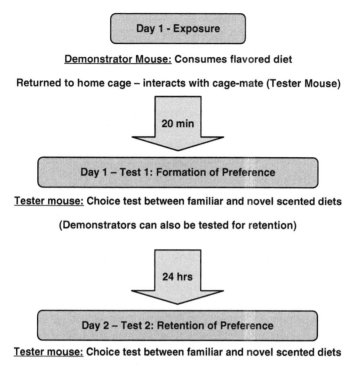

FIGURE 12.10 Social transmission of food preference. An appetitive task that takes advantage of the naturally occurring propensities of mice to learn about novel food sources from interactions with social partners. The social transmission of food preference can be used to examine acquisition processes or short- and long-term memory.

free-feeding weights. In preparation for testing, a small dot of commercially available hair dye (diluted with water) is placed on the back or rump of one mouse from each pair. This mark serves to designate the animal as the "demonstrator" in the test. The unmarked animal becomes the tester. Food is removed from mice 16 to 24 h before testing, with water available *ad libitum*. On the first day of testing, the demonstrator is removed from the home cage and placed into a clean mouse cage with a single bowl containing a flavored diet (Figure 12.10). After being allowed to eat for 30 min, the demonstrator is returned to the home cage, where the tester approaches the demonstrator and may interact with it for 20 min. During this time, interactions between the two animals are monitored for frequency of contacts, muzzle sniffs by the tester, and any aversive postures or responses by the demonstrator, including aversive posturing, clawing or biting the tester, or attacks. After 20 min the tester mouse is placed into a clean mouse cage with two small bowls fixed to the floor at opposite ends of the cage, creating a two-choice test. One bowl contains the diet eaten by the demonstrator; the other bowl contains a novel diet. The tester mouse is allowed to eat for 30 min, and consumption (weight) of the novel and familiar foods is recorded. The tester mice are reexamined one day later for retention using the same familiar–novel diet comparison. The preference for diet is scored in each test as the amount of novel diet eaten subtracted from the amount of familiar diet consumed and divided by the total amount eaten [137]. Positive scores reflect preferences for the demonstrator diet, suggesting the tester animal learned and retained the food preference of its demonstrator or social partner. A negative score denotes a preference for the novel diet. It is important to note that although investigators typically look for familiar preferences in the tester mice, a preference for a novel diet that is maintained over 24 h still signifies a choice that is remembered by the tester mouse. A strong preference for novelty suggests that memory is intact in the animal. However, this response suggests the tester mouse may present a novelty-seeking phenotype.

A note should be made with regard to the selection of flavors in these tests. Our Core Facility has obtained best results using a flavored mouse mash made from equal amounts water mixed with standard mouse chow (Richmond Diet 5001, Lab Diet Inc., Richmond, IN) that is ground to create a paste that can be flavored. The paste controls spillage and permits accurate weighing of the bowls before and after testing. The most common flavors used in mouse testing are cinnamon or cocoa [44, 136, 138]. However, sampling of different flavors or scents should be performed in advance with naïve mice to ensure that flavors or scents selected for testing do not evoke any inherent bias in the mice. In our laboratory, we found that a number of mouse strains have a strong predilection to consume foods with vanilla, cocoa, and peanut butter flavors. By contrast, other mouse strains have aversions to flavors of cinnamon and similar spices, even when the concentrations of these flavors are reduced to less than 1%. In our studies with the dopamine transporter knockout mice [137], we found lemon and almond flavors to be equally attractive to the mice, and neither flavor promoted strong attraction or avoidance responses. This is an important consideration, particularly when flavors are counterbalanced across testing for pairs of animals. If an investigator finds a bimodal distribution in performance, naturally occurring biases have to be considered before final conclusions of cognitive ability can be made.

Nonspatial Transverse-Pattern Test

One of the most intriguing tests for nonspatial learning in mice is the transverse-pattern task [139, 140]. This test has advantages over the tests previously described, in that this task evaluates the ability of the mouse to make associations between cues and to learn from these representations while also applying these associations to new contingencies. One disadvantage is that the test is labor intensive and can take considerable time to complete. However, the data can be analyzed from a variety of perspectives to study cognitive abilities in mice.

In preparation for the transverse-pattern task, mice are on food restriction for at least 1 week before training starts and are maintained at approximately 95% of their free-feeding weight during testing. The mouse is first trained to dig in a small bowl that contains sand or other small objects (beads or metal ball bearings) to uncover a food reward. Shaping typically takes 5 days to complete, some mutant strains require longer training to dig with the same efficiency as wild-type controls. Testing is conducted in clean mouse cages, with two identical food bowls mounted to the

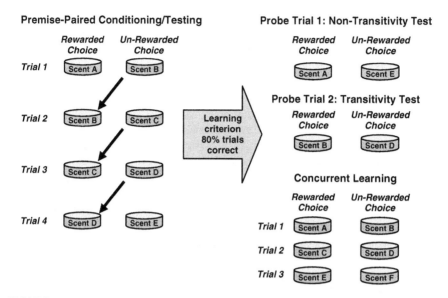

FIGURE 12.11 Nonspatial transverse-pattern testing. Transverse-pattern testing assesses the ability of a mouse to make associations between cues and to learn from these representations while applying the knowledge of these associations to new contingencies. The mice are first trained and tested using a transverse learning pattern, where the presence of a food reward in scented food is contingent upon the location of the food reward in the immediately preceding test trial. Once the mouse correctly obtains the food reward on 80% of the test trials, the animal is tested in two probe tests: the nontransitivity test and the transitivity test. Finally, the animal is assessed for the ability to make simple discriminations in concurrent learning trials.

floor on opposite ends of the cage. Five different odors designated as A, B, C, D, and E are used for testing and are created using various dry spices [139]; scents from sources such as talcum power, perfume, and essential oils are avoided (Figure 12.11). The spices are mixed with clean sand so that each spice constitutes only 1% of the total weight. For food rewards, 20-mg sucrose pellets for mice, or equivalent amounts of dark chocolate or dry cereal, can be used. In our laboratory we obtained the best results with small chips of dark chocolate, which are highly palatable to the mice and do not interfere or compromise the ability of the animals to discriminate the different scented sands.

Testing commences with an initial phase that is termed "premise-paired" conditioning and involves presenting the five different odors in pairs (Figure 12.11). Each pair of odors creates a two-choice test for the mouse. For premise-paired conditioning, the pairing is based on an ordered representation of the four odorant pairs, with scent A rewarded over scent B in pair 1, and scent B rewarded over scent C in pair 2, and scent C rewarded over D in pair 3, and finally scent D rewarded over E in pair 4. In the initial days of training, the mice are presented with the four pairs, with pair 1 being presented first, followed by pair 2, then pair 3, and finally pair 4. This series of trials is repeated three times on each training day for a total of 12 test trials. To ensure that spatial cues do not interfere with performance, the locations of the scents and food rewards are counterbalanced over test trials. Performance on each test trial is assessed by recording the bowl the mouse first approaches and sniffs, the bowl where the mouse subsequently digs, and whether the correct bowl is selected. Trials are scored as "correct" if the mouse picks the rewarded scent to dig in first or as a "failure" if the mouse selects the nonrewarded bowl first. As soon as the food reward is earned, the mouse is returned to its home cage.

Mice need to reach a learning criterion of at least 80% over three consecutive days before the sequence of premise pairs is randomized and interspersed with two different types of probe trials (Figure 12.11). The first probe trial, called the test of transitivity, involves scent B being rewarded over scent D. In this pair, two nonadjacent elements are selected, scents B and D, but by rewarding scent B, the ability of the mouse to perform transitive inference can be determined. The second probe trial is a novel nontransitive pairing, involving scents A and E. The nontransitive component comes from the fact that scent A is always rewarded and scent E is never rewarded. These two probe trials are essential and provide a powerful assessment for confirming whether the mice are capable of learning transitive inference using different pairs of odors.

In probe trials, mice that are deficient in transitive inference will typically demonstrate good performance when scents A and E are paired. Control groups should perform well on both probe trials. Additional confirmation of transitive inference can be assessed when two new pairs of odors are introduced, where scent W is rewarded over X, and scent Y is rewarded over Z. Typically, neither control nor mutant mice are able to solve this task, as the pairs lie outside of the established ordered-pair representation learned by the mice. Finally, once testing is completed, additional mice can be trained and tested over several days for concurrent pattern learning (Figure 12.11). In concurrent testing the pairs of scents are not overlapping as they are in transverse learning. Consequently, scent A is rewarded over B, and scent C is rewarded over scent D, and scent E is rewarded over scent F. Regardless of the order in which these paired scents are presented, odors A, C, and E are always rewarded. Moreover, discrimination of the rewarded scent from the nonrewarded scent in concurrent testing is not contingent upon which odor was rewarded in a previous test trial, as is the case in transverse testing. The concurrent test serves as a control parameter, assuring the investigator that the mouse can acquire and remember simple discriminations. If the mice are unable to make these simple discriminations, then tests of basic learning and short-term memory are required. Hence, comparison of the concurrent and transverse protocols allows the investigator to differentiate simple association learning from more demanding transverse-pattern learning.

SPATIAL AND CONTEXTUAL LEARNING

Spatial learning refers to the ability of the animal to learn the location of a reward [139, 144, 145]. However, in the original distinction between spatial and nonspatial learning, O'Keefe and Nadel [146] proposed that spatial learning involves the acquisition of cognitive maps that provide representation of the environment in terms of distances and directions. These maps allow animals to navigate to locations beyond their immediate perception to locate appetitive or aversive stimuli. It should be emphasized that nonspatial memory tasks do not require these maps. Cognitive maps require an intact hippocampus [146, 147], as animals with hippocampal damage exhibit abnormal exploratory activity in tests of spatial learning and experience difficulty navigating mazes to obtain food rewards [139, 148]. An analogous phenomenon is also found in humans, where hippocampal lesions result in errors in the use and organization of spatial information [147]. Although hippocampectomized patients are capable of learning new motor skills, these individuals are unable to learn new rules for applying these skills in novel or adaptive ways [149]. These findings demonstrate that the hippocampus plays a critical role in the acquisition of new information. The fact that the hippocampus is involved in explicit memory is further verified when rats are tested in multiarm mazes for both working and reference memory [150]. In these experiments, reference memory is defined by the ability of the animal to remember the location of food within the maze. By contrast, working or episodic memory refers to the ability of the mouse to use information acquired between test trials, such as which maze arms the animal had just visited and whether a food reward was found in that location. In these tests, Olton and colleagues [150] demonstrated that working memory was controlled in areas outside of the hippocampus, most notably in the frontal cortex and striatum. The relationship between reference and working memory in animals has been compared with semantic and episodic memory functions in humans [151]. This similarity between species serves to further strengthen rodent studies as tools to investigate basic mechanisms that underlie memory formation, and they also provide a foundation for understanding cognitive dysfunction in various psychiatric conditions [139].

Radial Arm Maze

The radial arm maze is used to study various aspects of learning and memory in rodents, including spatial and nonspatial attributes [145] as well as working and reference memory [152, 153]. The maze consists of an octagonal hub or central area, from which eight arms of identical size radiate outward. The original maze was elevated above the ground and had open arms so that the rat or mouse could use extramaze cues to navigate the maze [150]. More recently, intermaze cues have been employed to examine nonspatial learning [145]. In this case, specific patterns or textures are placed at the opening of each arm from the central hub or only at openings of particular arms during testing [145]. This strategy, in conjunction, with the traditional radial arm maze test, has allowed investigators to study both spatial mapping as well as working and reference memory.

In its traditional format, the radial arm maze test (also called the win-shift or win-stay paradigm) has food reward placed in some or all arms, and the animals have to remember the locations of the food or where they have just gone to retrieve the food [154, 155]. Over trials, animals learn to avoid re-entry into arms where food has been retrieved. Behavior is measured as the number of arms the animal enters before repeating an entry and the time required to retrieve all food rewards [34, 152, 156]. Animals with learning impairments typically make fewer arm visits before reentering an arm and may take longer to retrieve all food rewards. The radial arm maze has also been used to examine perseveration and impulsivity in cognitive dysfunction, which can interfere with cognitive performance [156].

Before testing in the radial arm maze, mice are subjected to daily handling and placed on food restriction to 90% of their free-feeding weight. Following five days on this regimen, animals are placed in a clear cup in the middle of the apparatus (e.g., the hub) and are given 5 min to eat Fruit-Loops food reward. Although some investigators shape the animals by scattering food along the arms of the maze [153], we have found that, once the mice learn to eat in the central hub of the maze, they have little difficulty in finding the rewards in the arms. Once the mouse has consumed food (equivalent to baiting all eight arms of the maze) reliably over three consecutive days within 300 sec, radial arm maze testing begins. In our laboratory, we utilize a protocol for mice based upon that designed for the rat by Addy and Levin [154]. At the start of the test trial, all arms or certain arms of the maze are baited, and the mouse is placed in the center of the maze inside a small cylinder. After 10 sec the cylinder is removed, allowing the mouse unimpeded access to the maze. Each arm entry is reinforced only once during the test trial. An arm entry is recorded when all four feet of the mouse cross into the arm. The animal remains in the maze until all food rewards are retrieved or until 300 sec has elapsed. Performance is assessed by the entries to repeat. In addition to this measure, the total time required for the animal to retrieve all food rewards is recorded. Aside from cognitive performance, rudimentary assessments of motivation can be made by noting the amount of food consumed. Perseveration can also be evaluated as the propensity of an animal to exit one arm and enter another arm repeatedly [156]. If behavior in the maze is video-recorded, the images can later be evaluated for the speed with which the mouse traverses the maze. High speeds with numerous errors may suggest impulsivity.

Morris Water Maze

The Morris water maze is an apparatus where mice learn to escape from water by swimming to a hidden platform located just below the surface of the water. Control animals learn this task in a relatively short time — only a few days [157, 158] or, under special training conditions, within a single day [159]. The water maze is advantageous, as it does not require food or water deprivation and takes advantage of the natural swimming behavior of the animals. Moreover, various studies have shown that an intact hippocampus is required for this task and that hippocampal-cortical pathways, as well as the dorsal striatum, play critical roles in the regulation of spatial memory in this test [160, 161].

The water-maze test requires a fairly large area of lab space where cues outside the maze can remain stable over the entire testing period. The maze uses a large circular pool of water that is made opaque by the addition of a white dye or nontoxic white poster paints. Although powdered milk can be used, the milk becomes rancid over days, and this feature increases the workload on the investigator. The behaviors of mice in the maze are video-taped with a camera mounted directly over the center of the pool. These data can be analyzed by several tracking programs such as Noldus Ethovision (Noldus Information Technology, Blacksburg, VA), SMART (San Diego Instruments, San Diego, CA), Water for Windows (HVS Image, Hampton, UK), or

Wintrack (University of Zurich, Switzerland). Each program provides basic measurements, including path length, swim time, and velocity as well as a number of analyses for path-length data [162]. Lighting the pools can be tricky, but placement of the lights behind diffusers in the ceiling or having the lights mounted on the walls at water level and pointing upward can produce adequate illumination without glare from the water surface — a critical consideration when using light-based automated tracking systems. The lighting should be even and maintained at 100–600 lux when measured from the surface of the water. A moveable platform, large enough for the mouse to stand comfortably with all four feet, should rest 1 to 2 cm below the surface of the water. It is important that the platform not be too far beneath the surface of the water; otherwise the mouse will have difficultly locating the platform. Optimum water temperatures are approximately 23 to 25°C.

The pool is divided into quadrants, designated as northeast, southeast, northwest, and southwest (Figure 12.12). One quadrant is designated as the location of the hidden

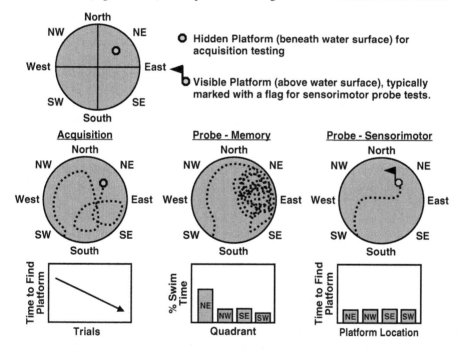

FIGURE 12.12 Morris Water Maze. The Morris water maze assesses spatial learning in the mouse. The test is conducted in a large circular pool of water, where a platform has been hidden beneath the water surface in one quadrant of the pool. Mice must use spatial cues to navigate and find the hidden platform. As the mice learn the location of the platform, the swim time to the platform will decrease. Periodically, mice are given probe trials where the platform is removed from the pool. If the mice have acquired the task, they should spend a higher proportion of swim time in the quadrant where the platform was located previously. After all testing is completed, the animals are given visible platform trials which assess sensorimotor capabilities. In these trials the platform is made visible by the placement of a flag on or over the platform. Regardless of which quadrant the platform is placed during these "flag" trials, the mice should rapidly swim to the platform.

platform, and this position is maintained throughout testing. Release points for the mice occur at seven different locations, spaced equally apart. It is important that release points be randomized across testing, and the same release points should be used for all animals. It is also important that the investigator be hidden from view of the mice during testing, as we have found that most mice preferentially swim toward the individual who removes them from the pool rather than the platform.

Before commencing testing, mice should be handled daily for one week. Testing typically consists of three phases: acquisition trials, where the animal uses extramaze cues to find the location of the hidden platform; retention or probe trials, where the platform is removed and the animal uses the cues to swim to the previous location of the platform; and cued navigation tasks or "flag trials," where a flag is placed over a movable platform to assess visual and sensorimotor function of the mice (Figure 12.12) [162]. On the day before testing, mice are trained to sit on the platform for 30 sec. They are allowed to swim freely around the maze for 60–90 sec and are then guided to the platform. This is repeated several times for each animal. If mice attempt to jump from the platform, they are returned to the platform for an additional 30 sec. Following each trial, mice are gently dried and placed back with cagemates. To avoid stressing the wet animals, it is important that the cage be kept out of drafts and, if needed, that a heating pad be set on a low setting and placed below the cage. The use of red lights or heat lamps is discouraged, as the brightness or heat from these lamps may be stressful to the animals.

On the first day of testing, the mice are released from different locations around the perimeter of the pool and allowed to swim until they find the platform or until 60 sec elapses. All mice are given four to six test trials per day, where pairs of trials are separated by an intertrial interval of 20 to 30 min. The acquisition phase of testing continues for 5 days for a total of 20 to 30 test trials. Variables used to establish learning during acquisition testing include path length (cm), the amount of time (sec) required to locate the platform, and swim velocity (cm/sec). Probe trials, where the platform is removed from the water maze, are given twice during acquisition training. We typically administer them midway and again at the end of testing. During the probe trial, the mouse is allowed to swim for 1 min with no platform present. In this time, the animal will use spatial cues to navigate and search the quadrant of the pool where the platform was previously located. Path length, swim time, and velocity are again examined, but this time they are measured for each quadrant of the pool. As the mouse acquires the task, it should demonstrate increased swim time in the quadrant on the probe trial where the platform was located previously. If no differences occur between quadrants, it is likely that the mouse has not learned to use spatial cues to find the platform. Increased swim time or path lengths in areas of the maze adjacent to the platform may indicate that the animal is confused about the relevant spatial cues or that knowledge concerning the position of the platform is still undergoing consolidation.

Although some investigators test noncued and cued navigation in the same animals, we have found that more-reliable data can be collected if naïve mice are used in both tests. On cued navigation trials, a "flag" is attached to the platform so that its location can be easily seen. We usually run 10 to 20 trials where the location of the platform is randomized across trials. This test is critical to ensure that the sensorimotor capabilities of the mice are intact. If the mice fail to swim to the visible platform, it is important that vision be examined more closely in the animals, particularly if a neurophysiological screen was not conducted. In addition, poor performance on the flag trials may necessitate examining locomotor and coordination abilities of the mice.

Mice have been shown to display distinct strategies and patterns in swimming navigation during learning [163]. Analyses of swim paths can provide some insights into these behaviors. In a detailed examination of swim paths from numerous animals, Wolfer and Lipp [163] found that in the initial phases of acquisition, mice usually adopt a pattern of thigmotaxis, or swimming along the perimeter of the maze. However, this pattern is replaced rather quickly to search behaviors referred to as scanning. These search patterns consist of the mouse engaging in swimming random paths across all quadrants of the pool or may include circling in wide loops systematically through the different areas of the pool. Scanning, in turn, will give way to a more focused search strategy, particularly after the mouse has inadvertently encountered the platform. After the mouse has acquired that task, it will swim directly to the platform. On the other hand, some mice will show random floating or will swim in very tight concentric circles that may not appear as part of any systematic search strategy. We may remove these mice from the study.

NEUROLOGICAL MECHANISMS UNDERLYING LEARNING AND MEMORY

O'Keefe and Nadel [146] first observed that the firing rates of hippocampal cells were correlated with specific locations in the test environment. These cells were later termed "place cells" [164]. Place cells were identified rapidly upon first exposure to the novel environment [165]. Once these patterns of activity in the hippocampus were established, the firing patterns persisted; even when spatial cues were removed [166, 167]. Because place cells only fired when a specific area of the environment was encountered, it was postulated that new comprehensive maps called "place fields" were constructed in the hippocampus every time the animal entered a new environment. These data suggested that activity of hippocampal neurons was not only long lasting, but also malleable over time.

Despite numerous studies conducted to characterize the hippocampal place fields and spatial memory, it was also known that the neural activity in the hippocampus could regulate nonspatial information and sequences of behavior that occur at regular intervals [168, 169]. Hence, neural activity in the hippocampus reflects a broad spectrum of responses, with some cells encoding unique events; some encoding certain stimuli, behaviors, and locations of events; and others regulating sequences of events or specific common features across different events.

The brain areas typically associated with the hippocampal memory system involve the cerebral association cortex, the hippocampus proper, and the parahippocampal region, which includes the perirhinal and entorhinal areas. The cortical areas provide perceptual and motor information to the hippocampus through the parahippocampus [139]. The hippocampus proper is composed of several cell layers and is organized in a distinct manner that allows information to be sequentially processed through defined circuits [170]. Three circuits are well defined: the *perforant pathway* includes projections from the entorhinal cortex to the granule cells of the dentate gyrus; the *mossy fiber pathway* runs from the granule cells to the CA3 pyramidal cells; and the *Schaffer collaterals* represent excitatory collaterals of the CA3 region that project to CA1 pyramidal cells. Transfer of information within the hippocampus is thought to occur along two glutaminergic pathways: the mossy fibers and Schaffer collaterals [114]. These pathways are under intense investigation, as long-term potentiation (LTP) and other processes — associated with plasticity and the formation of long-term memories — are easily studied in these neurons [171]. LTP was first described in the hippocampus, where it was observed that application of high-frequency electrical stimulation or tetanus to a particular neural pathway would augment excitatory synaptic potentials when the same pathway was restimulated with a single electrical pulse [172, 173]. Initially, studies in LTP were limited to the perforant pathway [174]; however, LTP was later observed in the mossy fibers and the Schaffer collaterals [175].

One of the first demonstrations that the glutamatergic NMDA receptors were important for spatial learning and synaptic plasticity was by Morris and colleagues [29]. In these experiments, intraventricular infusion of AP5 (an NMDA receptor antagonist) or direct infusion of this drug into hippocampus resulted in loss of spatial learning by rats in the hidden-platform version of the Morris water maze. Additional pharmacological and genetic experiments continued to support the idea that NMDA receptors, particularly those in the CA1 region of hippocampus, were necessary for the acquisition of spatial memory [129]. This idea was confirmed with homozygous mutant mice that specifically lacked NMDA receptors in the CA1 region of hippocampus [18,176]. These mice were deficient only in the hidden-platform version of the Morris water maze. While multielectrode recordings in freely moving mice showed the CA1 pyramidal cells to have place-related neural activity, spatial specificity for the individual place fields was perturbed. Intriguingly, induction of LTP in the Schaffer collaterals was also blocked. Although these studies provide strong evidence for the importance of the CA1 region and NMDA receptors in spatial learning and LTP, subsequent studies reveal that nonspatial learning is also disrupted (see Fear Conditioning section below) [44, 140].

Although NMDA receptors in CA1 hippocampus appear important for the acquisition or induction of memory, recall from long-term memory seems to be dependent upon other mechanisms [129]. As individual life experiences are unique moment-to-moment occurrences, it has been proposed that similar experiences only serve to reactivate certain aspects of stored memories rather than the complete memory [177]. Hence, this process should require that only certain neurons in a particular brain region become activated. Since CA3 hippocampus contains highly modifiable synapses, this brain region was hypothesized to be important for retrieval of hippocampal-dependent memories. As anticipated, CA3 NR1 knockout mice were deficient in recall that was dependent upon pattern completion [129]. These mutants could locate the hidden platform in the Morris water maze under full-cue conditions, but when probe trails were implemented with partial visual cues present, the knockout animals were unable to locate the platform. By contrast, mice with intact CA3 NR1 receptors were capable of finding the platform under both conditions [129]. Place-cell recordings from the CA3 NR1 receptor mutants revealed that responses from these cells to the degraded cued condition were severely impaired. Collectively, results with the CA1 and CA3 NR1 receptor knockout animals suggest that NMDA receptors in the CA1 region are intimately involved in the acquisition of memory, whereas recall is highly dependent upon NMDA receptors in CA3 hippocampus [129].

Despite this evidence, other laboratories have demonstrated that areas outside of CA1 and CA3 hippocampus are also critical for learning and memory. For instance, performance in the Morris water maze requires both spatial and nonspatial learning and memory [178], and some forms of learning appear to be independent of NMDA-mediated LTP in hippocampus [129]. Other brain areas that demonstrate LTP and are implicated in the induction and maintenance of memory processes include the prefrontal cortex [179, 180] and amygdala [181]. For example, enhancement of NMDA-receptor function in NR2B transgenic mice produces improved LTP in forebrain [182]. The transgenic mice also show enhanced retention in the novel-object recognition task, the Morris water maze, and fear conditioning.

Although a number of investigators have shown that glutaminergic transmission within hippocampal-prefrontal pathways plays a key role in LTP, other neurotransmitters including dopamine [180], norepinephrine [183], and serotonin [184] also affect LTP and long-term memory. For instance, stimulation of the ventral tegmental area provokes dopamine overflow in the prefrontal cortex and leads to a long-lasting enhancement in the magnitude of LTP [185]. By contrast, depletion of cortical dopamine levels dramatically reduces LTP [180]. The importance of dopamine in LTP is further emphasized by the fact that an optimal level of dopamine D1/D5 receptor expression is necessary for adequate LTP in the hippocampus [186, 187] as well as other brain regions, including the prefrontal cortex [188] and striatum [189]. Noradrenergic contributions to synaptic plasticity and LTP in both cortical and hippocampal areas have also been documented [190, 191]. Noradrenergic pathways, which originate in the locus coeruleus and innervate both the cerebral cortex and hippocampus, have been found to modulate glutamatergic activity and synaptic plasticity at perforant synapses in dentate gyrus [190], at mossy fiber synapses [192], and in CA1 hippocampus [193, 194]. In addition to catecholamines, serotonin also controls synaptic plasticity [195, 196]. Acute systemic administration of a serotonin reuptake inhibitor (fluvoxamine) increased the efficacy of synaptic transmission in the hippocampal-prefrontal cortical pathway [195]. When the rats were treated for 21 days with fluvoxamine, a marked enhancement in LTP was observed. Alternatively, destruction of the serotonin neurons in this pathway also enhanced LTP [196]. Both effects were attributed to disinhibition of the serotonin 1A receptor that regulates both NMDA-receptor function and LTP in the prefrontal cortex and hippocampus [197]. These findings are intriguing because they suggest that regulation of learning and memory involves mechanisms beyond glutamatergic control. In this regard, the atypical antipsychotic clozapine, which can bind to norepinephrine, dopamine D4, and serotonin receptors, has been found to reverse stress-induced impairments in LTP and cognitive functioning [180]. These studies provide critical links between mouse models of cognitive dysfunction and analogous symptoms observed in human psychiatric patients, suggesting an important role for these brain areas in the pathophysiology of various psychiatric conditions.

CONDITIONED EMOTIONAL RESPONSES

The conditioned fear response has become a common and powerful paradigm with which to study the neurological basis of emotional responses as well as learning and memory. This Pavlovian paradigm involves exposing the animal to a neutral conditioning stimulus (CS) and pairing this to an aversive unconditioned stimulus (UCS). Parenthetically, the CS is usually a light or tone, and it usually also cues within the same chamber where the animal was conditioned. After pairing the CS and UCS, the CS alone often elicits a defensive or "fearful" response that may involve immobility or freezing behavior.

Conditioned Taste Aversion (CTA)

CTA was discovered when investigators realized that irradiated rats avoided solutions or food that had been present during radiation treatments [198]. When rats encountered a novel taste (the CS) and this was followed by transient gastrointestinal distress caused by low-dose radiation (the UCS), CTA developed. This response results in a diminished intake of saccharin upon subsequent presentation. Later studies found that CTA could develop following exposure to a variety of other illness-producing agents, including chemotherapeutic agents, high doses of apomorphine or amphetamine, and lithium chloride [199]. For CTA to develop, the animal must be able to detect the CS; it must be able to become ill from UCS exposure; it must be able to form an association between the US and CS; and, finally, it must be able to avoid the CS.

CTA is a relatively simple test to conduct, and it typically requires two days of combined training and testing. Mice are placed on food restriction prior to testing to ensure that they will consume an adequate amount of the novel food during conditioning and to guarantee that an association between the CS and UCS will develop. Alternatively, water restriction can be used so that lithium chloride treatment can be paired with the consumption of a saccharin solution. In our laboratory we have found that both procedures work equally well with mice. The advantage of using a novel flavored food is that it is highly palatable to the mice, the amount of food consumed is easy to measure, and testing can be conducted over a brief period of time. By comparison, saccharin-flavored solutions typically require longer consumption periods to ensure that an adequate amount of the solution has been consumed. In addition, when using a liquid CS, intake can take several days to stabilize, and analyses of fluid intake typically require more elaborate and precise measurements, such as those provided by eating and drinking chambers available for mice (see Columbus Instruments, Columbus, OH). Readers interested in using saccharin-flavored water in CTA for mice should consult Cannon and colleagues [200].

Mice are individually housed the day before testing and placed on food restriction. On the test day, each mouse is placed into a clean test cage with a flavored diet located in a small feeding dish mounted in the center of the cage (Figure 12.13). The ground chow can be flavored with either 1% vanilla or almond extract and sweetened with a $0.25\text{-}M$ saccharin solution. Mice should be counterbalanced across flavors, so that half the animals are conditioned with vanilla-flavored chow and the

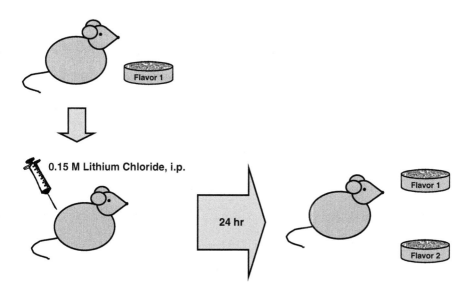

FIGURE 12.13 Conditioned taste aversion. A simple and rapid test to assess the ability of a mouse to develop a conditioned emotional response. Following food restriction, the mice are allowed to consume flavored chow. Immediately following consumption, the animals are injected with 0.15 M lithium chloride so as to pair the flavored chow with gastrointestinal distress. Twenty-four hours later, the mice are reexamined for a food preference between the flavored food (paired with the lithium chloride) and an alternative flavored diet. Mice that can develop a conditioned emotional response will not consume the diet paired with the lithium chloride.

other half with almond-flavored chow. The bowl with flavored diet is weighed before conditioning and again after 30 and 60 min of exposure. Because rodents typically restrict their initial consumption of novel food sources [201], it is important to weigh the bowl at both 30 and 60 min, allowing the mouse to consume an adequate amount of flavored diet. Mice on food restriction or food deprived the day before testing typically consume 0.3 to 1.0 g of food during the 1-h free-feeding period. After 1 h the mice are removed from the test chamber and injected with either 0.15 M lithium chloride or sterile water. The mice are returned to their home cage, maintained on food restriction, and tested 24 h later for retention. CTA testing involves a two-choice test between the two flavored chows, almond and vanilla. Mice conditioned to almond may be expected to prefer vanilla and those conditioned to vanilla should prefer almond. If control animals fail to demonstrate a preference between the two flavored chows, then several factors should be investigated, including increasing the lithium dose or increasing the saccharin concentration relative to the flavoring concentration to ensure that the animals are cued to flavor and not sweetness of the diet. It is also important that the mice be naïve, with no previous exposure to either almond or vanilla flavoring/scents or saccharin solutions before testing. Any of these experiences can create latent inhibition in the mice and weaken the CS-UCS association during conditioning [202].

A number of research studies have shown that the amygdala appears to be an important brain region for regulation and expression of CTA [199]. Although rats with lesions of the medial or central nucleus of the amygdala fail to demonstrate deficiencies in CTA, lesions of the basolateral amygdaloid (BLA) nucleus induce profound disruptions of CTA. The effect of BLA lesions has also been attributed to a reduction in neophobia. The BLA receives afferent information from all sensory modalities and relays this information to the central nucleus. The central nucleus is reciprocally connected to the hypothalamus and nucleus accumbens through stria terminalis and to the dorsal medial nucleus of the thalamus, the rostral cingulate cortex, orbital frontal cortex, and brain stem through the amygdalofugal pathway [203]. Given that lesions of the central nucleus fail to disrupt CTA, it can be assumed that the locus of control for this behavior rests within the BLA [199].

Fear-Potentiated Startle (FPS)

FPS is a test of Pavlovian conditioned fear where a reflexive acoustic startle response can be augmented when a startle stimulus is presented with an aversive stimulus that elicits a fear response [204, 205]. FPS has a number of features that make it attractive to investigators [206]. For instance, the reflexive startle response requires no learning on the part of the animal. The test is fully automated, and the potentiated startle response is long lasting and allows examination of long-term memory and extinction [207]. Unlike other tests of conditioned emotional responses, FPS can be examined following fear conditioning, permitting the investigator multiple opportunities to examine learning and memory under different paradigms [208]. Finally, the neural circuits that regulate emotional learning and memory are well characterized from the molecular to neurological levels [209].

Typically, FPS is conducted across four days (Figure 12.14). On each day, mice are acclimated to the chambers for 5 min before testing. On the first day, baseline startle responses are measured across a range of acoustic startle stimuli. On the next day, startle responses are paired with the CS. On the third day, mice are conditioned to the CS and scrambled electric shock, and on the fourth day, potentiation of the startle response to the CS is evaluated without shock.

Baseline testing on day 1 involves administering 40-msec bursts of white noise at several different intensities (100, 105, and 110 dB). Mice should demonstrate moderate startle responses to these stimuli. If the startle responses are too high (e.g., over 800 mA), mice are re-examined at lower intensities beginning at 90 dB. High baseline startle responses can result in a "ceiling" effect where potentiation of the response is more difficult to measure on subsequent test days. It is also important to evaluate mice on only 10 to 15 trials, where each stimulus is presented three to five times each, because the mice can habituate to these acoustic stimuli if they are administered too frequently [210]. On day 2, mice are placed back into the test chambers and given nine trials of baseline startle stimuli (three presentations/intensity). They are then presented with 18 test trials with these same acoustic stimuli (six presentations/intensity). On half the trials, the startle stimuli are administered immediately following a 30-sec, 12-kHz, 70-dB tone (CS); on the other half no tone is given. The tone+startle trials are interspersed in a pseudorandom order with the

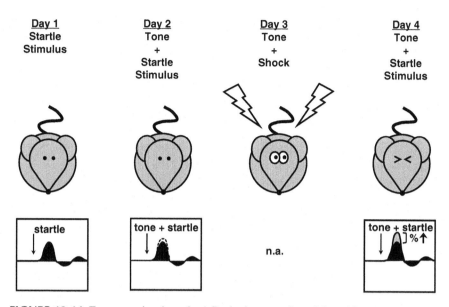

FIGURE 12.14 Fear-potentiated startle. A Pavlovian test of conditioned fear that utilizes the reflexive acoustic-startle response. Mice are examined for baseline startle responses on the first day. On the second day, animals are presented the startle stimuli and the CS (tone). On the third day, the mice are given the CS and shock pairing; 24 h later they are tested for retention. Mice that develop a conditioned emotional response exhibit potentiation of the startle response to the CS on the final test day, in absence of the shock.

startle-only trials. The magnitude of response to the CS (e.g., preconditioning potentiated response) is calculated by subtracting the response to startle-stimulus-only trials from the tone+startle-stimulus trials, dividing this number by the startle-stimulus-only responses, and multiplying the final score by 100. Positive values indicate that the tone augments the baseline startle response, whereas negative values suggest a reduction in startle response to the tone. If an increase or decrease of the response exceeds 15%, then the animals will have to be re-examined under various intensities of tone and white-noise startle stimuli until more desirable levels are obtained.

On day 3, the tone is presented and, at its termination, a 0.25-sec 0.3- to 0.4-mA scrambled shock is administered. The mice receive ten CS-UCS pairings, separated by an intertrial interval of 90 to 180 sec. Twenty-four hours later, the mice are tested for FPS under the same procedures used for day 2. Potentiation is measured by the same formula used to calculate the preconditioning potentiated startle. For animals that acquire FPS, the percent potentiation should be increased following conditioning compared with values obtained before conditioning. Parenthetically, C57BL/6 mice typically demonstrate FPS of 50 to 100% when postconditioned potentiation is compared with preconditioned responses [206]. In our laboratory we have generally found that control groups of unaffected WT mice or inbred C57BL/6 mice produce similar levels of FPS.

The FPS paradigm has a distinctive advantage over other tests of conditioned emotional memory in that it can be evaluated in humans of all ages with parameters similar to those used for rodents [211, 212]. Adult human patients diagnosed with posttraumatic stress disorder, depression, or bipolar disorder, and children with anxiety or temperament abnormalities, demonstrate enhanced and abnormally persistent FPS responses [213–215]. In addition, FPS in animals has been proposed to parallel the development of pathological fears and phobias in humans, where persistent and exaggerated responses to fear-provoking stimuli have been documented [216]. Anxiolytics block FPS in rodents, whereas anxiogenics can potentiate the response [217]. FPS is quite sensitive to drugs known to modulate states of fear and anxiety in both rodents and humans, including norepinephrine antagonists, benzodiazepines, opioids, and atypical anxiolytics [218, 219].

Although hippocampal or amygdala lesions interfere with expression of FPS [220, 221], the amygdala is considered the key site for FPS regulation [222]. For instance, the central and lateral amygdala are stimulated in response to the CS, and this activation appears to promote enhanced and prolonged neural responsiveness [223]. Besides these regions, the BLA also appears to be involved in FPS, as pharmacological lesions of the pathway between the central and basolateral nucleus impair expression of FPS following conditioning [207]. Moreover, acquisition of the conditioned emotional response for FPS is blocked when competitive NMDA antagonists are injected directly into the BLA; these agents also disrupt or prolong extinction of FPS. Hence, these findings indicate that the various regions of the amygdala serve different functions in FPS and that the basolateral amygdala is critical for the formation and plasticity of emotional memories that underlie conditioned emotional responses. For these reasons, FPS testing can provide important insights into the neural mechanisms that underlie associations between fear and anxiety found in human anxiety disorders [185, 209].

Fear Conditioning

The idea behind fear conditioning is that a fearful experience establishes an emotional memory that can result in long-term behavioral changes and, in some cases, these changes can become part of the permanent behavioral repertoire of the individual [224, 225]. In this paradigm, mice are conditioned by pairing a tone (CS) with foot shock (UCS). The animals are later examined in two tests; one evaluates contextual fear and the other examines fear responses elicited by the CS (cued fear conditioning). Fear in both tests is signified by immobility or freezing behavior (Figure 12.15). This response is a direct reflection of the conditioned emotional response in animals [226]. By examining freezing behavior in the different test conditions, dual mechanisms underlying conditional emotional memory can be examined. Freezing during the context test is attributed to hippocampal or temporal lobe processes. Deficits in freezing during the context and cued tests are indicative of amygdala dysfunction.

The strengths of the fear-conditioning paradigm are several-fold. For instance, conditioning only requires a single session. Additionally, the stimuli are under direct control of the investigator, and the behavioral responses have been operationally defined, validated, and are simple to measure. Further, the neural substrates for fear conditioning in animals have been identified. It is noteworthy that abnormalities in

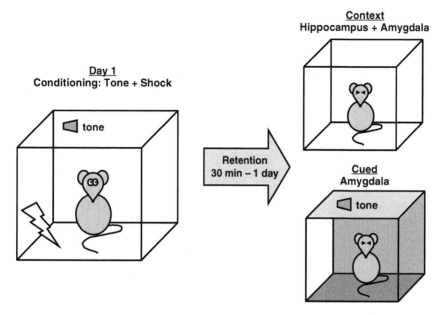

FIGURE 12.15 Fear conditioning. In Pavlovian fear conditioning, a fearful experience establishes a memory that can result in long-term behavioral changes. On the first day, mice are placed into a chamber and conditioned with a tone paired with foot shock. On the second day, the mice are returned to the chamber, and the incidences of freezing are examined in the absence of tone and foot shock (context test). Alternatively, the mice can be placed in a novel chamber, and freezing behavior is noted following the presentation of the tone without foot shock (cued test). In both tests, immobility or freezing behavior is scored.

the hippocampus and amygdala have been implicated in psychopathologies such as schizophrenia [227] and anxiety disorders [228, 229], including obsessive-compulsive disorder [230] and posttraumatic stress disorder [231–233].

Fear conditioning and testing can be conducted over two to three days, depending upon whether the same animals are used for the context and cued tests or whether different groups of mice are used in each test. In practice, we have found no differences between these two conditioning protocols. Typically, fear-conditioning testing is conducted either in a single chamber where the contextual features of the apparatus (e.g., lighting, floor texture, wall shape and texture, and visual cues) can be altered [13] or in two to three different chambers [231]. Behavioral responses are videotaped for later analyses with ethological scoring programs such as the Noldus Observer (Noldus Technologies, Blacksburg, VA) or are scored live by trained observers. Automated fear-conditioning chambers are also available, which permit simultaneous videotaping of mice with an automated threshold-detection system that monitors movements of the mice (Med-Associates, St. Albans, VT). The investigators can set detection thresholds so that freezing behaviors fall beneath one threshold and reactive behaviors such as jumping or running register above another. Activity between the two thresholds may constitute grooming, sitting, or low-intensity exploratory activity. When using these automated systems for fear conditioning, it is critical that the investigator first observe a pilot group of animals for freezing and general exploratory behaviors so that thresholds can be set to accurately differentiate these behaviors. In our laboratory we use both ethological and automated measures simultaneously. More recently, a fully integrated system has become available (Med-Associates, St. Albans, VT) that controls stimuli presented to the mice and records all behavioral responses through threshold-detection systems and video analyses.

During testing, particular precautions must be taken so that the observed emotional responses can be attributed directly to conditioning and not other influences. Novel scents can be powerful manipulators of behavior in rodents and, under particular circumstances, can provoke freezing behaviors [234]. Hence, it is important that the test chambers be cleaned adequately between animal testing. We use unscented Anlage spray (VWR International, West Chester, PA) or NPD-128 (VWR). The use of alcohol or ammonia-based cleaners between testing is discouraged, as these agents produce lingering odors that can be distracting to the animals during subsequent tests. Animals should be maintained in group housing before testing, as it has been shown that isolation can induce abnormal behaviors and interfere with attention [235] and learning [236]. During conditioning and testing, it is important that the mice be housed or maintained in rooms separate from the test room. If animals are to be evaluated in both contextual and cued fear conditioning, then modes of transit to the test rooms should be changed so that subtle but salient cues are not transmitted to the animal prior to testing. The investigator should also wear different lab coats and gloves for conditioning and testing.

In our lab, the mouse is placed in the test chamber for 2 min. After this time, a 30-sec, 65-dB, 2900-Hz tone is given, and 2 sec before termination of the tone, a 0.4-mA scrambled foot shock is administered where the tone and shock terminate simultaneously. The mouse remains in the apparatus for 30 sec before removal to the home cage. Parenthetically, while we find a single conditioning trial is often sufficient for conditioning, two to three CS-UCS pairings can provide more robust results when the pairings are separated by 30 to 60 sec. During the entire conditioning session, all incidences of spontaneous freezing behaviors are recorded. It may also be informative to observe whether the mice orient to the CS and whether they respond to the UCS during conditioning (e.g., running, vocalizing, or jumping). Mice that do not show these responses may have hearing loss [133] and be insensitive to the level of foot shock (see Appendix to this chapter).

Following conditioning, the mouse is returned to its home cage and tested 24 h later. Retention is measured in both the context and cued tests (Figure 12.15). For context testing, the mouse is returned to the original conditioning chamber and allowed 5 min of free exploration in the absence of tone. The next day the animal is evaluated in cued fear conditioning. Cued testing involves placing the mouse in a new test chamber or in the conditioning chamber where previous sensory cues have been disguised with novel floor textures, changes in wall patterns or colors, etc. In the cued test, the animal is allowed to explore the new surroundings for 2 min, after which the CS is presented for 3 min. In each test, freezing behavior is scored and expressed as the total sec of freezing, the percent test time freezing, or the percent time spent immobile per min [13]. The latter option is advantageous in that the pattern of behavior can be examined across testing for delays in freezing following exposure to the chamber or tone, or for early extinction of the behavior. Titration of the interval between conditioning and testing can be used to examine the acquisition and consolidation of emotional memory [181]. Alternatively, strength of the emotional memory can be evaluated through repeated apparatus exposure to follow extinction [237].

In fear conditioning two CS exist; one related to the contextual fear and the other associated with the auditory cue. As the contextual CS comprises the testing environment, the CS is more generalized and is present continuously throughout conditioning and testing. By comparison, the cued CS is usually restricted to a single sensory modality (e.g., tone) that is precisely timed to coincide with the presentation of the foot shock. Due to these distinctions, the contextual CS is more plastic and readily amenable to extinction compared with the cued CS. These observations suggested to Phillips and LeDoux [238] that different neural pathways control the expression of conditioned emotional responses in these two forms of fear conditioning. While the amygdala is critical for expression of conditioned fear in both the cued and contextual tests, the hippocampus is involved in contextual fear. Lesion studies have confirmed these hypotheses [238].

Conditioned fear is mediated by brain areas that can detect, process, and direct behavioral responses to perceived danger [225, 226]. Various investigators have consistently demonstrated that the amygdala is critical for the induction, BLA, and maintenance of conditioned fear [181, 239]. More specifically, the basolateral is a locus of sensory input that is proposed to be the location where the CS and UCS are initially associated. Neurons in the ventral hippocampus, subiculum, and CA1 region project to BLA, and damage to these areas can interfere with the development of contextual fear conditioning [224]. In addition, the central nucleus of the amygdala sends projections to the hypothalamus and brainstem to support the behavioral responses. Although lesions to the central nucleus can disrupt the expression of conditioned fear responses, damage to the specific projection areas can interrupt individual conditioned responses. For example, lesions to the periaqueductal gray can perturb freezing behavior while leaving blood pressure unaffected. Additionally, disruptions to the stria terminalis also interfere with conditioned release of adrenocorticotropin while leaving freezing and blood pressure responses intact [224].

Both glutamate and dopamine have been shown to be important for fear conditioning. For instance, NMDA receptors in the dorsal hippocampus and CA1 region play an important role in controlling freezing responses in contextual, but not cued, fear conditioning [44, 240]. Dopamine is also an important regulator of fear conditioning [225]. Dopamine D1 and D2 receptors are highly expressed in the amygdala [241]. Fear-arousing stimuli lead to the activation of dopamine neurons and the enhancement of dopamine neurotransmission in the amygdala [242]. Specific roles of the dopamine D1 and D2 receptors in fear conditioning have also been elucidated with mutant mice and drugs. When the D1 receptor mutants are examined in fear conditioning, they exhibit conditioned fear responses but fail to show extinction when repeatedly reexposed to the conditioning context [130]. Although D2 receptor homozygous mutants have not been examined in fear conditioning, administration of a selective D2 receptor antagonist (raclopride) into the amygdala impairs the acquisition and retention of the conditioned emotional response [243]. Bilateral intra-amygdala infusion of another D2 receptor antagonist (eticlopride) also disrupts formation and consolidation of emotional memory [244]. The roles of both the dopamine and glutamate in conditioned fear are particularly interesting because aberrations in these neurotransmitter systems are implicated in the pathologies of several psychiatric disorders where inappropriate fear responses are reported [225].

SUMMARY AND COMMENTS

The purpose of this chapter was to provide a general overview of tests designed to evaluate cognitive function in mice. An ancillary goal was to make the researcher aware that neurological and psychiatric dysfunction can also perturb cognition and that these complex behaviors have symbiosis. Given these relationships, it is important to evaluate mice on multiple dimensions of behavior to identify their phenotype. In cognitive testing, multiple behavioral and physiological deficiencies can confound cognitive performance. Finally, since the genetic background of mice can significantly influence behavior [245, 246], it is critical that the appropriate controls be included in the study so that any behavioral dysfunction can be more clearly ascribed to the gene(s) under study rather than representing other influences.

To date, animal models that use mutant mice have provided new insights into behavioral function. Although these models have given invaluable information, it is important to remember that these animals only provide approximations of the symptomologies and deficiencies of human patients. Nevertheless, subjecting mice to a comprehensive battery of tests provides a better framework for understanding not only the overall behavioral phenotype of the mutant, but also for more fully recognizing the limitations of the specific animal model. Despite this precaution, it should be emphasized that the goal of animal research is not to mimic precisely the human diseases or disorders under study, but to provide greater insights into basic genetic and molecular mechanisms involved in expression of the behavior. Once these mechanisms are better understood, new therapeutic strategies can be explored and developed for treating human patients.

APPENDIX: SHOCK-THRESHOLD TESTING

To evaluate sensitivity to foot shock, animals are exposed to different intensities of scrambled foot shock, and their behavioral responses are videotaped. In our laboratory, we use seven different intensities of foot shock (0, 0.05, 0.1, 0.15, 0.2, 0.3, and 0.4 mA) and present them in a random order, where each intensity is presented three times over 2 sec. Behavioral responses are scored by hand or with a computerized behavioral scoring program (Noldus Observer, Noldus Information Technology, Leesburg, VA). Responses are placed on a rating scale, with the lowest level of response scored as zero, indicating no overt response to the foot shock. Low-level responses include freezing, face wiping or self-grooming, shaking, or rapid forward departures; whereas moderate responses include tail rattling or retreating from foot shock. Moderately reactive responses include kicking, vocalization, and locomotor reactivity such as darting and leaping. The highest level of response includes jumping against the walls or ceiling of the chamber. Behavioral scores are summed based on the type of response and analyzed as a function of shock intensity. For behavioral conditioning, we select a level of foot shock that promotes only moderate responses. Foot shocks that induce continuous vocalizations, darting, or jumping may traumatize the animal and lead to inconclusive results.

REFERENCES

1. Provost, G.S., Kretz, P.L., Hamner, R.T., Matthews, C.D., Rogers, B.J., Lundberg, K.S., Dycaico, M.J., and Short, J.M., Transgenic systems for *in vivo* mutation analysis, *Mutation Res.*, 288, 133, 1993.
2. Reeves, R.H., Irving, N.G., Moran, T.H., Wohn, A., Kitt, C., Sisodia, S.S., Schmidt, C., Bronson, R.T., and Davisson, M.T., A mouse model for Down syndrome exhibits learning and behaviour deficits, *Nat. Genet.*, 11, 177, 1995.
3. Vitaterna, M.H., King, D.P., Chang, A.M., Kornhauser, J.M., Lowrey, P.L., McDonald, J.D., Dove, W.F., Pinto, L.H., Turek, F.W., and Takahashi, J.S., Mutagenesis and mapping of a mouse gene, *Clock*, essential for circadian behavior, *Science*, 264, 719, 1994.
4. Cairns, R.B., MacCombie, D.J., and Hood, K.E., A developmental-genetic analysis of aggressive behavior in mice, I: behavioral outcomes, *J. Comp. Psychol.*, 97, 69, 1983.
5. Shen, E.H., Harland, R.D., Crabbe, J.C., and Phillips, T.J., Bidirectional selective breeding for ethanol effects on locomotor activity: characterization of FAST and SLOW mice through selection generation 35, *Alcohol. Clin. Exp. Res.*, 19, 1234, 1995.
6. Van Oortmerssen, G.A. and Sluyter, F., Studies on wild house mice, V: aggression in lines selected for attack latency and their Y-chromosomal congenics, *Behav. Genet.*, 24, 73, 1994.
7. Antoch, M.P., Song, E.J., Chang, A.M., Vitaterna, M.H., Zhao, Y., Wilsbacher, L.D., Sangoram, A.M., King, D.P., Pinto, L.H., and Takahashi, J.S., Functional identification of the mouse circadian *Clock* gene by transgenic BAC rescue, *Cell*, 89, 655, 1997.
8. King, D.P., Zhao, Y., Sangoram, A.M., Wilsbacher, L.D., Tanaka, M., Antoch, M.P., Steves, T.D., Vitaterna, M.H., Kornhauser, J.M., Lowrey, P.L., Turek, F.W., and Takahashi, J.S., Positional cloning of the mouse circadian *Clock* gene, *Cell*, 89, 655, 1997.
9. Lin, L., Faraco, J., Li, R., Kadotani, H., Rogers, W., Lin, X., Qui, X., de Jong, P.J., Nishino, S., and Mignot, E., The sleep disorder canine narcolepsy is caused by a mutation in the hypocretin (orexin) receptor 2 gene, *Cell,* 98, 365, 1999.
10. Kim, H., Krege, J.H., Kluckman, K.D., Hagaman, J.R., Hodgin, J.B., Best, C.F., Jennette, C., Coffman, T.M., Maeda, N., and Smithies, O., Genetic control of blood pressure and the angiotensinogen locus, *Proc. Natl. Acad. Sci. U.S.A.*, 92, 2735, 1995.
11. Lowell, B.B., S-Susulic, V., Hamann, A., Lawitts, J.A., Himms-Hagen, J., Boyer, B.B., Kozak, L.P., and Flier, J.S., Development of obesity in transgenic mice after genetic ablation of brown adipose tissue, *Nature*, 366, 740, 1993.
12. Mohn, A.R., Gainetdinov, R.R., Caron, M.G., and Koller, B.H., Mice with reduced NMDA receptor expression display behaviors related to schizophrenia, *Cell*, 98, 427, 1999.

13. Pillai-Nair, N., Panicker, A.K., Rodriguiz, R.M., Foti, S., Huang, J., Wetsel, W.C., and Manness, P.F., NCAM-secreting transgenic mice display abnormalities in interneurons and behaviors related to schizophrenia, *J. Neurosci.*, 25, 4659, 2005.

14. Silva, A.J., Paylor, R., Wehner, J.M., and Tonegawa, S., Impaired spatial learning in alpha-calcium-calmodulin kinase II mutant mice, *Science,* 257, 206, 1992.

15. Usiello, A., Baik, J.H., Rouge-Pont, F., Picetti, R., Dierich, A., LeMeur, M., Piazza, P.V., and Borrelli, E., Distinct functions of the two isoforms of dopamine D2 receptors, *Nature*, 408, 199, 2000.

16. Ribar, T.J., Rodriguiz, R.M., Khiroug, L., Wetsel, W.C., Augustine, G.J., and Means, A.R., Cerebellar defects in Ca^{2+}/calmodulin kinase IV-deficient mice, *J. Neurosci.*, 20 (R107), 1, 2000.

17. Mayford, M., Bach, M.E., Huang, Y.Y., Wang, L., Hawkins, R.D., and Kandel, E.R., Control of memory formation through regulated expression of a CaMKII transgene, *Science*, 274, 1678, 1996.

18. Tsien, J.Z., Chen, D.F., Gerber, D., Tom, C., Mercer, E.H., Anderson, D.J., Mayford, M., Kandel, E.R., and Tonegawa, S., Subregion- and cell-type-restricted gene knockout in mouse brain, *Cell*, 87, 1317, 1996.

19. Lewandoski, M., Conditional control of gene expression in the mouse, *Nat. Rev. Genetics*, 2, 743, 2001.

20. Xia, H., Mao, Q., Eliason, S.L., Harper, S.Q., Martins, I.H., Orr, H.T., Paulson, H.L., Yang, L., Kotin, R.M., and Davidson, B.L., RNAi suppresses polyglutamine-induced neurodegeneration in a model of spinocerebellar ataxia, *Nat. Med.*, 10, 816, 2004.

21. Sora, I., Hall, F.S., Andrews, A.M., Itokawa, M., Li, X.F., Wei, H.B., Wichems, C., Lesch, K.P., Murphy, D.L., and Uhl, G.R., Molecular mechanisms of cocaine reward: combined dopamine and serotonin transporter knockouts eliminate cocaine place preference, *Proc. Natl. Acad. Sci. U.S.A.*, 98, 5300, 2001.

22. Henderson, N.D., Turri, M.G., DeFries, J.C., and Flint, J., QTL analysis of multiple behavioral measures of anxiety in mice, *Behav. Genet.*, 34, 267, 2004.

23. Lazarov, O., Robinson, J., Tang, Y.P., Hairston, I.S., Korade-Mirnics, Z., Lee, V.M., Hersh, L.B., Sapolsky, R.M., Mirnics, K., and Sisodia, S.S., Environmental enrichment reduces Aβ levels and amyloid deposition in transgenic mice, *Cell*, 120, 701, 2005.

24. Yerkes, R.M., *The Dancing Mouse: A Study in Animal Behavior*, Macmillan Press, Oxford, 1907.

25. Coburn, C.A., Heredity of wildness and savageness in mice, *Behav. Mono.*, 4, 71, 1922.

26. Washburn, M.F., Hunger and speed of running as factors in maze learning in mice, *J. Comp. Psychol.*, 6, 181, 1927.

27. Crawley, J.N. and Paylor, R., A proposed test battery and constellations of specific behavioral paradigms to investigate the behavioral phenotypes of transgenic and knockout mice, *Hormone Behav.*, 31, 197, 1997.

28. Collingridge, G.L., Kehl, S.J., and McLennan, H., Excitatory amino acids in synaptic transmission in the Schaffer collateral-commissural pathway of the rat hippocampus, *Physiology*, 334, 33, 1983.

29. Morris, R.G., Anderson, E., Lynch, G.S., and Baudry, M., Selective impairment of learning and blockade of long-term potentiation by an N-methyl-D-aspartate receptor antagonist, AP5, *Nature*, 319, 774, 1986.

30. Forrest, D., Yuzaki, M., Soares, H.D., Ng, L., Luk, D.C., Sheng, M., Stewart, C.L., Morgan, J.I., Connor, J.A., and Curran, T., Targeted disruption of NMDA receptor 1 gene abolishes NMDA response and results in neonatal death, *Neuron*, 13, 325, 1994.

31. Li, Y., Erzurumlu, R.S., Chen, C., Jhaveri, S., and Tonegawa, S., Whisker-related neuronal patterns fail to develop in the trigeminal brainstem nuclei of NMDAR1 knockout mice, *Cell*, 76, 427, 1994.
32. Cui, Z., Wang, H., Tan, Y., Zaia, K.A., Zhang, S., and Tsien, J.Z., Inducible and reversible NR1 knockout reveals crucial role of the NMDA receptor in preserving remote memories in the brain, *Neuron*, 41, 781, 2004.
33. Grove, M., Demyanenko, G., Rodriguiz, R.M., Quiroz, M.E., Martensen, S.A., Robinson, M.R., Wetsel, W.C., Maness, P.F., and Pendergast, A.M., Ablation of Abl-interactor 2 (Abi2), a novel component of early adherens junctions and dendritic spines, elicits defective cell morphology and migration in the eye and brain, *Mol. Cell. Biol.*, 24, 10905, 2004.
34. Gitler, D., Takagishi, Y., Feng, J., Ren, Y., Rodriguiz, R.M., Wetsel, W.C., Greengard, P., and Augustine, G.A., Different presynaptic roles of synapsins at excitatory and inhibitory synapses, *J. Neurosci.*, 24, 11368, 2005.
35. Shepherd, J.K., Grewal, S.S., Fletcher, A., Bill, D.J., and Dourish, C.T., Behavioural and pharmacological characterization of the elevated "zero-maze" as an animal model of anxiety, *Psychopharmacology*, 116, 5664, 1994.
36. Kliethermes, C.L., Cronise, K., and Crabbe, J.C., Anxiety-like behavior in mice in two apparatuses during withdrawal from chronic ethanol vapor inhalation, *Alcoholism: Clin. Exp. Res.*, 28, 1012, 2004.
37. Pogorelov, V.M., Rodriguiz, R.M., Inscol, M.L., Caron, M.G., and Wetsel, W.C., Novelty seeking and stereotypic activation of behavior in mice with disruption of the *Dat1* gene, *Neuropsychopharmacology*, 30, 1818, 2005.
38. Randall, C.L., Becker, H.C., and Middaugh, L.D., Effect of prenatal ethanol exposure on activity and shuttle avoidance behavior in adult C57 mice, *Alcohol Drug Res.*, 6, 351, 1985.
39. Chiu, C.S., Brickley, S., Jensen, K., Southwell, A., Mckinney, S., Cull-Candy, S., Mody, I., and Lester, H.A., GABA transporter deficiency causes tremor, ataxia, nervousness, and increased GABA-induced tonic conductance in cerebellum, *J. Neurosci.*, 25, 3234, 2005.
40. Rogers, D.C., Fisher, E.M., Brown, S.D., Peters, J., Hunter, A.J., and Martin, J.E., Behavioral and functional analysis of mouse phenotype: SHIRPA, a proposed protocol for comprehensive phenotype assessment, *Mamm. Genome*, 8, 711, 1997.
41. Irwin, S., Comprehensive observational assessment, 1A: a systematic, quantitative procedure for assessing the behavioural and physiologic state of the mouse, *Psychopharmacologia*, 13, 222, 1968.
42. Ralph, R.J., Varty, G.B., Kelly, M.A., Wang, Y.M., Caron, M.G., Rubinstein, M., Grandy, D.K., Low, M.J., and Geyer, M.A., The dopamine D2, but not D3 or D4, receptor subtype is essential for the disruption of prepulse inhibition produced by amphetamine in mice, *J. Neurosci.*, 19, 4627, 1999.
43. Geyer, M.A., Krebs-Thompson, K., Braff, D.L., and Swerdlow, N.R., Pharmacological studies of prepulse inhibition models of sensorimotor gating deficits in schizophrenia: a decade in review, *Psychopharmacology*, 156, 117, 2001.
44. Rampon, C., Tang, Y.P., Goodhouse, J., Shimizu, E., Kyin, M., and Tsien, J.Z., Enrichment induces structural changes and recovery from nonspatial memory deficits in CA1 NMDAR1-knockout mice, *Nat. Neurosci.*, 3, 238, 2000.
45. Harrison, A.A., Everitt, B.J., and Robbins, T.W., Central serotonin depletion impairs both the acquisition and performance of a symmetrically reinforced go/no-go conditional visual discrimination, *Behav. Brain Res.*, 100, 99, 1999.

46. Bodyak, N. and Slotnick, B., Performance of mice in an automated olfactometer: odor detection, discrimination and odor memory, *Chem. Senses*, 24, 637, 1999.
47. Robbins, T.W., Muir, J.L, Killcross, A.S., and Pretsell, D., Methods for assessing attention and stimulus control in the rat, in *Behavioral Neuroscience: A Practical Approach*, Vol. I, Sahgal, A., Ed., Oxford University Press, New York, 1993, p. 13.
48. Killcross, A.S., Dickinson, A., and Robbins, T.W., Amphetamine-induced disruptions of latent inhibition are reinforcer mediated: implications for animal models of schizophrenic attentional dysfunction, *Psychopharmacology*, 115, 185, 1994.
49. Lubow, R.E., *Latent Inhibition and Conditioned Attention Theory*, Cambridge University Press, New York, 1989.
50. Posner, M.I., Attention in cognitive neuroscience, in *The Cognitive Neurosciences*, Gazzaniga, M.S., Ed., MIT Press, Cambridge, 1995, p. 615.
51. Venables, P.H., The effect of auditory and visual stimulation on the skin potential responses of schizophrenics, *Brain*, 83, 77, 1960.
52. McGhie, A. and Chapman, J., Disorders of attention and perception in early schizophrenia, *Br. J. Med. Psychol.*, 34, 102, 1961.
53. Hoffman, H.S. and Searle, J.L., Acoustic variables in the modification of the startle reaction in the rat, *J. Comp. Physiol. Psych.*, 60, 53, 1965.
54. Braff, D.L, Geyer, M.A., and Swerdlow, N.R., Human studies of prepulse inhibition of startle: normal subjects, patient groups, and pharmacological studies, *Psychopharmacology*, 156, 234, 2001.
55. Braff, D.L. and Freedman, R., The importance of endophenotypes in studies of the genetics of schizophrenia, in *Neuropsychopharmacology, the 5th Generation of Progress*, Davis, K.L., Charney, D., Coyle, J.T., and Nemeroff, C., Eds., Lippincott, Williams & Wilkins, Baltimore, 2002, p. 703.
56. Swerdlow, N.R., Shoemaker, J.M., Auerbach, P.P., Pitcher, L., Goins, J., and Platten, A., Heritable differences in dopaminergic regulation of sensorimotor gating: temporal, pharmacological and generational analyses of apomorphine effects on prepulse inhibition, *Psychopharmacology*, 174, 452, 2004.
57. Swerdlow, N.R., Platten, A., Shoemaker, J., Pitcher, L., and Auerbach, P., Effects of pergolide on sensorimotor gating of the startle reflex in rats, *Psychopharmacology*, 158, 230, 2001.
58. Bushnell, P.J., Behavioral approaches to the assessment of attention in animals, *Psychopharmacology*, 138, 231, 1998.
59. Beane, M. and Marrocco, R.T., Norepinephrine and acetylcholine mediation of the components of reflexive attention: implications for attention deficit disorders, *Prog. Neurobiol.*, 74, 167, 2004.
60. Berger, A., Henik, A., and Rafal, R., Competition between endogenous and exogenous orienting of visual attention, *J. Exper. Psychol. Gen.*, 134, 207, 2005.
61. Gallagher, M. and Holland, P.C., The amygdala complex: multiple roles in associative learning and attention, *Proc. Natl. Acad. Sci. U.S.A.*, 91, 11771, 1994.
62. Aston-Jones, G., Chiang, C., and Alexinsky, T., Discharge of noradrenergic locus coeruleus neurons in behaving rats and monkeys suggest a role in vigilance, *Prog. Brain Res.*, 88, 501, 1991.
63. Nobre, A.C., Orienting attention to instants in time, *Neuropsychologia*, 39, 1317, 2001.
64. Chudasama, Y., Nathwani, F., and Robbins, T.W., d-Amphetamine remediates attentional performance in rats with dorsal prefrontal lesions, *Behav. Brain Res.*, 158, 97, 2005.
65. Ranganath, C. and Rainer, G., Neural mechanisms for detecting and remembering novel events, *Nat. Rev. Neurosci.*, 4, 193, 2003.

66. Hubel, D.H. and Wiesel, T.N., *Brain and Visual Perception*, Oxford University Press, Oxford, 2005.
67. Ross, R.G., Olincy, A., Harris, J.G., Radant, A., Adler, L.E., Compagnon, N., and Freedman, R., The effects of age on a smooth pursuit tracking task in adults with schizophrenia and normal subjects, *Biol. Psychiatry*, 46, 383, 1999.
68. Gooding, D.C., Miller, M.D., and Kwapil, T.R., Smooth pursuit eye tracking and visual fixation in psychosis-prone individuals, *Psychiatry Res.*, 93, 41, 2000.
69. Gooding, D.C. and Tallent, K.A., Spatial, object, and affective working memory in social anhedonia: an exploratory study, *Schizophr. Res.*, 63, 247, 2003.
70. Smyrnis, N., Kattoulas, E., Evdokimidis, I., Stefanis, N.C., Avramopoulos, D., Pantes, G., Theleritis, C., and Stefanis, C.N., Active eye fixation performance in 940 young men: effects of IQ, schizotypy, anxiety and depression, *Exp. Brain Res.*, 156, 1, 2004.
71. Clementz, B.A., Farber, R.H., Lam, M.N., and Swerdlow, N.R., Ocular motor responses to unpredictable and predictable smooth pursuit stimuli among patients with obsessive-compulsive disorder, *J. Psychiatry Neurosci.*, 21, 21, 1996.
72. Munoz, D.P., Armstrong, I.T., Hampton, K.A., and Moore, K.D., Altered control of visual fixation and saccadic eye movements in attention-deficit hyperactivity disorder, *J. Neurophysiol.*, 90, 503, 2003.
73. Crevits, L., Vandierendonck, A., Stuyven, E., Verschaete, S., and Wildenbeest, J., Effect of intention and visual fixation disengagement on prosaccades in Parkinson's disease patients, *Neuropsychologia*, 42, 624, 2004.
74. Davidson, M.C. and Marrocco, R.T., Local infusion of scopolamine into intraparietal cortex slows covert orienting in rhesus monkeys, *J. Neurophysiol.*, 83, 536, 2000.
75. Posner, M.I., Orienting of attention, *Q. J. Exp. Psychol.*, 32, 3, 1980.
76. Honey, R.C., Watt, A., and Good, M., Hippocampal lesions disrupt an associative mismatch process, *J. Neurosci.*, 18, 2226, 1998.
77. Eisenberg, N., Guthrie, I.K., Fabes, R.A., Shepard, S., Losoya, S., Murphy, B.C., Jones, S., Poulin, R., and Reiser, M., Prediction of elementary school children's externalizing problem behaviors from attentional and behavioral regulation and negative emotionality, *Child Dev.*, 71, 1367, 2000.
78. Ellis, L.K., Rothbart, M.K., and Posner, M.I., Individual differences in executive attention predict self-regulation and adolescent psychosocial behaviors, *Ann. N.Y. Acad. Sci.*, 1021, 337, 2004.
79. Dehaene, S., Artiges, E., Naccache, L., Martelli, C., Viard, A., Schurhoff, F., Recasens, C., Martinot, M.L., Leboyer, M., and Martinot, J.L., Conscious and subliminal conflicts in normal subjects and patients with schizophrenia: the role of the anterior cingulate, *Proc. Natl. Acad. Sci. U.S.A.*, 100, 13722, 2003.
80. Spessot, A.L., Plessen, K.J., and Peterson, B.S., Neuroimaging of developmental psychopathologies: the importance of self-regulatory and neuroplastic processes in adolescence, *Ann. N.Y. Acad. Sci.*, 1021, 86, 2004.
81. Barkley, R.A., Behavioral inhibition, sustained attention, and executive functions: constructing a unifying theory of ADHD, *Psychol. Bull.*, 121, 65, 1997.
82. Wurtz, R.H. and Albano, J.E., Visual-motor function of the primate superior colliculus, *Annu. Rev. Neurosci.*, 3, 189, 1980.
83. Posner, M.I. and Dehaene, S., Attentional networks, *Trends Neurosci.*, 17, 75, 1994.
84. Aston-Jones, G., Rajkowski, J., Kubiak, P., and Akaoka, H., Acute morphine induces oscillatory discharge of noradrenergic locus coeruleus neurons in the waking monkey, *Neurosci. Lett.*, 140, 219, 1992.
85. Robbins, T.W., The 5-choice serial reaction time task: behavioural pharmacology and functional neurochemistry, *Psychopharmacology*, 163, 362, 2002.

86. Dalley, J.W., Cardinal, R.N., and Robbins, T.W., Prefrontal executive and cognitive function in rodents: neural and neurochemical substrates, *Neurosci. Biobehav. Rev.*, 28, 771, 2004.

87. McGaughy, J., Dalley, J.W., Morrison, C.H., Everitt, B.J., and Robbins, T.W., Selective behavioral and neurochemical effects of cholinergic lesions produced by intra-basalis infusions of 192 IgG-saporin on attentional performance in a five-choice serial reaction time task, *Neurosci.*, 22, 1905, 2002.

88. Carli, M., Robbins, T.W., Evenden, J.L., and Everitt, B.J., Effects of lesions to ascending noradrenergic neurones on performance of a 5-choice serial reaction task in rats; implications for theories of dorsal noradrenergic bundle function based on selective attention and arousal, *Behav. Brain Res.*, 9, 361, 1983.

89. Logan, G.D., On the ability to inhibit thought and action: a users' guide to the stop signal paradigm, in *Inhibitory Processes in Attention, Memory and Language*, Dagenbach, D. and Carr, T.H., Eds., Academic Press, San Diego, 1994, p. 189.

90. Nelson, J.K., Reuter-Lorenz, P.A., Sylvester, C-Y.C., Jonides, J., and Smith, E.E., Dissociable neural mechanisms underlying response-based and familiarity-based conflict in working memory, *Proc. Natl. Acad. Sci. U.S.A.*, 100, 11171, 2003.

91. Eagle, D.M. and Robbins, T.W., Inhibitory control in rats performing a stop-signal reaction time task: effects of lesions of the medial striatum and d-amphetamine, *Behav. Neurosci.*, 117, 1302, 2003.

92. Marston, H.M., Sahgal, A., and Katz, J.L., Signal-detection methods, in *Behavioral Neuroscience: A Practical Approach*, Vol. II, Sahgal, A., Ed., Oxford University Press, New York, 1993, p. 188.

93. Aggleton, J.P., Behavioural tests for the recognition of non-spatial information by rats, in *Behavioral Neuroscience: A Practical Approach*, Vol. I, Sahgal, A., Ed., Oxford University Press, New York, 1993, p. 81.

94. Logan, G.D. and Cowan, W.B., On the ability to inhibit thought and action: a theory of an act of control, *Psychol. Rev.*, 91, 295, 1984.

95. Schachar, R.J., Tannock, R., Marriott, M., and Logan, G., Deficient inhibitory control in attention-deficit hyperactivity disorder, *J. Abnorm. Child Psychol.*, 23, 411, 1995.

96. Rubia, K., Smith, A.B., Brammer, M.J., Toone, B., and Taylor, E., Abnormal brain activation during inhibition and error detection in medication-naive adolescents with ADHD, *Am. J. Psychiatry*, 162, 1067, 2005.

97. Brown, J.W. and Braver, T.S., Learned predictions of error likelihood in the anterior cingulate cortex, *Science*, 307, 1118, 2005.

98. Carter, C.S., Braver, T.S., Barch, D.M., Botvinick, M.M., Noll, D., and Cohen, J.D., Anterior cingulate cortex, error detection, and the online monitoring of performance, *Science*, 280, 747, 1998.

99. Rushworth, M.F., Walton, M.E., Kennerley, S.W., and Bannerman, D.M., Action sets and decisions in the medial frontal cortex, *Trends Cogn. Sci.*, 8, 410, 2004.

100. Semrud-Clikeman, M., Steingard, R.J., Filipek, P., Biederman, J., Bekken, K., and Renshaw, P.F., Using MRI to examine brain-behavior relationships in males with attention deficit disorder with hyperactivity, *J. Am. Acad. Child. Adolesc. Psychiatry*, 39, 477, 2000.

101. Ferry, A.T., Ongur, D., An, X., and Price, J.L., Prefrontal cortical projections to the striatum in macaque monkeys: evidence for an organization related to prefrontal networks, *J. Comp. Neurol.*, 425, 447, 2000.

102. Slovin, H., Abeles, M., Vaadia, E., Haalman, I., Prut, Y., and Bergman, H., Frontal cognitive impairments and saccadic deficits in low-dose MPTP-treated monkeys, *J. Neurophysiol.*, 81, 858, 1999.

103. Sommer, W., Leuthold, H., and Schubert, T., Multiple bottlenecks in information processing? An electrophysiological examination, *Psychon. Bull. Rev.*, 8, 81, 2001.
104. Clarke, H.F., Dalley, J.W., Crofts, H.S., Robbins, T.W., and Roberts, A.C., Cognitive inflexibility after prefrontal serotonin depletion, *Science*, 304, 878, 2004.
105. Pantelis, C., Harvey, C.A., Plant, G., Fossey, E., Maruff, P., Stuart, G.W., Brewer, W.J., Nelson, H.E., Robbins, T.W., and Barnes, T.R., Relationship of behavioural and symptomatic syndromes in schizophrenia to spatial working memory and attentional set-shifting ability, *Psychol. Med.*, 34, 693, 2004.
106. Purcell, R., Maruff, P., Kyrios, M., and Pantelis, C., Cognitive deficits in obsessive-compulsive disorder on tests of frontal-striatal function, *Biol. Psychiatry*, 43, 348, 1998.
107. Meyer, U., Chang, de L.T., Feldon, J., and Yee, B.K., Expression of the CS- and US-pre-exposure effects in the conditioned taste aversion paradigm and their abolition following systemic amphetamine treatment in C57BL6/J mice, *Neuropsychopharmacology*, 29, 2140, 2004.
108. Rimer, M., Barrett, D.W., Maldonado, M.A., Vock, V.M., and Gonzalez-Lima, F., Neuregulin-1 immunoglobulin-like domain mutant mice: clozapine sensitivity and impaired latent inhibition, *Neuroreport*, 16, 271, 2005.
109. Harrell, A.V. and Allan, A.M., Improvements in hippocampal-dependent learning and decremental attention in 5-HT(3) receptor overexpressing mice, *Learn. Mem.*, 10, 410, 2003.
110. Mason, S.T. and Lin, D., Dorsal noradrenergic bundle and selective attention in the rat, *J. Comp. Physiol. Psychol.*, 94, 819, 1980.
111. Clark, A.J., Feldon, J., and Rawlins, J.N., Aspiration lesions of rat ventral hippocampus disinhibit responding in conditioned suppression or extinction, but spare latent inhibition and the partial reinforcement extinction effect, *Neuroscience*, 48, 821, 1992.
112. Yee, B.K., Feldon, J., and Rawlins, J.N., Latent inhibition in rats is abolished by NMDA-induced neuronal loss in the retrohippocampal region, but this lesion effect can be prevented by systemic haloperidol treatment, *Behav. Neurosci.*, 109, 227, 1995.
113. Schiller, D., Zuckerman, L., and Weiner, I., Abnormally persistent latent inhibition induced by lesions to the nucleus accumbens core, basolateral amygdala and orbitofrontal cortex is reversed by clozapine but not by haloperidol, *J. Psychiatr. Res.*, 40, 167, 2005.
114. Kandel, E.R., Kupfermann, I., and Iversen, S., Learning and memory, in *Principles of Neural Science*, Kandell, E.R., Schwartz, J.H., and Jessell, T.M., Eds., McGraw Hill, New York, 2000, p. 1225.
115. Kimble, G.A., *Hilgard and Marquis' Conditioning and Learning*, 2nd ed., Appleton-Century-Crofts, New York, 1961, p. 44.
116. Baddeley, A.D. and Hitch, G., The recency effect: implicit learning with explicit retrieval? *Mem. Cognit.*, 21, 146, 1993.
117. Squire, L.R. and Zola, S.M., Memory, memory impairment, and the medial temporal lobe, *Cold Spring Harb. Symp. Quant. Biol.*, 61, 185, 1996.
118. Schacter, D.L., Norman, K.A., and Koutstaal, W., The cognitive neuroscience of constructive memory, *Annu. Rev. Psychol.*, 49, 289, 1998.
119. Raaijmakers, J.G. and Shiffrin, R.M., Models for recall and recognition, *Annu. Rev. Psychol.*, 43, 205, 1992.
120. Haberlandt, K., Thomas, J.G., Lawrence, H., and Krohn, T., Transposition asymmetry in immediate serial recall, *Memory*, 13, 274, 2005.
121. Wietrzych, M., Meziane, H., Sutter, A., Ghyselinck, N., Chapman, P.F., Chambon, P., and Krezel, W., Working memory deficits in retinoid X receptor gamma-deficient mice, *Learn. Mem.*, 12, 318, 2005.

122. Frankland, P.W., Josselyn, S.A., Anagnostaras, S.G., Kogan, J.H., Takahashi, E., and Silva, A.J., Consolidation of CS and US representations in associative fear conditioning, *Hippocampus*, 14, 557, 2004.

123. Laxmi, T.R., Stork, O., and Pape, H.C., Generalisation of conditioned fear and its behavioural expression in mice, *Behav. Brain Res.*, 145, 89, 2003.

124. Weinberger, S.B., Koob, G.F., and Martinez, J.L., Jr., Differences in one-way active avoidance learning in mice of three inbred strains, *Behav. Genet.*, 22, 177, 1992.

125. Clincke, G.H. and Wauquier, A., Pharmacological protection against hypoxia-induced effects on medium-term memory in a two-way avoidance paradigm, *Behav. Brain. Res.*, 14, 139, 1984.

126. Thomas, S.A. and Palmiter, R.D., Disruption of the dopamine β-hydroxylase gene in mice suggests roles for norepinephrine in motor function, learning, and memory, *Behav. Neurosci.*, 111, 579, 1997.

127. Anisman, H., Differential effects of scopolamine and D-amphetamine on avoidance: strain interactions, *Pharmacol. Biochem. Behav.*, 3, 809, 1975.

128. Tulving, E., Episodic memory: from mind to brain, *Ann. Rev. Psychol.*, 53, 1, 2002.

129. Nakazawa, K., McHugh, T.J., Wilson, M.A., and Tonegawa, S., NMDA receptors, place cells and hippocampal spatial memory, *Nat. Rev. Neurosci.*, 5, 361, 2004.

130. El-Ghundi, M., O'Dowd, B.F., and George, S.R., Prolonged fear responses in mice lacking dopamine D1 receptor, *Brain Res.*, 892, 86, 2001.

131. Ennaceur, A. and Delacour, J., A new one-trial test for neurobiological studies of memory in rats, I: behavioral data, *Behav. Brain Res.*, 31, 47, 1988.

132. Nagy, Z.M., Porada, K.J., and Monsour, A.P., Ontogeny of short- and long-term memory capacities for passive avoidance training in undernourished mice, *Dev. Psychobiol.*, 13, 373, 1980.

133. Carlson, S. and Willott, J.F., The behavioral salience of tones as indicated by prepulse inhibition of the startle response: relationship to hearing loss and central neural plasticity in C57BL/6J mice, *Hear. Res.*, 15, 168, 1996.

134. Chen, T.H., Wang, M.F., Liang, Y.F., Komatsu, T., Chan, Y.C., Chung, S.Y., and Yamamoto, S., A nucleoside-nucleotide mixture may reduce memory deterioration in old senescence-accelerated mice, *J. Nutr.*, 130, 3085, 2000.

135. Tang, Y.P., Shimizu, E., Dube, G.R., Rampon, C., Kerchner, G.A., Zhuo, M., Liu, G., and Tsien, J.Z., Genetic enhancement of learning and memory in mice, *Nature*, 401, 63, 1999.

136. Kogan, J.H., Frankland, P.W., Blendy, J.A., Coblentz, J., Morowitz, Z., Schutz, G., and Silva, A.J., Spaced training induces normal long-term memory in CREB mutant mice, *Curr. Biol.*, 7, 1, 1996.

137. Rodriguiz, R.M., Chu, R., Caron, M.G., and Wetsel, W.C., Aberrant responses in social interaction of dopamine transporter knockout mice, *Behav. Brain Res.*, 148, 185, 2004.

138. Wrenn, C.C., Harris, A.P., Saavedra, M.C., and Crawley, J.N., Social transmission of food preference in mice: methodology and application to galanin-overexpressing transgenic mice, *Behav. Neurosci.*, 117, 21, 2003.

139. Eichenbaum, H. and Cohen, N.J., *From Conditioning to Conscious Recollection: Memory Systems of the Brain*, Oxford University Press, New York, 2001.

140. Rondi-Reig, L., Libbey, M., Eichenbaum, H., and Tonegawa, S., CA1-specific N-methyl-D-aspartate receptor knockout mice are deficient in solving a nonspatial transverse patterning task, *Proc. Natl. Acad. Sci. U.S.A.*, 98, 3543, 2001.

141. Smith, D.R., Striplin, C.D., Geller, A.M., Mailman, R.B., Drago, J., Lawler, C.P., and Gallagher, M., Behavioural assessment of mice lacking D1A dopamine receptors, *Neuroscience*, 86, 135, 1998.

142. Strupp, B.J. and Levitsky, D.A., Early brain insult and cognition: a comparison of malnutrition and hypothyroidism, *Dev. Psychobiol.*, 16, 535, 1983.

143. Alvarez, P., Lipton, P.A., Melrose, R., and Eichenbaum, H., Differential effects of damage within the hippocampal region on memory for a natural, nonspatial odor-odor association, *Learn. Mem.*, 8, 79, 2001.

144. Silva, A.J., Giese, K.P., Fedorov, N.B., Frankland, P.W., and Kogan, J.H., Molecular, cellular, and neuroanatomical substrates of place learning, *Neurobiol. Learn Mem.*, 70, 44, 1998.

145. Schwegler, H. and Crusio, W.E., Correlations between radial-maze learning and structural variations of septum and hippocampus in rodents, *Behav. Brain Res.*, 67, 29, 1995.

146. O'Keefe, J. and Nadel, L., *The Hippocampus as a Cognitive Map*, Oxford University Press, New York, 1979.

147. Cave, C.B. and Squire, L.R., Equivalent impairment of spatial and nonspatial memory following damage to the human hippocampus, *Hippocampus*, 1, 329, 1991.

148. O'Keefe, J. and Conway, D.G., Hippocampal place units in the freely moving rat: why they fire when they fire, *Exp. Brain Res.*, 31, 573, 1978.

149. Scoville, W.B. and Milner, B., Loss of recent memory after bilateral hippocampal lesions, *J. Neurochem.*, 20, 11, 1957.

150. Olton, D.S., Becker, J.T., and Handlemann, G.E., Hippocampus, space and memory, *Brain Behav. Sci.*, 2, 313, 1979.

151. Tulving, E., Episodic and semantic memory, in *Organization of Memory*, Tulving, E. and Donaldson, W., Eds., Academic Press, New York, 1972, p. 382.

152. Levin, E.D., Christopher, N.C., Lateef, S., Elamir, B.M., Patel, M., Liang, L-P., and Crapo, J.D., Extracellular superoxide dismutase overexpression protects against age-induced cognitive impairments in mice, *Behav. Genet.*, 32, 119, 2002.

153. Rawlins, J.N., Lyford, G.L., Seferiades, A., Deacon, R.M., and Cassaday, H.J., Critical determinants of nonspatial working memory deficits in rats with conventional lesions of the hippocampus or fornix, *Behav. Neurosci.*, 107, 420, 1993.

154. Addy, N. and Levin, E.D., Nicotine interactions with haloperidol, clozapine and risperidone and working memory function in rats, *Neuropsychopharmacology*, 27, 534, 2002.

155. DiMattia, B.V. and Kesner, R.P., Serial position curves in rats: automatic versus effortful information processing, *J. Exp. Psychol. Anim. Behav. Process.*, 10, 557, 1984.

156. Gainetdinov, R.R., Wetsel, W.C., Jones, S.R., Levin, E.D., Jaber, M., and Caron, M.G., Role of serotonin in the paradoxical calming effect of psychostimulants on hyperactivity, *Science*, 283, 397, 1999.

157. Stewart, C.A. and Morris, R.G.M., The watermaze, in *Behavioral Neuroscience: A Practical Approach*, Vol. I, Sahgal, A., Ed., Oxford University Press, New York, 1993, p. 105.

158. Lipp, H.P. and Wolfer, D.P., Genetically modified mice and cognition, *Curr. Opin. Neurobiol.*, 8, 272, 1998.

159. Tsai, G., Ralph-Williams, R.J., Martina, M., Bergeron, R., Berger-Sweeney, J., Dunham, K.S., Jiang, Z., Caine, S.B., and Coyle, J.T., Gene knockout of glycine transporter 1" characterization of the behavioral phenotype, *Proc. Natl. Acad. Sci. U.S.A.*, 22, 8454, 2004.

160. Buhot, M.C., Wolff, M., Savova, M., Malleret, G., Hen, R., and Segu, L., Protective effect of 5-HT1B receptor gene deletion on the age-related decline in spatial learning abilities in mice, *Behav. Brain Res.*, 142, 135, 2003.

161. Wolff, M., Savova, M., Malleret, G., Segu, L., and Buhot, M.C., Differential learning abilities of 129T2/Sv and C57BL/6J mice as assessed in three water maze protocols, *Behav. Brain Res.*, 136, 463, 2002.

162. Wolfer, D.P., Madani, R., Valenti, P., and Lipp, H.P., Extended analyses of path data from mutant mice using the public domain software Wintrack, *Physiol. Behav.*, 73, 745, 2001.

163. Wolfer, D.P. and Lipp, H.P., Dissecting the behavior of transgenic mice: is it the mutation, the genetic background, or the environment, *Exper. Physiol.*, 85, 627, 2000.

164. O'Keefe, J. and Dostrovsky, J., The hippocampus as a spatial map: preliminary evidence from unit activity in the freely moving rat, *Brain Res.*, 34, 171, 1971.

165. Wilson, M.A. and McNaughton, B.L., Dynamics of the hippocampal ensemble code for space, *Science*, 261, 1055, 1993.

166. Thompson, L.T. and Best, P.J., Long-term stability of place-field activity of single units recorded from the dorsal hippocampus of freely behaving rats, *Brain Res.*, 509, 299, 1990.

167. O'Keefe, J. and Speakman, A., Single unit activity in the rat hippocampus during a spatial memory task, *Exp. Brain Res.*, 68, 1, 1987.

168. Olton, D.S., Wenk, G.L., Church, R.M, and Meck, W.H., Attention and the frontal cortex as examined by simultaneous temporal processing, *Neuropsychologia*, 26, 307, 1988.

169. Eichenbaum, H., The topography of memory, *Nature*, 402, 597, 1999.

170. Andersen, P., Bliss, T.V.P., and Skrede, K.K., Lamellar organization of hippocampal excitatory pathways, *Exp. Brain Res.*, 13, 222, 1971.

171. Rosenzweig, E.S., Redish, A.D., McNaughton, B.L., and Barnes, C.A., Hippocampal map realignment and spatial learning, *Nat. Neurosci.*, 6, 609, 2003.

172. Andersen, P. and Lomo, T., Control of hippocampal output by afferent volley frequency, *Prog. Brain Res.*, 27, 400, 1967.

173. Bliss, T.V. and Lomo, T., Long-lasting potentiation of synaptic transmission in the dentate area of the anaesthetized rabbit following stimulation of the perforant path, *J. Physiol.*, 232, 331, 1973.

174. Lomo, T., The discovery of long-term potentiation, *Philos. Trans. R. Soc. London, B. Biol. Sci.*, 358, 617, 2003.

175. Chen, C. and Tonegawa, S., Molecular genetic analysis of synaptic plasticity, activity-dependent neural development, learning, and memory in the mammalian brain, *Ann. Rev. Neurosci.*, 20, 157, 1997.

176. McHugh, T.J., Blum, K.I., Tsien, J.Z., Tonegawa, S., and Wilson, M.A., Impaired hippocampal representation of space in CA1-specific NMDAR1 knockout mice, *Cell*, 87, 1339, 1996.

177. Marr, D.A., Simple memory: a theory for archicortex, *Philos. Trans. R. Soc. London, B. Biol. Sci.*, 202, 437, 1971.

178. Saucier, D. and Cain, D.P., Spatial learning without NMDA receptor-dependent long-term potentiation, *Nature*, 378, 186, 1995.

179. Herry, C. and Garcia, R., Prefrontal cortex long-term potentiation, but not long-term depression, is associated with the maintenance of extinction of learned fear in mice, *J. Neurosci.*, 22, 577, 2002.

180. Jay, T.M., Rocher, C., Hotte, M., Naudon, L., Gurden, H., and Spedding, M., Plasticity at hippocampal to prefrontal cortex synapses is impaired by loss of dopamine and stress: importance for psychiatric diseases, *Neurotox. Res.*, 6, 233, 2004.

181. Rodrigues, S.M., Schafe, G.E., and LeDoux, J.E., Molecular mechanisms underlying emotional learning and memory in the lateral amygdala, *Neuron*, 44, 75, 2004.

182. Tang, Y.P., Wang, H., Feng, R., Kyin, M., and Tsien, J.Z., Differential effects of enrichment on learning and memory function in NR2B transgenic mice, *Neuropharmacology*, 41, 779, 2001.

183. Harley, C., Noradrenergic and locus coeruleus modulation of the perforant path-evoked potential in rat dentate gyrus supports a role for the locus coeruleus in attentional and memorial processes, *Prog. Brain Res.*, 88, 307, 1991.

184. Tachibana, K., Matsumoto, M., Togashi, H., Kojima, T., Morimoto, Y., Kemmotsu, O., and Yoshioka, M., Milnacipran, a serotonin and noradrenaline reuptake inhibitor, suppresses long-term potentiation in the rat hippocampal CA1 field via 5-HT1A receptors and alpha 1-adrenoceptors, *Neurosci. Lett.*, 357, 91, 2004.

185. Martinez, J.L. and Kesner, R.P., *Neurobiology of Learning and Memory*, Academic Press, San Diego, 1998.

186. Otmakhova, N.A. and Lisman, J.E., D1/D5 dopamine receptor activation increases the magnitude of early long-term potentiation at CA1 hippocampal synapses, *J. Neurosci.*, 16, 7478, 1996.

187. Yang, H.W., Lin, Y.W., Yen, C.D., and Min, M.Y., Change in bi-directional plasticity at CA1 synapses in hippocampal slices taken from 6-hydroxydopamine-treated rats: the role of endogenous norepinephrine, *Eur. J. Neurosci.*, 16, 1117, 2002.

188. Gurden, H., Takita, M., and Jay, T.M., Essential role of D1 but not D2 receptors in the NMDA receptor-dependent long-term potentiation at hippocampal-prefrontal cortex synapses *in vivo*, *J. Neurosci.*, 20, RC106, 2000.

189. Kerr, J.N. and Wickens, J.R., Dopamine D-1/D-5 receptor activation is required for long-term potentiation in the rat neostriatum *in vitro*, *J. Neurophysiol.*, 85, 117, 2001.

190. Stanton, P.K. and Sarvey, J.M., Depletion of norepinephrine, but not serotonin, reduces long-term potentiation in the dentate gyrus of rat hippocampal slices, *J. Neurosci.*, 5, 2169, 1985.

191. Gray, R. and Johnston, D., Noradrenaline and beta-adrenoceptor agonists increase activity of voltage-dependent calcium channels in hippocampal neurons, *Nature*, 327, 620, 1987.

192. Huang, Y.Y. and Kandel, E.R., Modulation of both the early and the late phase of mossy fiber LTP by the activation of beta-adrenergic receptors, *Neuron*, 16, 611, 1996.

193. Katsuki, H., Izumi, Y., and Zorumski, C.F., Noradrenergic regulation of synaptic plasticity in the hippocampal CA1 region, *J. Neurophysiol.*, 77, 3013, 1997.

194. Izumi, Y. and Zorumski, C.F., Norepinephrine promotes long-term potentiation in the adult rat hippocampus *in vitro*, *Synapse*, 31, 196, 1999.

195. Ohashi, S., Matsumoto, M., Otani, H., Mori, K., Togashi, H., Ueno, K., Kaku, A., and Yoshioka, M., Changes in synaptic plasticity in the rat hippocampo-medial prefrontal cortex pathway induced by repeated treatments with fluvoxamine, *Brain Res.*, 949, 131, 2002.

196. Ohashi, S., Matsumoto, M., Togashi, H., Ueno, K., and Yoshioka, M., The serotonergic modulation of synaptic plasticity in the rat hippocampo-medial prefrontal cortex pathway, *Neurosci. Lett.*, 342, 179, 2003.

197. Staubli, U. and Otaky, N., Serotonin controls the magnitude of LTP induced by theta bursts via an action on NMDA-receptor-mediated responses, *Brain Res.*, 643, 10, 1994.

198. Garcia, J., Kimeldorf, D.J., and Koelling, R., Conditioned aversion to saccharin resulting from exposure to gamma radiation, *Science*, 122, 157, 1955.

199. Reilly, S.A. and Bornovalova, M., Conditioned taste aversion and amygdala lesions in the rat: a critical review, *Neurosci. Biobehav. Rev.*, 29, 1067, 2005.

200. Cannon, C.M., Scannell, C.A., and Palmiter, R.D., Mice lacking dopamine D1 receptors express normal lithium chloride-induced conditioned taste aversion for salt but not sucrose, *Eur. J. Neurosci.*, 21, 2600, 2005.

201. Corey, D.T., The determinant of exploration and neophobia, *Neurosci. Biobehav. Rev.*, 2, 235, 1978.

202. De la Casa, L.G. and Lubow, R.E., Delay-induced super-latent inhibition as a function of order of exposure to two flavours prior to compound conditioning, *Q. J. Exp. Psychol. Sect. B. Compar. Physiol. Psychol.*, 58, 1, 2005.

203. Swanson, L.W., The projections of the ventral tegmental area and adjacent regions: a combined fluorescent retrograde tracer and immunofluorescence study in the rat, *Brain Res. Bull.*, 9, 321, 1982.

204. Davis, M. and Astrachan, D.I., Conditioned fear and startle magnitude: effects of different footshock or backshock intensities used in training, *J. Exp. Psychol. Anim. Behav. Process.*, 4, 95, 1978.

205. Brown, J.S., Kalish, H.I., and Farber, I.E., Conditioned fear as revealed by the magnitude of startle response to an auditory stimulus, *J. Exper. Psychol.*, 41, 317, 1951.

206. Falls, W.A., Fear-potentiated startle in mice, in *Current Protocols in Neuroscience*, 8.11B1, 2002.

207. Davis, M., Falls, W.A., Campeau, S., and Kim, M., Fear-potentiated startle: a neural and pharmacological analysis, *Behav. Brain Res.*, 58, 175, 1993.

208. Gewirtz, J.C., Falls, W.A., and Davis, M., Normal conditioned inhibition and extinction of freezing and fear-potentiated startle following electrolytic lesions of medical prefrontal cortex in rats, *Behav. Neurosci.*, 111, 712, 1997.

209. Fendt, M. and Fanselow, M.S., The neuroanatomical and neurochemical basis of conditioned fear, *Neurosci. Biobehav. Rev.*, 23, 743, 1999.

210. Sasaki, A., Wetsel, W.C., Rodriguiz, R.M., and Meck, W.H., Timing of acoustic startle response in mice: habituation and dishabituation as a function of the interstimulus interval, *Int. J. Comp. Psychol.*, 14, 258, 2002.

211. Grillon, C., Ameli, R., Foot, M., and Davis, M., Fear-potentiated startle: relationship to the level of state/trait anxiety in healthy subjects, *Biol. Psychiatry*, 33, 566, 1993.

212. Grillon, C. and Davis, M., Fear-potentiated startle conditioning in humans: explicit and contextual cue conditioning following paired versus unpaired training, *Psychophysiology*, 34, 451, 1997.

213. Grillon, C., Ameli, R., Goddard, A., Woods, S.W., and Davis, M., Baseline and fear-potentiated startle in panic disorder patients, *Biol. Psychiatry*, 35, 431, 1994.

214. Grillon, C., Morgan, C.A., Davis, M., and Southwick, S.M., Effects of experimental context and explicit threat cues on acoustic startle in Vietnam veterans with posttraumatic stress disorder, *Biol. Psychiatry*, 44, 1027, 1998.

215. Schmidt, L.A. and Fox, N.A., Fear-potentiated startle responses in temperamentally different human infants, *Dev. Psychobiol.*, 32, 113, 1998.

216. Silva, J.A., Giese, K.P., Fedorev, N.B., Frankland, P.W., and Kogan, J.H., Molecular, cellular and neuroanatomical substrates of place learning, *Neurobiol. Learn. Mem.*, 70, 44, 1998.

217. Patrick, C.J., Berthot, B.D., and Moore, J.D., Diazepam blocks fear-potentiated startle in humans, *J. Abnorm. Psychol.*, 105, 89, 1996.

218. Miczek, K.A., Weerts, E.M., Vivian, J.A., and Barros, H.M., Aggression, anxiety and vocalizations in animals: GABAA and 5-HT anxiolytics, *Psychopharmacology*, 121, 38, 1995.

219. Joordens, R.J., Hijzen, T.H., and Olivier, B., The effects of 5-HT1A receptor agonists, 5-HT1A receptor antagonists and their interaction on the fear-potentiated startle response, *Psychopharmacology*, 139, 383, 1998.

220. Heldt, S.A., Coover, G.D., and Falls, W.A., Posttraining but not pretraining lesions of the hippocampus interfere with feature-negative discrimination of fear-potentiated startle, *Hippocampus*, 12, 774, 2002.

221. Heldt, S., Sundin, V., Willott, J.F., and Falls, W.A., Posttraining lesions of the amygdala interfere with fear-potentiated startle to both visual and auditory conditioned stimuli in C57BL/6J mice, *Behav. Neurosci.*, 114, 749, 2000.

222. Rosen, J.B. and Davis, M., Enhancement of acoustic startle by electrical stimulation of the amygdala, *Behav. Neurosci.*, 102, 195, 1988.

223. Romanski, L.M., Clugnet, M.C., Bordi, F., and LeDoux, J.E., Somatosensory and auditory convergence in the lateral nucleus of the amygdala, *Behav. Neurosci.*, 107, 444, 1993.

224. LeDoux, J.E., The emotional brain, fear, and the amygdala, *Cell Mol. Neurobiol.*, 23, 727, 2003.

225. Pezze, M.A. and Feldon, J., Mesolimbic dopaminergic pathways in fear conditioning, *Prog. Neurobiol.*, 74, 301, 2004.

226. LeDoux, J.E., *The Emotional Brain: The Mysterious Underpinnings of Emotional Life*, Simon & Schuster, New York, 1996.

227. Schauz, C. and Koch, M., Blockade of NMDA receptors in the amygdala prevents latent inhibition of fear-conditioning, *Learn. Mem.*, 7, 393, 2000.

228. Kent, J.M. and Rauch, S.L., Neurocircuitry of anxiety disorders, *Curr. Psychiatry Rep.*, 5, 266, 2003.

229. Rauch, S.L., Shin, L.M., and Wright, C.I., Neuroimaging studies of amygdala function in anxiety disorders, *Ann. N.Y. Acad. Sci.*, 985, 389, 2003.

230. Cannistraro, P.A. and Rauch, S.L., Neural circuitry of anxiety: evidence from structural and functional neuroimaging studies, *Psychopharmacol. Bull.*, 37, 8, 2003.

231. Debiec, J. and LeDoux, J.E., Disruption of reconsolidation but not consolidation of auditory fear conditioning by noradrenergic blockade in the amygdala, *Neuroscience*, 129, 267, 2004.

232. Jha, S.K., Brennan, F.X., Pawlyk, A.C., Ross, R.J., and Morrison, A.R., REM sleep: a sensitive index of fear conditioning in rats, *Eur. J. Neurosci.*, 21, 1077, 2005.

233. Pawlyk, A.C., Jha, S.K., Brennan, F.X., Morrison, A.R., and Ross, R.J., A rodent model of sleep disturbances in posttraumatic stress disorder: the role of context after fear conditioning, *Biol. Psychiatry*, 57, 268, 2005.

234. Blanchard, D.C., Griebel, G., and Blanchard, R.J., Conditioning and residual emotionality effects of predator stimuli: some reflections on stress and emotion, *Prog. Neuropsychopharmacol. Biol. Psychiatry*, 27, 1177, 2003.

235. Powell, S.B., Swerdlow, N.R., Pitcher, L.K, and Geyer, M.A., Isolation rearing-induced deficits in prepulse inhibition and locomotor habituation are not potentiated by water deprivation, *Physiol. Behav.*, 77, 55, 2002.

236. Gariepy, J.-L., The mediation of aggressive behavior in mice: a discussion of approach/withdrawal processes in social adaptations, in *Behavioral Development: Concepts of Approach/Withdrawal and Integrative Levels*, Hood, K.E., Greenberg, G., and Tobach, E., Eds., Garland Publishing, New York, 2005, p. 231.

237. Waddell, J., Dunnett, C., and Falls, W.A., C57BL/6J and DBA/2J mice differ in extinction and renewal of extinguished conditioned fear, *Behav. Brain Res.*, 154, 567, 2004.

238. Phillips, R.G. and LeDoux, J.E., Differential contribution of amygdala and hippocampus to cued and contextual fear conditioning, *Behav. Neurosci.*, 106, 274, 1992.

239. LeDoux, J.E., Emotion circuits in the brain, *Ann. Rev. Neurosci.*, 23, 155, 2000.

240. Bast, T., Zhang, W.N., and Feldon, J., Dorsal hippocampus and classical fear conditioning to tone and context in rats: effects of local NMDA-receptor blockade and stimulation, *Hippocampus*, 13, 657, 2003.

241. Boyson, S.J., McGonigle, P., and Molinoff, P.B., Quantitative autoradiographic localization of the D1 and D2 subtypes of dopamine receptors in rat brain, *J. Neurosci.*, 6, 3177, 1986.

242. Inglis, F.M. and Moghaddam, B., Dopaminergic innervation of the amygdala is highly responsive to stress, *J. Neurochem.*, 72, 1088, 1999.

243. Greba, Q., Gifkins, A., and Kokkinidis, L., Inhibition of amygdaloid dopamine D2 receptors impairs emotional learning measured with fear-potentiated startle, *Brain Res.*, 899, 218, 2001.

244. Guarraci, F.A., Frohardt, R.J., Falls, W.A., and Kapp, B.S., The effects of intra-amygdaloid infusions of a D2 dopamine receptor antagonist on Pavlovian fear conditioning, *Behav. Neurosci.*, 114, 647, 2000.

245. Crabbe, J.C., Wahlsten, D., and Dudek, B.C., Genetics of mouse behavior: interactions with laboratory environment, *Science*, 284, 1670, 1999.

246. Crawley, J.N., *What's Wrong with My Mouse? Behavioral Phenotyping of Transgenic and Knockout Mice*, John Wiley & Sons, New York, 2000.

Section IV

Model Applications and Future Developments

13 Cognitive Pharmacology in Aging Macaques

Jerry J. Buccafusco
Medical College of Georgia
Veterans Administration Medical Center
and
Prime Behavior Testing Laboratories

CONTENTS

COGNITIVE CHANGES WITH AGE IN MACAQUES

Various species of macaques, particularly *Macaca mulatta* (rhesus), have played an important role in our current understanding of the anatomical sites and in the behavioral and molecular mechanisms involved in cognition and memory. These versatile animals have been taught to perform a broad landscape of behavioral paradigms, many of which are relevant to human learning and memory. The rhesus monkey shares 92 to 95% genetic homology with the human species, and these animals enjoy a cognitive domain that allows for evaluation of higher-order cognitive and executive function. Nonverbal human cognitive tasks have been adapted for nonhuman primates, and vice versa, without reservation that a form of declarative memory is readily accessible in these animals. Numerous operant paradigms have been designed to evaluate aspects of reference memory, attention, and working memory (both short term and long term). Despite the various macaque species employed, and the variety of testing modalities applied, the largely consistent finding is that, with age, there is a decrement in memory-related task performance [1–4]. Without doubt there are macaques (and other nonhuman primate families) within the aging population that do not appear to exhibit age-dependent decrements in memory-related task performance. The potential for successful cognitive aging in these animals obviously has a human counterpart. The ability to study the differences at anatomical, cellular, neurochemical, and functional levels between animals that are successful vs. nonsuccessful with regard to cognitive aging has largely been the domain of rodent studies. Very few labs have used this approach with regard to nonhuman primates [e.g., 5, 6]. Part of the difficulty resides in the relatively long lifespan of macaques. Most often it is not possible to know whether an aged subject that displays cognitive impairment is that way because of true age-related impairment, or because the animal was below the range of normal intelligence for these animals from early on.

Do Old Monkeys Get Alzheimer's(-like) Disease?

Aged rhesus monkeys can succumb to many of the same diseases that afflict the human elderly [4], with the outward appearance of aging apparent over a time course commensurate with the approximately three-fold difference in human–monkey lifespan. Though the consequences of aging on the macaque brain have been the subject of numerous studies, there continues to be controversy as to whether declining cognition is related to cortical neuronal loss [7, 8]. Perhaps more convincing is the finding of the loss of specific types of subcortical neurons, such as the cholinergic cells arising from the medial septal nucleus [9] or the nucleus basalis [8]. Indeed, the loss of presynaptic cholinergic activity with age in the basal ganglia of rhesus monkeys, as measured by positron emission tomography (PET), could partly explain the loss of cognitive function with age [10]. These PET studies also revealed that a minority of older subjects exhibited near-normal cholinergic activity — a finding in consort with the minority of aged animals that fail to show cognitive deficits with age.

Since advanced age continues to be the most potent risk factor for Alzheimer's disease, the nature of the changes underlying age-related cognitive deficits in human and nonhuman primates remains relevant. But most nonhuman primates, and particularly macaques, do not appear to acquire an Alzheimer's-like syndrome. Though most aged rhesus monkeys (for example) are found to be cognitively impaired relative to their young counterparts, their impairment is relatively constant throughout the remainder of their lives. They do not normally exhibit a rapidly degenerating cognitive deterioration as would be encountered in Alzheimer's disease. However aged monkeys do often present with amyloid plaques on autopsy. These plaques can be visualized by using human anti-$A\beta$ protein antibodies, by human anti-apoenzyme E antibodies, and by human anti-ubiquitin antibodies [11, 12]. As with human Alzheimer's disease, plaque densities in aged rhesus monkeys are not well correlated with antemortem cognitive power. Also, plaque densities found in aged primate brains are not as severe or extensive as is commonly encountered in advanced Alzheimer's disease. One other interspecies difference, with some rare exceptions [e.g., 13], is the absence of human-type neurofibrillary tangles in the aged nonhuman primate brain. Neurofibrillary tangles represent the other hallmark pathological entity of Alzheimer's disease. Whether the absence of tangles in the aged macaque brain is a result of the underlying tau protein being expressed with a different amino acid composition from human form [14], or whether the animals simply do not live long enough to develop tangles [15], is as yet unresolved. So the answer to the question posed by the subtitle for this section must be that aged monkeys as they are commonly used in biomedical research do not develop true Alzheimer's disease.

DELAYED MATCHING-TO-SAMPLE AS A PARADIGM FOR ASSESSING COGNITION

Matching-to-sample performance was first described in laboratory animals over 40 years ago [16]. The delayed matching-to-sample (DMTS) task allows for the determination of components of reference and working memory in monkeys, i.e., delay-dependent effects can be distinguished from delay-independent effects on working memory. Increasing mnemonic burden decreases performance efficiency, and the resultant retention curve can provide information regarding the effect of a drug treatment or other manipulation on the components of memory [16]. Proficiency by macaques in the DMTS task depends upon unimpaired hippocampal and inferotemporal cortical function [17, 18], whereas acquisition of the "matching" component of the task is dependent upon frontotemporal cortical interactions [19, 20]. It is not too surprising, therefore, that the DMTS task is capable of detecting the decrease in cognitive proficiency that occurs with advanced age both in humans [21] and in monkeys [22]. Versions of the DMTS task have been utilized in numerous clinical studies as an instrument for evaluating working memory and reaction time as well as the effectiveness of cognition-enhancing drugs [23–28]. DMTS paradigms have been computer automated [21, 22], allowing for greater testing throughput, particularly when the animals are tested in their home cages. The trade-off for the latter convenience is the potentially added distraction of other animals in the vicinity of testing, although the stress of chairing and isolating the subject is avoided. Computer automation also obviates potential experimenter bias that can be associated with face-to-face testing.

A COMPUTER-AUTOMATED PROCEDURE

A touch-sensitive screen (15-in. AccuTouch LCD Panelmount TouchMonitor) with pellet dispenser units (Med Associates) mounted in a lightweight aluminum chassis is attached to the home cage. The stimuli include red, blue, and yellow rectangles arranged so that the sample stimulus is presented above and to the center of the choice stimuli. A trial is initiated by presentation of a sample rectangle composed of one of the three colors. The sample rectangle remains in view until the monkey touches within its borders to initiate a preprogrammed delay (retention) interval. Following the delay interval, the two choice rectangles located below where the sample had been are presented. One of the two choice rectangles is presented with its color matching the stimulus color, whereas the other (incorrect) choice rectangle is presented as one of the two remaining colors. A correct (matching) choice is reinforced (flavored food pellet dispensed). Nonmatching choices are neither reinforced nor punished. The intertrial interval is 5 sec, and each session consists of 96 trials. The presentation of stimulus color, choice colors, and choice position (left or right on the screen) is fully counterbalanced so as to relegate nonmatching strategies to chance levels of accuracy. Five different presentation sequences are rotated through each daily session to prevent the subjects from memorizing the first several trials. Delay intervals are established during numerous nondrug or vehicle sessions prior to initiating the study. The duration for each delay interval is adjusted for each subject until three levels of performance accuracy are approximated: zero delay interval (85 to 100% of trials answered correctly); short delay interval (75 to 84% correct); medium delay interval (65 to 74% correct); and long delay interval (55 to 64% correct). The assignment of retention intervals based upon an individual's baseline task accuracy is necessary to avoid ceiling effects in the most proficient animals during drug studies, while also serving to ensure that each animal begins testing at relatively the same level of task difficulty. In addition to session accuracy, two response latencies also are measured: the "sample latency," which is the time between presentation of the stimulus rectangle and the animal touching within the sample rectangle, and the "choice latency," which is the time between presentation of the choice stimuli and the animal touching within one of the choice rectangles.

Another advantage of the computer-assisted DMTS task is the long-term stability of baseline performance, permitting longitudinal study designs in which the subject can serve as his own control, thus enhancing statistical power. Computer automation allows for the precise measurement of task latencies that can provide information with regard to the effect of treatments on psychomotor speed and motivation. Analysis of choice preference can provide information related to compulsive behavior, i.e., perseveration, important for the detection of potential side effects associated with drug treatments. Most importantly, DMTS testing in macaques has proven to be a reliable preclinical method for predicting clinical effectiveness or clinical failure of novel cognition-enhancing drugs [29].

SEVENTEEN YEARS OF TESTING COGNITION-ENHANCING DRUGS

Since our first report of the positive mnemonic action of nicotine in macaques in their performance of the DMTS task [30], we have had the great fortune to work in collaborative studies with our colleagues in the pharmaceutical industry (and through our own modest drug discovery program) in the evaluation of a wide variety of potential drugs for the treatment of the cognitive impairment associated with neurodegenerative diseases and other psychiatric conditions. Though some of these data continue to be proprietary, a comprehensive database has been established that has allowed us to address several questions regarding aging, cognition, and the effects of drugs on these factors. One aspect of this work that has been of greatest surprise is the diversity of neurotransmitter substances involved in cognition. This diversity could reflect the concept that memory is represented by several distinct processes, and that different types of memory are relegated to different (but sometimes overlapping) brain regions. Even within working memory or episodic memory, there appear to exist separable and interacting components that may include acquisition (attention), consolidation, and retention (short and long term) and, alternatively, encoding, retrieval, storage, and consolidation. Using adult and aged macaques trained to perform the DMTS task, we have noted acute efficacy in improving task accuracy by drugs whose primary target is acetylcholinesterase, subtypes of nicotinic and muscarinic acetylcholine receptors, α-adrenergic receptors, 5-HT_{1A} receptors (antagonist), $5HT_{2A}$ receptors (antagonist), 5-HT_3 receptors (antagonist), 5-HT_4 receptors, glutamate AMPA (α-Amino-3-hydroxy-5-methylisoxazole-4-propionic acid) receptors, and angiotensin-1 receptors (antagonist). No particular pharmacological target appears to stand out as to providing the greatest degree of improvement in task performance, and within a specific pharmacological class, there can be wide differences in efficacy among analogs (Figure 13.1). This is true even among the cholinesterase inhibitors, perhaps because even they have differing selectivity for their molecular targets, e.g., acetylcholinesterase versus butyrylcholinesterase [31].

AGE-RELATED DIFFERENCES IN THE ACTIONS OF COGNITION-ENHANCING DRUGS

One other surprising finding to evolve from these studies was the rather similar degree of effectiveness of many of the cognitive-enhancing drugs in both young and aged animals. Figure 13.1 indicates the relative effectives upon acute administration of several classes of potential memory-enhancing drugs in both young and aged monkeys. Though not every compound is represented in both age groups, it is clear that the maximal efficacy (about 18% trials correct) was attained in both. For two classes of drugs that we have studied more comprehensively than others, nicotinic receptor agonists and acetylcholinesterase inhibitors, we have made head-to-head comparisons of their effectiveness in the two age groups [32–35]. When these compounds were titrated in to maximal effects in the two age groups, rather similar increases in task accuracies were obtained. In our study with donepezil, aged monkeys required higher doses than young animals to exhibit equivalent degrees of improvement in task performance [35]. One apparent exception was the AMPA modulator IDRA 21, which appeared to be more effective in young than in aged rhesus monkeys [36]. Perhaps one general point to take away from these still-premature observations is that it is unlikely that significant animal testing in the preclinical phase of drug discovery can be avoided, particularly when it comes to the effects of drugs on the central nervous system. The use of nonhuman primates in studies of cognitive pharmacology continues to provide a level of clinical predictability difficult to obtain with other animal models.

As discussed at the outset of this chapter, aged macaques are generally impaired cognitively with respect to their younger counterparts. In the context of cognitive pharmacology, it is important to note how these deficits are manifested. Under baseline conditions, aged monkeys require longer periods of DMTS training, and once they have established a stable baseline performance, their retention curves are composed of shorter delay intervals than for young subjects. That is, aged monkeys require shorter delay intervals (time between extinguishing the sample stimulus and the presentation of the choice stimuli) than young animals to achieve the same accuracy levels.

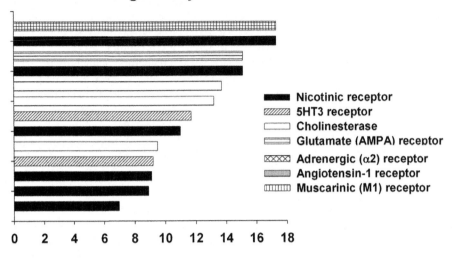

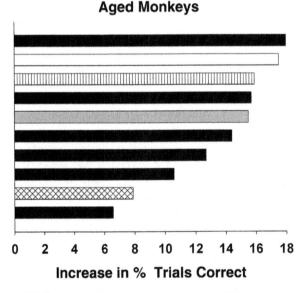

Increase in % Trials Correct

FIGURE 13.1 The increase in performance efficiency by young and aged (>19 yr) macaques well trained in the performance of the DMTS task initiated after the administration of potential memory-enhancing agents. Each data point indicates the increase in delayed matching-to-sample accuracy relative to baseline (vehicle) produced by the most effective dose of drug acting at the indicated drug target, and during the most improved delay interval.

Another aspect of aging in primates is their susceptibility to distraction. Distractor stimuli are presented in a nonpredictable manner to the test subject on 24 of the 96 trials completed during distractor DMTS sessions. The stimuli are initiated

1 sec into the delay interval and remain active for 3 sec. They consist of a random pattern of three colored rectangles (identical to the rectangles that served as sample and choice stimuli in the standard DMTS task) that are caused to flash in an alternating manner. The total duration of onset for a given colored rectangle is 0.33 sec. Immediately as one colored rectangle is extinguished, a different colored rectangle is presented. Thus, during presentation of the distractor, each color is presented in random order three separate times. Distractor stimuli are present an equal number of times on trials with short, medium, and long delay intervals. The remaining trials are considered nondistractor trials, and the data for distractor and nondistractor trials can undergo separate statistical treatments.

When a distractor 3 sec in duration is placed near the start of the delay interval (distractor stimuli placed near the end of the delay interval are not as effective), subsequent choice accuracy decreases considerably, often to near chance (50% correct) performance. The degree of distractor-induced decrement in task accuracy is dependent upon distractor duration. A distractor 1 sec in duration has virtually no effect on young monkeys but significantly impairs short-delay accuracy in aged monkeys [37]. Moreover, drugs that improve attention such as methylphenidate can reverse distractor-impaired accuracies in young monkeys but not in aged monkeys. In this laboratory, we routinely use young macaques for studies of drugs expected to reverse distractibility. Efficacy in this model has relevance to the treatment of attention-deficit disorders. However, the more difficult task of pharmacologically reversing distractibility in aged monkeys could partly explain the reduced potency of cognition-enhancing drugs in aged monkeys. Attention is an indispensable component of working memory, and it is conceivable that drugs capable of improving attention (in addition to the other components of memory) in aged or impaired subjects would enjoy greater effectiveness relative to those that act primarily on memory consolidation, retention, or recall. One other potentially interesting aspect of our work with the distractor-DMTS model is that it is not uncommon to find, for a particular compound, that the doses (or dose) that are optimal for improving accuracies during nondistractor trials are quite different from those that are optimal for reversing distractibility, even within the same session [38]. Thus dose sensitivity for one task does not always predict the same effectiveness in another, even closely related task. Again, this finding has implications for the design of clinical studies.

COGNITIVE CHANGES WITH AGE IN MACAQUES AND MAN

Macaque Data

The decline in cognition and in memory ability with age is an unquestioned tenet of human biology. Yet it is equally clear that not all humans exhibit the same rate of cognitive decline. Our observations over the years have confirmed that this scenario is also true for macaques. It was of interest, therefore, to determine whether any of the components of the computer-assisted DMTS task were correlated with age in these animals. We are fortunate to have the records of representatives of nearly the entire lifespan of macaques that, either formerly or extant at our facility, have

TABLE 13.1

Comparison of the Number of Delayed Matching-to-Sample (DMTS) Sessions Required for Training to a Long-Delay Retention Interval of 20 sec by Young and Aged Macaques

	Young Monkeys		Aged Monkeys	
	Mean	S.E.M.[a]	Mean	S.E.M.[a]
Age at 1st session (yr)[b]	7.6	0.75	21.9[c]	1.11
No. trials to 20-sec-long delay interval	207.4	39.0	488.8[c]	94.2
Duration of delay interval				
Short delay (% trials correct)	7.4	0.42	5.2	0.30
Medium delay (% trials correct)	12.5	0.42	10.6	0.37
Long delay (% trials correct)	20.0	0.0	20.0	0.0
N	33	—	21	—

[a] Standard error of the mean.

[b] The first session of DMTS acquisition began after achieving the criterion of at least 75% trials correct during 0-sec delay interval.

[c] Statistical significance ($p < 0.01$) with respect to mean for young monkeys.

contributed to a database repository. The database houses every trial of every session for each subject maintained at our facility beginning in 1990.

As an initial approach to quantify the effect of age on task acquisition, animals were divided into two groups: those younger than 20 years old, and those 20 years and older. A total of 54 monkeys were included in this analysis. The number of trials to achieve a long delay (retention interval) of 20 sec was compared (Table 13.1). As a group, aged monkeys required more than twice the number of sessions to achieve this milestone as compared with the young cohort. An alternative means of assessing task acquisition is to note the duration of retention intervals after achieving a fixed number of sessions, in this case 100 sessions. Sixty-one monkeys were included in this analysis. Data are presented for the assigned delay intervals termed zero, short, medium, and long. The zero delay interval represents near simultaneous matching. Short, medium, and long delay intervals are individually adjusted to achieve the idealized retention curve, as discussed above. If aged animals show impairment in task acquisition, it follows that for a fixed number of sessions, their assigned delay intervals would be shorter in duration than those assigned to young subjects. As a group, this appeared to be the case, since assigned delay intervals were uniformly twice as long for young subjects as compared with aged subjects (Table 13.2). Even after 100 sessions of DMTS testing, aged subjects did not attain the idealized accuracy for each delay interval, although the younger cohort had. These group differences were reflected to some extent when the assigned delay interval was correlated with age (Figure 13.2). Though the relationships between the durations of each of the delay intervals and age were statistically significant, the slopes of the regression lines were not that impressive. The reason for the less than

TABLE 13.2
Comparison of the Durations of the Assigned Delay Intervals (sec) and the Delayed Matching-to-Sample (DMTS) Task Accuracies (% C) Achieved after 100 Sessions by Young and Aged Macaques

	Age (yr)	Duration of Delay Interval (sec)			DMTS Task Accuracy (% C)[a]				No.
		Short	Medium	Long	Zero	Short	Medium	Long	
				Young Monkeys					
Mean	9.8	5.4	9.0	14.2	88.2	75.5	67.1	60.0	34
S.E.M.[b]	0.83	1.07	1.70	2.91	1.78	1.90	1.45	1.51	—
				Aged Monkeys					
Mean	22.2[c]	2.9[c]	5.0[c]	7.8[c]	77.8[c]	67.9[c]	60.9[c]	58.7	27
S.E.M.[b]	0.36	0.39	0.69	1.25	2.06	1.94	1.53	1.83	—

[a] The first session of DMTS acquisition began after achieving the criterion of at least 75% trials correct (% C) during 0-sec delay interval.
[b] Standard error of the mean.
[c] Statistical significance ($p < 0.05$) with respect to mean for young monkeys.

perfect concordance between delay interval duration and age is perhaps better illustrated in the relationship between overall task accuracy and age (Figure 13.2). Again, though statistically significant, the slope of the regression line was not particularly impressive. The relationship was skewed by the presence of five or six aged individuals that exhibited unusually high task accuracies, and by two or three young subjects that exhibited unusually low task accuracies.

When aged subjects are well trained in the DMTS task, i.e., after they have completed at least 200 sessions, because their performance is normalized by adjusting the duration of the delay intervals, they exhibit virtually identical retention curves as the young monkeys. The normalization procedure therefore obscures the quite marked differences in starting baseline accuracies that would otherwise have occurred. This is important for drug studies, wherein it would difficult to see improvement in animals with high predrug task accuracies or decrements in accuracy produced by amnestic agents in animals with low predrug baselines. One final point of potential interest is that recently four of our rhesus monkeys who have been maintained at our institution for several years have become old enough to qualify as aged subjects (>19 years). Over the past few years during this transition, there was essentially no change in their DMTS performance in any of the task categories. This maintenance of task performance during the years of transition from young to aged by well-trained animals might not seem that surprising. But some consideration might also be given to the contention that has been put forth recently that cognitive training inhibits the onset or severity of mental components of neurodegenerative diseases in humans [39]. Our observation might prove to be a nonhuman primate analog of the human concept: use it, or lose it.

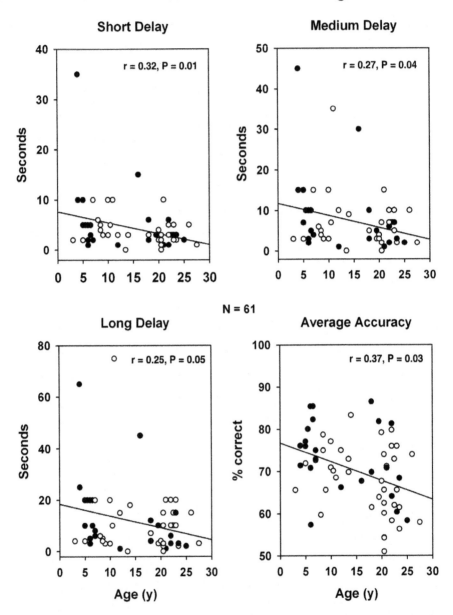

FIGURE 13.2 The delayed matching-to-sample (DMTS) performance by macaques during task acquisition plotted as a function of age. Task acquisition was initiated after each animal had reached the criterion of performing 0-sec delay trials with >75% correct accuracy. Each value represents the data obtained on the 100th DMTS session. Note that both aged (>19 yr) and young monkey groups included individuals that deviated from the general pattern that aged animals are impaired relative to their younger counterparts. (Filled circles indicate males; open circles indicate females.)

HUMAN DATA

Without doubt, the closer the relationship or relevance between nonhuman and human testing instruments, the more effective the preclinical data will be in predicting clinical efficacy. As discussed above, many versions of delayed response or delayed matching tasks have been used to assess human working memory. However, until the DMTS task enjoys wide clinical use, it will be necessary to relate the data from these operant tasks in animals to the more common neuropsychological instruments in current use. We use a fixed battery of these standard instruments within our human Alzheimer's research program. As with our nonhuman primate program, the human test findings also are maintained in a standard database. It was of interest, therefore, to compare the relationship between subject age and task performance as we had done for the nonhuman primate studies discussed above. To avoid the confound of disease, we retrieved only data derived from healthy elderly control subjects.

Figure 13.3 presents the data obtained from 54 healthy adult and elderly participants ages 44 through 84 years. The scores derived from three separate test instruments are plotted as a function of age:

1. *Mini-Mental-State Examination (MMSE)*: The most widely used measure of cognitive function. The MMSE has a maximum score of 30 points, assessing the subject in the domains of orientation to time and place, registration of three words, attention and calculation, recall of three words, language, and visual construction.
2. *Clinical Dementia Rating (CDR)*: Used as a global measure of dementia.
3. *Repeatable Battery for the Assessment of Neuropsychological Status (RBANS)*: Helps determine the neuropsychological status of adults aged 20 to 89 years; used to detect and characterize dementia in the elderly.

Included within the data set is the mean value for 42 participants with Alzheimer's disease for comparison. Though the data do not represent the entire human life span, the relationship between age and cognitive status for the normal adults and the elderly is highly reminiscent of the DMTS task parameters by macaques as correlated with age. The similarities include the mostly statistically significant correlation values, although again the slopes of the regression lines were not that impressive. Indeed, the regression values were quite similar to those we obtained for the monkey data, i.e., $r \approx 0.3$. In fact, the RBANS scores, like the DMTS accuracy values in the monkey data set, show that the reason for the less than impressive regression slope is the presence of several elderly participants with particularly strong RBANS scores, as well as a few younger subjects with particularly poor scores. Whereas it could be that other tests of working memory in nonhuman primates would provide similar performance vs. age relationships, it appears that the computer-assisted DMTS task provides a measure of cognitive proficiency in macaques that is analogous to standard neuropsychological evaluations in humans.

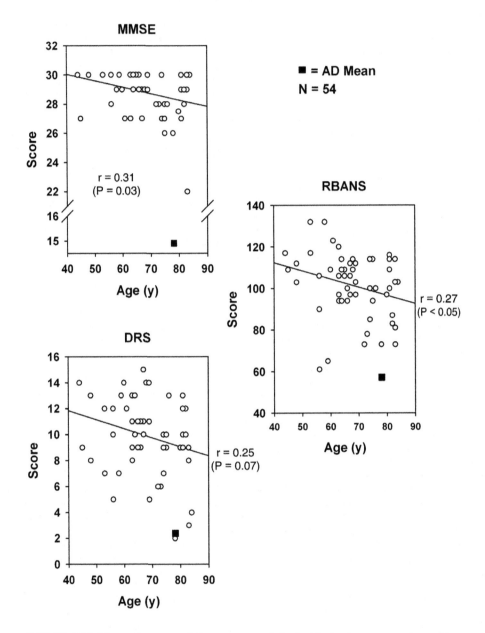

FIGURE 13.3 The performance of three neuropsychological tests of memory and cognition by 54 healthy elderly human participants plotted as a function of age. MMSE (mini mental state exam); DRS (dementia rating scale); RBANS (repeatable battery for the assessment of neuropsychological status). The filled square in each graph indicates the mean value derived for each test by a cohort of 42 individuals with probable Alzheimer's disease (AD).

FUTURE DIRECTIONS

Thus far, the evaluation of novel memory-enhancing drugs in nonhuman primates in advance of clinical trials has been rather limited, usually relegated to dose-response analysis during acute administration regimens. Because computer automation and home-cage monitoring have greatly improved testing efficiency in these animals (two technicians can test up to 50 animals per day by using 18 test panels operating from six work stations), preclinical studies can be performed in a timely manner, providing important information regarding potency, effectiveness, and potential side effects. With a large database of standard test compounds, new drugs can be compared to provide an even greater level of clinical predictability. One factor common for many drugs that improve cognition is the induction of a protracted degree of task improvement that occurs long after the compound has been metabolized or eliminated [40, 41]. This lack of pharmacokinetic–pharmacodynamic concordance can lead to inappropriate dosing regimens during clinical trials. Estimation of the true pharmacodynamic response to a new compound is a relatively simple matter during standard DMTS procedures in monkeys. Also, as nonhuman primate testing becomes a more common model for the preclinical evaluation of new cognition-enhancing drugs, additional information would be gained from the evaluation of chronic drug administration regimens. The data from such studies would be of enormous help in designing the expensive clinical trials that almost always require at least subacute dosing paradigms. Now that most patients with newly diagnosed Alzheimer's disease are placed on one or more of the available treatment regimens, it will be difficult to recruit nontreated Alzheimer's subjects for clinical trials. The testing of new compounds in individuals with Alzheimer's disease might require patients to cross over from their current medication(s) to the novel regimen. Therefore, similar scenarios should be considered during preclinical testing situations, particularly in nonhuman primates.

Computer-automated testing methods also can be applied to the preclinical evaluation of other classes of central nervous system-active drugs. Reversible pharmacological models that use the acute administration of amnestic agents such as scopolamine or psychotomimetic drugs such as ketamine or MK-801 can be used to evaluate novel compounds for cognition enhancement under impaired conditions. The use of psychotomimetic drugs could help in the study of novel structures that could improve the cognitive status of individuals with schizophrenia. The lack of effectiveness of new drugs recently developed for the treatment of stroke and other cerebrovascular insults to the brain has been partly attributed to lack of relevance of the rodent models of stroke to the human syndrome. New stroke models developed for nonhuman primates should prove more predictive for novel drug testing.

In the recent past, nonhuman primate models for human diseases of the central nervous system have not been fully exploited as preclinical predictors for effectiveness and potential side effects. This has largely been a reflection of the cost, regulatory concerns, expertise available, and most importantly, slow throughput. Now that most testing paradigms can be computer automated, and with the use of touch-sensitive video screens permitting tremendous flexibility, the issues pertaining to trained personnel and slow throughput can be minimized, if not eliminated.

ACKNOWLEDGMENTS

The author would like to acknowledge his long-term colleague Dr. Alvin V. Terry, Jr., for his valued participation in the many studies described in this chapter. He also would like to thank: primate technicians Nancy Kille and Daniel Martin for their skillful contributions; Scott Webster and Aarti Arun for their database analysis; and Rosanne Schade, study coordinator for the Alzheimer's Research Center. Much of the work cited for the author's studies was supported by: the Office of Research and Development, Medical Research Service, Department of Veterans Affairs; the Institute for the Study of Aging; and Philip Morris USA Inc. and Philip Morris International.

REFERENCES

1. Herndon, J.G. et al., Patterns of cognitive decline in aged rhesus monkeys, *Behav. Brain Res.*, 87, 25, 1997.
2. Bartus, R.T., On neurodegenerative diseases, models, and treatment strategies: lessons learned and lessons forgotten a generation following the cholinergic hypothesis, *Exp. Neurol.*, 163, 495, 2000.
3. Voytko, M.L. and Tinkler, G.P., Cognitive function and its neural mechanisms in nonhuman primate models of aging, Alzheimer disease, and menopause, *Frontiers Biosci.*, 9, 1899, 2004.
4. Roth, G.S. et al., Aging in rhesus monkeys: relevance to human health interventions, *Science*, 305, 1423, 2004.
5. Rapp, P.R. and Amaral, D.G., Individual differences in the cognitive and neurobiological consequences of normal aging, *Trends Neurosci.*, 15, 340, 1992.
6. Calhoun, M.E. et al., Reduction in hippocampal cholinergic innervation is unrelated to recognition memory impairment in aged rhesus monkeys, *J. Comp. Neurol.*, 474, 238, 2004.
7. Merrill, D.A., Roberts, J.A., and Tuszynski, M.H., Conservation of neuron number and size in entorhinal cortex layers II, III, and V/VI of aged primates, *J. Comp. Neurol.*, 422, 396, 2000.
8. Smith, D.E. et al., Memory impairment in aged primates is associated with focal death of cortical neurons and atrophy of subcortical neurons, *J. Neurosci.*, 24, 4373, 2004.
9. Stroessner-Johnson, H.M., Rapp, P.R., and Amaral, D.G., Cholinergic cell loss and hypertrophy in the medial septal nucleus of the behaviorally characterized aged rhesus monkey, *J. Neurosci.*, 12, 1936, 1992.
10. Voytko, M.L. et al., Cholinergic activity of aged rhesus monkeys revealed by positron emission tomography, *Synapse*, 39, 95, 2001.
11. Summers, J.B. et al., Localization of ubiquitin in the plaques of five aged primates by dual-label fluorescent immunohistochemistry, *Alzheimers Res.*, 3, 11, 1997.
12. Summers, J.B. et al., Co-localization of apolipoprotein E and beta-amyloid in plaques and cerebral blood vessels of aged non-human primates, *Alzheimer's Rep.*, 1, 119, 1998.
13. Kiatipattanasakul, W. et al., Abnormal neuronal and glial argyrophilic fibrillary structures in the brain of an aged albino cynomolgus monkey (*Macaca fascicularis*), *Acta Neuropathol.*, 100, 580, 2000.
14. Nelson, P.T. et al., Molecular evolution of tau protein: implications for Alzheimer's disease, *J. Neurochem.*, 67, 1622, 1996.

15. Nakayama, H., Uchida, K., and Doi, K., A comparative study of age-related brain pathology — are neurodegenerative diseases present in nonhuman animals? *Medical Hypotheses*, 63, 198–202, 2004.

16. Paule, M.G. et al., Symposium overview: the use of delayed matching-to-sample procedures in studies of short-term memory in animals and humans, *Neurotoxicol. Teratol.*, 20, 493, 1998.

17. Sahgal, A. and Iversen, S.D., Categorization and retrieval after selective inferotemporal lesions in monkeys, *Brain Res.*, 146, 3410, 1978.

18. Jackson, W.J., Regional hippocampal lesions alter matching by monkeys: an anorexiant effect, *Physiol. Behav.*, 32, 593, 1984.

19. Stern, C.E. et al., Medial temporal and prefrontal contributions to working memory tasks with novel and familiar stimuli, *Hippocampus*, 11, 337, 2001.

20. Bussey, T.J., Wise, S.P., and Murray, E.A., Interaction of ventral and orbital prefrontal cortex with inferotemporal cortex in conditional visuomotor learning, *Behav. Neurosci.*, 116, 703, 2002.

21. Robbins, T.W. et al., Cambridge neuropsychological test automated battery (CANTAB): a factor analytic study of a large sample of normal elderly volunteers, *Dementia*, 5, 266, 1994.

22. Buccafusco, J.J., Terry, A.V., Jr., and Murdoch, P.B., A computer assisted cognitive test battery for aged monkeys, *J. Mol. Neurosci.*, 19, 179, 2002.

23. Irle, E., Primate learning tasks reveal strong impairment in patients with presenile dementia of the Alzheimer type, *Brain Cognition*, 6, 429, 1987.

24. Squire, L.R., Zola-Morgan, S., and Chen, K.S., Human amnesia and animal models of amnesia: performance of amnesic patients on tests designed for the monkey, *Behav. Neurosci.*, 102, 210, 1988.

25. Riekkinen M. et al., Tetrahydroaminoacridine improves the recency effect in Alzheimer's disease, *Neuroscience*, 83, 471, 1998.

26. Fowler, K.S. et al., Paired associate performance in the early detection of DAT, *J. Int. Neuropsychol. Soc.*, 8, 58, 2002.

27. White, K.G. and Ruske, A.C., Memory deficits in Alzheimer's disease: the encoding hypothesis and cholinergic function, *Psychonomic Bull. Rev.*, 9, 426, 2002.

28. Barbeau, E. et al., Evaluation of visual recognition memory in MCI patients, *Neurology*, 62, 1317, 2004.

29. Buccafusco, J.J. and Terry, A.V., Jr., Multiple CNS targets for eliciting beneficial effects on memory and cognition, *J. Pharmacol. Exp. Ther.*, 295, 438, 2000.

30. Elrod, K., Buccafusco, J.J., and Jackson, W.J., Nicotine enhances delayed matching-to-sample performance by primates, *Life Sci.*, 43, 277, 1988.

31. Giacobini, E., Drugs that target cholinesterases, in *Cognitive Enhancing Drugs*, Buccafusco, J.J., Ed., Birkhäuser Verlag, Basel, 2004, chap. 2.

32. Buccafusco, J.J. and Jackson, W.J., Beneficial effects of nicotine administered prior to a delayed matching-to-sample task in young and aged monkeys, *Neurobiol. Aging*, 12, 233, 1991.

33. Terry, A.V., Jr., Buccafusco, J.J., and Jackson, W.J., Scopolamine reversal of nicotine enhanced delayed matching-to-sample performance by monkeys, *Pharmacol. Biochem. Behav.*, 45, 925, 1993.

34. Prendergast, M.A. et al., Improvement in accuracy on a delayed recall task by aged monkeys and mature monkeys after intramuscular or transdermal administration of the CNS nicotinic receptor agonist ABT-418, *Psychopharmacology*, 130, 276, 1997.

35. Buccafusco, J.J. and Terry, A.V., Jr., Donepezil-induced improvement in delayed matching accuracy by young and old rhesus monkeys, *J. Mol. Neurosci.*, 24, 85, 2004.

36. Buccafusco, J.J. et al., The effects of IDRA 21, a positive modulator of the AMPA receptor, on delayed matching performance by young and aged rhesus monkeys, *Neuropharmacology*, 46, 10, 2003.

37. Prendergast, M.A. et al., Age-related differences in distractibility and response to methylphenidate in monkeys, *Cerebral Cortex*, 8, 164, 1998.

38. Bain, J.N. et al., Enhanced attention in rhesus monkeys as a common factor for the cognitive effects of drugs with abuse potential, *Psychopharmacology*, 169, 150, 2003.

39. Scarmeas, N. and Stern, Y., Cognitive reserve and lifestyle, *J. Clin. Exp. Neuropsychol.*, 25, 625, 2003.

40. Buccafusco, J.J. et al., Cognitive effects of nicotinic cholinergic agonists in non-human primates, *Drug Dev. Res.*, 38, 196, 1996.

41. Buccafusco, J.J. et al., Drug-induced cognitive improvement: evidence for pharmacokinetic-pharmacodynamic discordance, *Trends Pharmacol. Sci.*, 26, 352, 2005.

14 Cognitive Impairment following Traumatic Brain Injury

Mark D. Whiting, Anna I. Baranova, and Robert J. Hamm
School of Medicine, Virginia Commonwealth University

CONTENTS

INTRODUCTION

Animal models of traumatically induced cognitive impairment are designed to reproduce those features of human traumatic brain injury (TBI) that result in long-term cognitive impairment and disability. The sequelae of behavioral impairments associated with human TBI include disruption along nearly every level of information processing. However, the most severely affected cognitive domains are memory and information-processing speed and efficiency.[1] Both retrograde and anterograde memory deficits are common following TBI. Memory function seems to be particularly vulnerable to head injury, and Levin and coworkers[2] suggest that long-term memory processes are disproportionately affected compared with other areas of cognitive functioning following TBI.

ANIMAL MODELS OF TRAUMATIC BRAIN INJURY

Experimental models of TBI have been developed to reproduce those aspects of TBI seen in the clinical setting. Although many different models of TBI have been developed and are in use today, this chapter focuses on three specific models: fluid percussion, weight drop/impact acceleration, and cortical impact. A review of the techniques and characteristics specific to each model follows.

Fluid Percussion

The rodent fluid percussion (FP) model of brain injury is the most commonly used model in brain injury research today. This injury model has been reproduced in several species, including cat, dog, mouse, rabbit, sheep, and swine. Injury is either applied centrally (CFP), over the sagittal suture midway between bregma and lambda, or laterally (LFP), over the parietal cortex. A brief fluid pressure pulse applied to the intact dural surface through a craniotomy produces brief displacement and deformation of neural tissue. By altering the magnitude of fluid pressure applied to the brain, a graded range of neurological, histopathological, and cognitive deficits have been reliably reproduced.[3] Traumatic pathology including hemorrhage, cavitation, vascular disruption,[4] alterations in metabolism, changes in ionic homeostasis and blood flow,[5] traumatic axonal injury leading to axonal swelling and disconnection,[6–8] and cell death have been reported following FP injury.[9,10] Deficits in spatial memory, working memory, and neurological motor function are common following FP injury, with spatial memory deficits persisting for as long as one year following injury.[11] The hippocampus, a structure important in learning and memory processes, is particularly vulnerable to FP injury. A number of hippocampus-dependent memory tasks are disrupted following CFP injury in rats.[12] However, Lyeth and coworkers[13] demonstrated that memory impairment may exist in the absence of hippocampal cell death following CFP injury. This suggests that memory impairment may be due not only to cell death, but also to a disruption of normal neuronal functioning, especially within the hippocampus.

Weight Drop/Impact Acceleration

Weight-drop injury models produce diffuse brain injury utilizing a simple device characterized by a free-falling weight through a Plexiglas tube, with varying weights and heights.[14] To prevent skull fractures, a small stainless steel helmet-disk is placed on the rodent skull while the animal is supported by the foam bed. Weight-drop injury induces blood-brain barrier disruption, edema formation, and hypertension. This injury model also produces widespread neuronal injury, axonal disruption, and microvasculature pathology.[15] Injury to the neurons mostly occurs bilaterally in the cerebral cortex as well as in the brainstem. In addition, weight-drop injury produces diffuse axonal injury that occurs secondary to trauma and is often observed in the clinical situation. This makes weight-drop injury an appropriate tool in studying neuronal, axonal, and vascular changes in the experimental setting.

Closed Cortical Impact

A major difference of closed cortical impact (CCI) from the previous two injury models is in the pneumatically driven controlled-impact device compared with the free-falling weight in the weight-drop models.[16] The advantage of the controlled injury allows for manipulation of variables such as duration, velocity, and depth of the impact. As well, a possible secondary hit or rebound of the free-falling weight is impossible, unlike in the weight-drop models. Similar to other TBI models, CCI produces edema, changes in the cerebral microvasculature, neuronal damage, and axonal damage ultimately leading to traumatic axonal injury.[17] Controlled cortical-impact injury also produces significant neurobehavioral deficits. Spatial memory impairments in rats tested in the Morris water maze (MWM) were observed up to 34 days postinjury.[18] More recently, Dixon and coworkers[19] report that spatial memory performance in the MWM may persist for up to one year following CCI injury. Motor deficits as well as learning and memory deficits postCCI were also observed in mice.[20,21] Such neurobehavioral impairment is analogous to human TBI, which makes CCI a useful technique in trauma research, especially if precise control of contact velocity and brain deformation is necessary in the experimental procedure.

INFORMATION PROCESSING: A BRIEF OVERVIEW

Although numerous experiments have demonstrated that experimental TBI produces memory deficits similar to those seen in humans, few researchers have addressed how TBI disrupts the flow of information through the brain. For example, is the disruption of memory processes following TBI due to a failure to store information in the brain, or a failure to retrieve information in storage? To address such questions, a basic understanding of contemporary information-processing theory is helpful. The most widely accepted information-processing theory is based on the model of Atkinson and Shriffin,[22] also known as stage theory. Information from the environment is processed in three stages: sensory memory, short-term or working memory, and long-term memory. Sensory memory processes are short-lived and involve the process of signal transduction. Very little information from sensory memory would pass to the next stage without the help of attentional processing, which aids in the creation of short-term or working memories. If information in working memory is thoroughly processed, it will be encoded in long-term memory. Long-term memories are then available for retrieval, leading to behavioral responses. Hypothetically, TBI could disrupt information processing at any of the stages in this model. Retrograde memory impairments could involve the inability to retrieve information already in storage or a failure of storage/consolidation, while anterograde memory impairments may involve a disruption in any of the stages of information processing.

MEMORY IMPAIRMENT AFTER TBI

RETROGRADE AMNESIA

Trauma-induced retrograde amnesia (RA) is a common consequence of brain injury. People who experience a head injury typically forget things that occurred from several minutes to years prior to the injury.[23–25] In one of the first experiments to investigate the cognitive consequences of TBI in animals, Ommaya and coworkers[26] tested animals for retention of a passive-avoidance task after TBI. In this study, rats received an escapable foot shock when they entered a black compartment. When the animal escaped to the white (nonshocked) compartment, it was removed and TBI was administered. When the rats were returned to the white (nonshocked) compartment, the latency to enter the black (previously shocked) compartment was recorded. Long latencies to enter the black compartment indicate good retention of the task. On the other hand, short latencies represent a retention deficit (i.e., amnesia). Results indicated that the step-through latency of the injured rats was significantly shorter than uninjured controls. The impairment of retention of a task learned prior to injury confirms that TBI does produce RA, as is observed in cases of human brain injury.

Zhou and Riccio[27] conducted a series of experiments to characterize the RA produced by TBI. In the first experiment, they also observed a trauma-induced retention deficit of a passive-avoidance (PA) task following a weight-drop model of TBI, replicating the findings of Ommaya and coworkers.[26] In the second experiment, they investigated the temporal gradient of TBI-induced RA. Other experimental treatments that produced RA[28] (e.g., electroconvulsive shock [ECS]) produce a time-dependent amnesia). That is, retention performance increases directly with the interval between learning and the amnesic treatment, with events close to the injury being most vulnerable to amnesia. In their study, the interval between PA training and TBI was varied from 1 min, 30 min, 6 h, 24 h, 3 days, and 5 days. Results indicated that TBI within 6 h of training produced a profound amnesia. Training-TBI intervals of 24 h and 3 days yielded less severe deficits, and normal retention was observed with a 5-day delay between training and TBI. This temporally graded RA is also observed in human TBI.

The last experiment conducted by Zhou and Riccio[27] investigated the underlying mechanism responsible for the RA produced by TBI. In TBI patients, Benson and Geschwind[29] proposed that RA was the result of a retrieval deficit and not the result of a storage or consolidation failure. Previous research has demonstrated that many forms of experimentally induced RA are the result of a retrieval failure, since giving the subjects an appropriate cue prior to retention testing can reduce the retention deficit. For example, Miller and Springer[30] found that giving the subject a noncontingent foot shock prior to retention testing could alleviate ECS-induced RA. The effectiveness of reminder treatments has been observed with a variety of amnesic treatments. These findings support the hypothesis that the RA produced by TBI may be the result of a retrieval deficit rather than impairment of consolidation or storage. To test the role of a retrieval deficit, three conditions were tested. The first condition was the typical training-TBI sequence with retention testing 3 days after TBI. The second condition investigated was the same as the first with the addition of a noncontingent foot shock administered in a different apparatus 2 min before retention testing (remainder treatment). The last condition was a control condition that received electric shock in novel environments (to control for any nonassociative effects of the foot shocks). The second condition produced the typical trauma-induced RA effect. However, the reminder foot-shock condition completely eliminated the RA usually seen following TBI. These findings suggest that a pretest reminder cue improved retention by facilitating the retrieval of the prior learning.

Using the LFP model, Smith and colleagues[31] investigated the effect of TBI on the retention of a previously learned Morris water maze (MWM) task. Rats were trained prior to injury and then tested for retention 42 h after LFP injury. A significant disruption in maze performance was observed, and the degree of the deficit was related to the severity of the injury. In another study following a similar procedure, Smith et al.[32] observed that a reduction in hippocampal dentate hilar neurons was related to the severity of the RA observed 2 weeks following injury. However, the cue-induced alleviation of RA found by Zhou and Riccio[27] suggests that neuronal cell loss and loss of memory storage may not be a necessary prerequisite for TBI-induced RA.

Using an operant testing procedure to investigate cognitive function after TBI, Gorman et al.[33] trained rats to press a bar for food. The front panel of the operant chamber box (Skinner box) had an opening for the delivery of food rewards, two retractable response levers (one on the right and one on the left) that the animal could press to receive food, and stimulus lights (one over each response lever). To examine performance on a visual discrimination task, the light over a particular lever was illuminated. If the animal pressed that lever, food was delivered (a response on the nonilluminated lever turned the light out and no food was delivered). After animals had learned this task (pressing only the illuminated lever), they were subjected to a brain injury. Following the injury, animals were tested again. Testing began 1 to 2 min after injury and continued for five additional days. Results revealed a transitory reduction in the percentage of correct lever presses in the first test session that rapidly disappeared on subsequent test sessions. The results of this experiment found that brain injury did not produce a long-lasting RA of this relatively simple task.

TBI reliably produces an RA effect that is temporally graded. While it is difficult to make generalizations because of differences in injury model and severity, it appears that the duration of RA following TBI may be related to difficulty of task to be remembered. Simple or overtrained tasks may produce a transitory RA (i.e., 1-day deficit on a visual operant task and several days on a passive-avoidance task). More complex tasks (such as the MWM) may produce a more enduring deficit.

The cognitive processes that account for RA after brain injury are a matter of speculation. One could propose that the memory attributes of the learning situation are stored almost immediately, but an elaboration process may continue for an extended period of time and produce a postacquisition state-dependent retention effect. Under these conditions, successful retrieval of a memory episode is a function of the similarity between the encoding context and the retrieval context. Thus, if animals are trained prior to TBI, encoding takes place in a "normal" neuronal context. When animals are tested for retention after TBI, numerous pathological changes produced by TBI provide the context under which retrieval is tested. If retention is tested soon after injury, the difference between the "normal" neuronal-encoding context may be very different from the neuronal-retrieval context that exists soon after injury. As postinjury interval increases, the neuronal-retrieval context becomes more similar to the "normal" encoding context, and retention performance improves.

This state-dependent type of explanation for RA may account for the temporal gradient of RA observed following TBI. Similarly, the effectiveness of retrieval cues in eliminating the TBI-induced RA also may be explained by a state-dependent learning. As indicated previously, the successful retrieval of a memory is a function of the similarity between the encoding and retrieval environments. Providing a cue prior to retention testing may reactivate the memory representation produced in acquisition of the memory and thus facilitate retention performance.

Working Memory

Lyeth and associates[13] evaluated working-memory function after CFP injury with the radial arm maze (RAM) test. The RAM typically has eight arms that radiate from a central start platform. Each arm ends with a goal box. To investigate working memory, six of the eight arms were baited with food. A hungry rat is placed in the central start area and allowed to freely enter any of the arms of the maze and eat the food in the goal box. A trial continues until the rat has entered all the baited arms. The measure of the animal's working-memory function is how many correct choices it makes (i.e., not reentering arms that it has already visited). Prior to TBI, rats were trained until they made very few working-memory errors. After reaching criterion, animals were subjected to a mild or moderate level of CFP injury. Testing after injury found that magnitude and duration of the deficits in working memory were related to the severity of injury. In the mild-injury condition, working memory returned to normal after 10 days. In the moderate-injury condition, the impairment of working memory was more robust and more enduring.

Working-memory function has also been examined after injury using an operant procedure, as described by Gorman et al.[33] On this task, one lever was presented to the rat. After the animal pressed the lever, the lever was retracted. After a variable delay (2 to 12 sec), both levers were presented. To receive a food reward, the animal was required to press the same lever that it pressed prior to the delay (in other words, the animal had to remember which lever it had pressed previously). Following acquisition of this task, animals were injured using CFP. Injured animals demonstrated a retention deficit on the 2- and 4-sec delays only on the first test session. No significant effects were observed with the 8-sec delay. However, as the delay increased to 12 sec, brain-injured animals performed poorly for 5 days after TBI. Thus, as the delay increased, the magnitude and duration of the working-memory deficit increased.

Working memory has also been studied using a modification of the MWM procedure.[34] This study was designed to examine working memory following CFP-induced TBI. Rats were injured at a moderate level of injury or received a sham injury. On days 11 to 15 postinjury, working memory was assessed. Each animal received eight pairs of trials per day. For each pair of trials, animals were randomly assigned to one of four possible starting points and one of four possible escape-platform positions. On the first trial of each pair (the information trial), rats were placed in the maze and given 120 sec to locate the hidden escape platform. After remaining on the goal platform for 10 sec, they were placed back into the maze for the second trial of the pair (test trial). The platform position and the start position remained unchanged on this trial. After the second trial, the animal was given a 4-min intertrial rest. Between pairs of trials, both the start position and the goal location were changed. Analysis of the latency to reach the goal platform indicated that the sham-injured animals performed significantly better on the second trial than on the first trial of each pair. However, injured animals did not significantly differ between first and second trial goal latencies on any day. In a similar study, Kline and coworkers[35] described working-memory deficits in an identical MWM task following CCI injury. These results indicate that injured animals have a profound and enduring deficit in spatial working-memory function on days 11 to 15 after TBI.

ANTEROGRADE AMNESIA

Experimental TBI in rodents produces anterograde memory deficits similar to those seen in human head injury. While retrograde memory tasks are designed to assess performance of a task learned prior to traumatic insult, anterograde memory tasks are designed to assess performance on a previously unlearned task. In terms of the information-processing model discussed above, anterograde memory deficits may be caused by any of the stages of information processing.

The MWM has been used extensively to assess anterograde memory deficits following both FP- and CCI-type injuries. Hamm and coworkers[18] used a CCI model of TBI to produce a moderate level (1.5- to 2.0-mm deformation) of injury in rats. Following injury, animals received motor-function assessment on the beam balance and beam walk tasks. This was done to ensure that cognitive assessment in the MWM was not confounded by motor deficits in the injured animals. Animals then received cognitive assessment in the MWM on postinjury days 11 to 15 and 30 to 34. CCI injury produced a significant disruption in maze performance at both time points. Cognitive impairment in injured animals was evidenced not only by increased goal latencies, but also by their search patterns during trials. Typically, uninjured control animals will swim in the center of the maze to find the hidden platform. CCI-injured animals, however, spent most of their time swimming around the perimeter of the maze. This suggests that injured animals failed to learn the spatial location of the hidden platform in relation to extramaze cues. Since the MWM task is particularly sensitive to hippocampal dysfunction,[36] the results from this experiment suggest that the hippocampus is sensitive to CCI injury. In another study, Scheff and coworkers[37] demonstrated that lateral CCI injury produces anterograde memory deficits in the MWM for 14 days postinjury. Animals were injured at either a mild (1-mm deformation) or moderate (2 mm) level of lateral CCI injury. In this experiment, animals received MWM assessment on either day 7 or 14 following injury, with all trials being completed in a single day. Injured animals had significantly increased escape latencies and spent less time in the target quadrant than sham-injured animals. Moreover, the anterograde memory impairment produced by this model appears to be dependent on the severity of injury. Moderately injured animals performed significantly worse than mildly injured animals, with both groups performing significantly worse than the control group.

Hamm and coworkers[12] provide further evidence that the anterograde memory deficits observed following experimental TBI are not the result of a generalized deficit in learning and memory ability, but rather are due to selective damage to the hippocampus. In this study, rats were injured at a moderate level (2.1 atm) of CFP injury or surgically prepared but not injured. Following CFP or sham injury, cognitive performance was evaluated using three tasks: passive avoidance, a constant-start version of the MWM, and a variable-start MWM task. In the constant-start version of the MWM, the animal's start location in the maze remains the same across all trials and all days. Previous research has demonstrated that the variable-start version of the MWM task is disrupted by damage to the hippocampus,[36,38,39] while both the passive-avoidance and constant-start MWM tasks do not depend on hippocampal processing. On day 9 postinjury, all animals received training on a single-trial passive-avoidance task. Retention of this task was tested 24 h later. Animals then received training in either a constant-start or variable-start version of the MWM on postinjury days 11 to 15. In accordance with the hypothesis that TBI would selectively impair performance on tasks that are hippocampally dependent, injured animals displayed a significant deficit in the variable-start version of the MWM but showed no impairment in the passive-avoidance and constant-start MWM tasks.

Anterograde memory deficits have also been reported in weight-drop and impact-acceleration models of brain injury. Isaksson and coworkers[40] report spatial learning deficits following severe weight-drop brain injury in rats. Following severe injury, animals had significantly increased escape latencies in the MWM on postinjury days 10 to 13 compared with sham-injury animals. Schmidt and coworkers[41] examined the effects of moderate and severe impact-acceleration injury on MWM performance in rats. Injury was delivered using a 500-g, 2.1-m weight drop, and animals were divided into moderate or severe injury based on righting times. MWM performance was assessed on days 5 to 7 postinjury. Severity of injury was strongly correlated with spatial learning and memory impairments 1 week postinjury. However, this impairment resolved within 3 to 5 weeks postinjury. Interestingly, weight-drop models of minimal traumatic brain injury produced long-term cognitive impairment similar to that seen in severe weight-drop brain injury.[42] In this study, mice were administered a noninvasive, closed-head weight-drop injury of 20, 25, or 30 g and then tested in the MWM 7, 30, 60, and 90 days postinjury. Escape latencies of control mice improved by up to 450%, while injured mice could only improve their scores by 50%. Importantly, the cognitive deficits in injured mice occurred in absence of any neurological impairment or anatomical damage to the brain.

Similar findings have been reported in mouse models of repeated mild brain injury (RMI). RMI models are designed to mimic the repeated head injuries often seen in sports such as soccer or football. Deford and coworkers[43] examined the effects of RMI in B6C3F1 mice. Using a noninvasive weight-drop model, masses of 50, 100, and 150 g were dropped from 40 cm to produce injury. Four injuries were administered 24 h apart. Cognitive performance was then evaluated in the MWM on days 7 to 11 postinjury. Mice injured with 100- and 150-g masses had significantly increased escape latencies in the MWM compared with sham-injury animals and those injured with a 50-g mass. Additionally, mice injured with a single mild injury (SMI) of 150 g did not exhibit cognitive impairment in the MWM. As with previously described studies, the cognitive deficits produced by RMI were observed in the absence of overt cell death. Creeley and coworkers[44] have also reported MWM deficits in mice following weight-drop RMI. However, in this study, behavioral impairment was observed in conjunction with a contra-coup injury involving ventral brain structures close to the skull.

Anterograde memory deficits following experimental TBI are common. However, the cognitive basis of anterograde memory impairment is still unknown. In terms of information processing, anterograde memory deficits involve a disruption in the attending to, encoding of, storage, or retrieval of information. However, when injured rats enter a water maze with a visible platform, deficits in escape latencies compared with shams disappear. This suggests that injured animals do not suffer deficits in sensory-attentional processing. Moreover, experimental evidence indicates that the hippocampus is selectively vulnerable to TBI. This brain region is especially important in memory processes and is likely involved in the encoding and storage of memories, suggesting that memory impairment is due to a failure to encode or store relevant information about the environment. Further progress in understanding the cognitive basis of anterograde memory impairment after TBI will depend on the development and application of more-sensitive behavioral outcome measures following TBI.

AGE EFFECTS ON MEMORY IMPAIRMENT AFTER TBI

Age is an important predictor of outcome following TBI. Animal models of TBI have demonstrated that both cognitive and motor deficits following injury increase with age. Hamm and coworkers[45] examined the effects of a low level (1.7 to 1.8 atm) of CFP injury in both aged (20 months) and young (3 months) rats. Following injury, motor function was assessed on days 1 to 5 postinjury. MWM performance was then assessed on days 11 to 15 after injury. While this magnitude of injury failed to produce motor deficits in 3-month-old rats, the aged rats exhibited significant motor deficits on the beam balance and beam walking tasks. Compared with young animals, aged rats exhibited increased cognitive deficits in the MWM as well. One study used impact acceleration to produce injury in both aged (20 to 23 months) and young (2 to 3 months) rats.[46] Following injury, aged rats had significantly impaired MWM performance compared with young rats up to 5 weeks following injury. Although some improvement was noted in aged injured rats from weeks 3 to 5, their performance was still significantly worse than young injured rats during the same period. These results indicate that both FP and impact-acceleration injury produce age-dependent cognitive deficits in rats.

TBI is the leading cause of injury-related deaths in children under age 15. Thus, it is important to develop experimental models of pediatric brain injury that address the concerns unique to TBI in children. Adelson and coworkers[47] have examined the effect of severe closed-head weight-drop injury in immature (postnatal 17 days) Sprague Dawley rats. Animals were administered injury with either a 100-g (severe) or 150-g (ultrasevere) mass from 2 m. Motor function was measured daily on beam balance, grip test, and inclined plane, and cognitive assessment was performed in the MWM on postinjury days 11 to 22. On the beam balance and inclined plane, motor deficits were evident in the severely injured rats that persisted for 4 days after injury. The ultraseverely injured group exhibited profound motor deficits compared with both shams and severely injured animals that persisted for up to 10 days after injury. Cognitive assessment indicated that only ultraseverely injured animals displayed MWM deficits for up to 22 days following injury. To further investigate the long-term cognitive deficits produced by the ultrasevere injury, Adelson et al.[48] conducted another study in which MWM performance was assessed in injured rats for 3 months postinjury. P17 rats injured with a 150-g mass from 2 m displayed significantly increased escape latencies in the MWM for 3 months following injury. Moreover, this injury level in P17 rats resulted in significantly lower body and brain weight gain in the immature rats as measured at 3 months postinjury. Motor deficits were similar to those reported in the previous study.

These studies suggest that severe weight-drop brain injury is capable of producing long-term cognitive impairment in immature rats that is associated with a reduction in body and brain weight gain compared with sham-injured animals. The effects of CFP injury, unilateral entorhinal cortex lesion (UEC), or CFP combined with UEC have also been examined in juvenile (P28) rats.[49] While neither CFP nor UEC alone produced cognitive impairment in the MWM task, the combined injury method produced significant MWM deficits in P28 rats. Taken together, the findings from these studies indicate that experimental models of TBI produce reversible motor deficits and persistent cognitive deficits in immature rats. While juvenile rats appear to be somewhat more resistant to long-term cognitive impairment than adult animals, further investigation is needed to address those aspects of TBI that are unique in developing animals.

The effect of age on cognitive outcome following experimental TBI is similar to that reported in the clinical setting. Numerous studies have shown that increasing age is an important predictor of morbidity and mortality in human TBI. Vollmer and coworkers[50] examined the outcome data for 661 TBI patients aged 15 and older at the time of injury. Based on multivariate analysis of factors such as age, severity of injury, prior systemic disease, and injury mechanism, age could not be ruled out as an independent predictor of poor outcome following injury. This suggests that poor outcome in older TBI patients may be due to the brain's impaired ability to respond to traumatic insult as age increases. While many studies have confirmed these findings in adult patients, Leurssen and coworkers[51] report that outcome following TBI in the pediatric population actually improves with increasing age, with the best outcome in children of ages 12 to 15. In the pediatric TBI population then, very young children may suffer the worst outcome. These results indicate that the effect of age on outcome following human TBI produces a U-shaped function, with the worst outcome in the youngest and oldest patients.

COMPARING EXPERIMENTAL MODELS

The cognitive impairment produced by experimental TBI is dependent on many factors, including severity of injury, the age of the experimental animals, and the injury model used. Hallam and coworkers[52] compared the behavioral deficits induced by both LFP and weight-drop brain injuries. In this study, LFP injury resulted in significant memory impairments in both the MWM and RAM tasks. However, weight-drop injury did not produce deficits on the two tasks. Research has also demonstrated that minor changes in methodology may produce differential effects on cognitive outcome in a single-injury model. For example, one study demonstrated that craniotomy position affects both MWM performance and hippocampal cell loss following LFP injury in rats.[53] LFP injury was delivered through craniotomies at four locations: rostral, caudal, medial, and lateral. Injury at the medial and caudal craniotomies produced significantly greater impairments in the MWM compared with the lateral and rostral injuries. Injuries at the caudal, medial, and lateral locations produced both cortical damage and significant cell loss in areas CA2 and CA3 ipsilateral to the injury, while the rostral injury produced cortical damage but little hippocampal damage. The importance of craniotomy position has also been reported by Vink and coworkers.[54] In this study, magnetic resonance imaging (MRI) was used to detect lesion development following moderate LFP injury in rats. With the craniotomy placed adjacent to the sagittal suture, both ipsilateral and contralateral lesion development was observed. Ipsilateral lesion development and cortical damage increased as the craniotomy site was shifted laterally away from the sagittal suture. Contralateral lesion development could be detected until the center of the craniotomy was more than 3.5 mm from the sagittal suture; beyond this point, MRI detected no contralateral damage. These studies indicate that different injury models, or even minor changes in methodology in a single model, may result in differential cognitive impairment and histological outcome following experimental TBI. Addressing factors such as these becomes especially important when studies are designed to understand injury pathology or test therapeutic treatments.

CONCLUSION

The cognitive processes that result in memory impairment following TBI are not fully understood. While numerous studies have demonstrated that TBI induces both retrograde and anterograde memory deficits in experimental animals, few have questioned the cognitive basis of these deficits. It is uncertain what types of learning or memory are affected following TBI, and many cognitive tasks, including the MWM, involve more than one type of learning or memory. It is also unclear which anatomical structures are responsible for the observed deficits. Although certain brain regions appear to be especially vulnerable to traumatic injury, no single brain region is responsible for the cognitive dysfunction observed following trauma. Furthermore, the role of anatomical damage in mediating cognitive dysfunction following injury is controversial. While many studies correlate behavioral impairment with anatomical markers of injury such as cell death or degeneration, many studies have reported long-term cognitive impairment in the absence of overt anatomical damage. Although cell death and other anatomical damage are likely sufficient to produce some types of cognitive impairment, a number of studies have demonstrated that such damage is not necessary to produce cognitive deficits after TBI.

Experimental models of TBI have successfully reproduced the cognitive deficits observed in cases of human traumatic brain injury, which makes them ideal for studying pathological mechanisms of injury or potential therapeutic treatments. However, factors such as age, severity of injury, and injury model may affect both pathological and cognitive outcomes following TBI. Even slight differences in methodology can produce differential outcomes in a single model. It is therefore important to address the methodological factors that result in differential cognitive impairment following experimental TBI. Lastly, to better understand the cognitive processes that mediate memory impairment following TBI, it is important to apply a variety of behavioral outcome measures that are sensitive to multiple types of memory dysfunction.

REFERENCES

1. Capruso, D.X. and Levin, H.S., Cognitive impairment following closed head injury, *Neurol. Clin.*, 10, 879, 1992.
2. Levin, H.S. et al., Disproportionately severe memory deficit in relation to normal intellectual functioning after closed head injury, *J. Neurol. Neurosurg. Psychiatry*, 51, 1294, 1988.
3. Dixon, C.E. et al., A fluid percussion model of experimental brain injury in the rat, *J. Neurosurg.*, 67, 110, 1987.
4. McIntosh, T.K. et al., Traumatic brain injury in the rat: characterization of a lateral fluid-percussion model, *Neuroscience*, 28, 233, 1989.
5. Hovda, D.A. et al., The increase in local cerebral glucose utilization following fluid percussion brain injury is prevented with kynurenic acid and is associated with an increase in calcium, *Acta. Neurochir. Suppl.*, 51, 331, 1990.
6. Cordobes, F. et al., Post-traumatic diffuse axonal brain injury: analysis of 78 patients studied with computed tomography, *Acta. Neurochir.*, 81, 27, 1986.
7. Adams, J.H. et al., Diffuse axonal injury in head injury: definition, diagnosis and grading, *Histopathology*, 15, 49, 1989.
8. Maxwell, W.L., Povlishock, J.T., and Graham, D.L., A mechanistic analysis of non-disruptive axonal injury: a review, *J. Neurotrauma*, 14, 419, 1997.
9. Rink, A. et al., Evidence of apoptotic cell death after experimental traumatic brain injury in the rat, *Am. J. Pathol.*, 147, 1575, 1995.
10. Yakovlev, A.G. et al., Activation of CPP32-like caspases contributes to neuronal apoptosis and neurological dysfunction after traumatic brain injury, *J. Neurosci.*, 17, 7415, 1997.
11. Pierce, J.E.S. et al., Enduring cognitive, neurobehavioral and histopathological changes persist for up to one year following severe experimental brain injury in rats, *Neuroscience*, 87, 359, 1998.
12. Hamm, R.J. et al., Selective cognitive impairment following traumatic brain injury in rats, *Behav. Brain Res.*, 59, 169, 1993.
13. Lyeth, B.G. et al., Prolonged memory impairment in the absence of hippocampal cell death following traumatic brain injury in the rat, *Brain Res.*, 526, 249, 1990.
14. Marmarou, A. et al., A new model of diffuse brain injury in rats, part I: pathophysiology and biomechanics, *J. Neurosurg.*, 80, 291, 1994.
15. Montasser, A., Foda, M.A., and Marmarou, A., A new model of diffuse brain injury in rats, part II: morphological characterization, *J. Neurosurg.*, 80, 301, 1994.
16. Lighthall, J.W., Controlled cortical impact: a new experimental brain injury model, *J. Neurotrauma*, 5, 1, 1988.
17. Smith, D.H. et al., A model of parasagittal controlled cortical impact in the mouse: cognitive and histopathologic effects, *J. Neurotrauma*, 12, 169, 1995.

18. Hamm, R.J. et al., Cognitive deficits following traumatic brain injury produced by controlled cortical impact, *J. Neurotrauma*, 9, 11, 1992.

19. Dixon, C.E. et al., One-year study of spatial memory performance, brain morphology, and cholinergic markers after moderate controlled cortical impact in rats, *J. Neurotrauma*, 16, 109, 1999.

20. Fox, G.B. et al., Effect of traumatic brain injury on mouse spatial and nonspatial learning in the Barnes circular maze, *J. Neurotrauma*, 15, 1037, 1998.

21. Fox, G.B. et al., Sustained sensory/motor and cognitive deficits with neuronal apoptosis following controlled cortical impact brain injury in the mouse, *J. Neurotrauma*, 15, 599, 1998.

22. Atkinson, R. and Shiffrin, R., Human memory: a proposed system and its control processes, in *The Psychology of Learning and Motivation: Advances in Research and Theory*, Vol. 2, Spence, K. and Spence, J., Eds., Academic Press, New York, 1968.

23. Russell, W.R., Amnesia following head injuries, *Lancet*, 2, 762, 1935.

24. Russell, W.R. and Nathan, P.W., Traumatic amnesia, *Brain*, 69, 280, 1946.

25. Wilson, B.A., *Rehabilitation of Memory*, Guilford Press, New York, 1987.

26. Ommaya, A.K., Geller, A., and Parsons, L.C., The effect of experimental head injury on one-trial learning in rats, *Int. J. Neurosci.*, 1, 371, 1971.

27. Zhou, Y. and Riccio, D.C., Concussion-induced retrograde amnesia in rats, *Physiol. Behav.*, 57, 1107, 1995.

28. Duncan, C.P., The retroactive effect of electroshock on learning, *J. Comp. Physiol. Psychol.*, 42, 32, 1949.

29. Benson, D.F. and Geshwind, N., Shrinking retrograde amnesia, *J. Neurol. Neurosurg. Psychiatry*, 30, 539, 1967

30. Miller, R.R. and Springer, A.D., Amnesia, consolidation and retrieval, *Psychol. Rev.*, 80, 69, 1973.

31. Smith, D.H. et al., Evaluation of memory dysfunction following experimental brain injury using the Morris water maze, *J. Neurotrauma*, 8, 259, 1991.

32. Smith, D.H. et al., Persistent memory dysfunction is associated with bilateral hippocampal damage following experimental brain injury, *Neurosci. Lett.*, 168, 151, 1994.

33. Gorman, L.K., Shook, B.L., and Becker, D.P., Traumatic brain injury produces impairments in long-term and recent memory, *Brain Res.*, 614, 29, 1993.

34. Hamm, R.J. et al., Working memory deficits following traumatic brain injury in the rat, *J. Neurotrauma*, 13, 317, 1996.

35. Kline, A.E. et al., Attenuation of working memory and spatial acquisition deficits after a delayed and chronic bromocriptine treatment regimen in rats subjected to traumatic brain injury by controlled cortical impact, *J. Neurotrauma*, 19, 415, 2002.

36. Morris, R.G.M. et al., Place navigation in rats with hippocampal lesions, *Nature*, 297, 681, 1982.

37. Scheff, S.W. et al., Morris water maze deficits in rats following traumatic brain injury: lateral controlled cortical impact, *J. Neurotrauma*, 14, 615, 1997.

38. Schenk, F. and Morris, R.G.M., Dissociation between components of spatial memory in rats after recovery from the effects of retrohippocampal lesions, *Exp. Brain. Res.*, 58, 11, 1985.

39. Sutherland, R.J., Kolb, B., and Whishaw, I.Q., Spatial mapping: definitive disruption by hippocampal or medial frontal cortical damage in the rat, *Neurosci. Lett.*, 31, 271, 1982.

40. Isaksson, J., Hillered, L., and Olsson, Y., Cognitive and histopathological outcome after weight-drop brain injury in the rat: influence of systemic administration of monoclonal antibodies to ICAM-1, *Acta. Neuropathol.*, 102, 246, 2001.

41. Schmidt, R.H., Scholten, K.J., and Maughan, P.H., Cognitive impairment and synaptosomal uptake in rats following impact acceleration injury, *J. Neurotrauma*, 17, 1129, 2000.

42. Zohar, O. et al., Closed-head minimal traumatic brain injury produces long-term cognitive deficits in mice, *Neuroscience*, 118, 949, 2003.

43. Deford, S.M. et al., Repeated mild brain injuries result in cognitive impairment in B6C3F1 mice, *J. Neurotrauma*, 19, 427, 2002.

44. Creeley, C.E. et al., Multiple episodes of mild traumatic brain injury result in impaired cognitive performance in mice, *Acad. Emerg. Med.*, 11, 809, 2004.

45. Hamm, R.J. et al., The effect of age on motor and cognitive deficits after traumatic brain injury in rats, *Neurosurgery*, 31, 1072, 1992.

46. Maughan, P.H., Scholten, K.J., and Schmidt, R.H., Recovery of water maze performance in aged versus young rats after brain injury with the impact acceleration model, *J. Neurotrauma*, 17, 1141, 2000.

47. Adelson, P.D. et al., Motor and cognitive functional deficits following diffuse traumatic brain injury in the immature rat, *J. Neurotrauma*, 14, 99, 1997.

48. Adelson, P.D., Dixon, C.E., and Kochanek, P.M., Long-term dysfunction following diffuse traumatic brain injury in the immature rat, *J. Neurotrauma*, 17, 273, 2000.

49. Prins, M.L., Povlishock, J.T., and Phillips, L.L., The effects of combined fluid percussion traumatic brain injury and unilateral entorhinal deafferentation on the juvenile rat brain, *Brain. Res. Dev. Brain. Res.*, 140, 93, 2003.

50. Vollmer, D.G. et al., Age and outcome following traumatic coma: why do older patients fare worse? *J. Neurosurg.*, 75, S37, 1991.

51. Luerssen, T.G. et al., Outcome from head injury related to patient's age, *J. Neurosurg.*, 68, 406, 1988.

52. Hallam, T.M. et al., Comparison of behavioral deficits and acute neuronal degeneration in rat lateral fluid percussion and weight-drop brain injury models, *J. Neurotrauma*, 21, 521, 2004.

53. Floyd, C.L. et al., Craniectomy position affects Morris water maze performance and hippocampal cell loss after parasagittal fluid percussion, *J. Neurotrauma*, 19, 303, 2002.

54. Vink, R. et al., Small shifts in craniotomy position in the lateral fluid percussion injury model are associated with differential lesion development, *J. Neurotrauma*, 18, 839, 2001.

15 Cognitive Impairment Models Using Complementary Species

Daniel T. Cerutti and Edward D. Levin
Duke University

CONTENTS

FISH MODELS

The causes of illness are many and problematic. Our first line of investigation is descriptive — the generally straightforward problem of discovering a reliable syndrome. The second step is generally epidemiological — discovering the environmental or genetic factors that comprise the disease mechanism. Cholera, lead poisoning, hypothyroidism, and Huntington's chorea are familiar examples of this two-step process.

Our ability to identify disease sources includes the number of causal variables that sum to produce an effect, the probability that they produce an effect, and the immediacy of their effects. Thus, it was easy to discover the toxic effects of lead but more difficult to understand the link between iodine deficiency and thyroid function. Among the most intractable disorders are those like schizophrenia and attention-deficit disorder, in which the syndrome also includes a loosely defined and oftentimes subtle cognitive impairment. These more difficult cases benefit greatly from animal models that can open doors to theoretical and applied investigations.

Animal modeling hastens the rate of discovery to the extent that it simplifies a problem and accelerates the search for critical causal variables.[1] The simplification arises from the improved control over critical dependent variables; the rate of discovery improves because variables can be screened more rapidly, thereby addressing the ever-present problem of experimental throughput. Both of these can be significant obstacles in the case of human cognitive impairments, which can be both subtle and may only appear once an individual reaches adulthood.

But the use of animal models is not without problems. The accuracy of a model — its external validity — depends on the match between the model and target brain-behavior processes. By definition, however, we develop a model because of our failure to understand the target process, a classical "bootstrapping" problem. How do we overcome this seemingly intractable predicament? The answer must be that we approach modeling by approximation. We first look for a fair match between the behavior of our model and the target and then balance that concern with others such as economics, throughput, and so forth. In this sense, our animal model is a product of expedience, but more importantly, it is a classical, testable "theory," one that we can support or refute by experimental trial.

The choice of model species is always justified in terms of its external validity. In the case of a cognitive impairment, we must also seek validity in the behavioral assay we employ. Classic animal models of cognitive impairment use rodent species in tests of learning, memory, and attention. Their justification is based on both the similarity of brain abnormality and the corresponding behavioral impairment. For example, research on Alzheimer's disease is supported by a mouse model that shows brain plaques with impaired spatial learning and memory[2-5]; rats with ventral hippocampal damage have been used as models of attentional disorders in schizophrenia[6]; and nonhuman primates have been used to model working-memory problems in schizophrenia.[7] Clearly, nonhuman primates offer much closer homologies to human neural and behavioral function, as described in Chapter 13 of this book.[8]

Mammalian models of cognitive impairment are now complemented by non-mammalian vertebrates and invertebrates — fish, flies, and worms — that are the subject of this chapter. It has become customary to think of these species as "alternative models," but the term "complementary model" is used instead because it better describes the use of these models. They do not replace the classical rodent and primate models, which remain quite valuable. Rather they complement them, offering some unique advantages but also drawbacks that justify the continued use of the classic models. Their validity can be high when questions are clearly delimited, for example, when the problem is to understand the impact of a particular gene, drug, or toxicant.[9-13] This chapter reviews work with zebrafish (*Danio rerio*), goldfish (*Carassius auratus*), fruit flies (*Drosophila melanogaster*), and nematodes

(*Caenorhabditis elegans*), with an emphasis on their contribution to the understanding of mechanisms of cognitive impairment.

Despite the obvious anatomical adaptations to their aquatic habitat, fish show just about all of the behavior seen in terrestrial species in some form or other, and even have special adaptations of their own. In addition to the five senses found in mammals, their lateral-line organ allows them to "see" the location, size, and features of submerged objects; and many fish generate and detect weak electrical currents, a sense that they use to detect predators and prey.[14] Although fish species differ dramatically in social behavior, there are some examples that show monogamous mating for life, e.g., angelfish,[15] which show individual recognition by sight or odor,[16] social learning,[17] complex mate-selection strategies,[18] ritualized aggression,[19] and communication of danger.[20,21] In terms of adaptive behavior and learning, fish show advanced abilities for spatial navigation,[22] nonassociative learning such as habituation,[23,24] precise timing abilities,[25-27] and Pavlovian conditioning of a variety of adaptive behaviors,[28] including operant behavior motivated by aversive stimuli such as shuttle-box behavior,[29] negatively reinforced avoidance,[30] and positively reinforced lever-pressing food responses.[31] Behavioral similarities with terrestrial species are thus just as obvious as their anatomical differences, making fish a promising model of vertebrate behavioral development and cognition.

Behavioral research with fish began with ethologists and comparative psychologists asking questions about the evolution of learning and cognition.[32-34] These questions about behavior have invariably accompanied those about the evolution of brain function[35-38]; it goes without saying that the understanding of the teleost brain has been driven in large part by the development of appropriate behavioral assays. The extent to which basic behavioral and brain processes in mammals and fish are analogous remains an open question; there are clear similarities and differences. As with all animal models, the validity of a fish model hinges on the particular research question. Many species of fish have been used in models of cognitive impairment; for example, the Japanese medaka (*Oryzias latipes*) is being used in toxicological studies on effects of the insecticide diazinon,[39] and walleye (*Stizostedion vitreum*) have been used to demonstrate the adverse impacts of insecticides on cholinergic systems.[40] This section emphasizes procedures and behavioral processes with two of the more commonly studied species, goldfish (*Carassius auratus*) and zebrafish (*Danio rerio*), that parallel those employed with rodents.

GOLDFISH (*CARASSIUS AURATUS*)

Cognitive studies of processes such as attention, memory, and choice have been conducted with many fish species, as reviewed by Reebs.[14] A favorite subject is the goldfish (*Carassius auratus*), a member of the cyprinid family that includes carp and zebrafish.[41,42] The goldfish has appeared as a model to study development, anatomy, brain and behavior evolution, pharmacology and toxicology, and the ecological determinants of cognitive behavior.[43] Goldfish models have taken advantage of both basic, instinctual behaviors, such as habituation to fearful stimuli, and more complex behavior such as maze learning.

Reflexes and Habituation

Reflexes provide animals with important adaptations to problems that would be too costly to learn through experience. For example, many animals freeze (or flee) in the presence of movements or noise[44–46]; the behavior is easy to understand because even a slight disturbance could be a sign of an approaching predator.[47] On the other hand, it pays to learn to ignore such stimuli if they do not signal danger. We speak of *habituation* when responses become less likely with repeated stimulation; habituation is seen when stimuli are not particularly harmful, such as when the startle response for rats decreases with repeated presentations of noise.[48] The opposite of habituation, *potentiation*, tends to be seen in cases where stimuli are painful (e.g., rats jump and squeal more vigorously with repeated electric shocks). A great deal of research into the physiology of habituation has been conducted with simpler organisms, for example, the gill-withdrawal response in aplysia;[49–52] see Marcus et al.[53] for a demonstration of variables that lead to potentiation versus habituation in aplysia. In these organisms, research suggests that habituation is a psychological process originating at very specific neural locations in the sensory systems.[50] Although superficially simple, habituation and potentiation are considered forms of learning to the extent that the effect of later stimuli depends on the memory of preceding stimuli. Both processes are temporary, with a reflex recovering its original magnitude some interval of time after stimuli are no longer presented.[49]

Goldfish show habituation in startle (tail flip) and arousal (erection of dorsal fin and movement of pectoral fins) responses following repeated exposure to a Plexiglas rod thrust into the water.[54] These responses show different dynamics to repeated stimulation, with startle responses habituating first and arousal showing an increase before habituating. As in other species,[55,56] habituation in goldfish is more rapid and more complete with shorter intervals between stimuli, as evidenced by the "rate-sensitive property" of habituation.[46,54–63] Several studies show that telencephalic ablation in the goldfish interferes with habituation of startle[46] without impairing simple escape responses,[64] probably because ablation disrupts memory functions. However, the brain processes involved in habituation remain contested[34–37,61,65]

Pavlovian Conditioning

Students of behavior are familiar with the basic facts of classical conditioning in which organisms come to anticipate biologically significant stimuli such as mates, food, and danger by learning about signals that precede them.[66] Pavlovian conditioning is among the most ubiquitous forms of learning: it is found in organisms ranging from invertebrates like *aplysia*[67,68] to humans, and it is involved in a broad range of processes from reproduction to feeding. A number of experiments have shown Pavlovian conditioning in goldfish.[29,69–74] For example, a light closely followed by shock — known as "short-delay conditioning" — will come to elicit startle responses,[70] but little conditioning to light is seen when light and shock appear simultaneously.[66] The reason is that a conditioned response will only be learned if the light is temporally *predictive* of shock; not if it is simply redundant. Although fear conditioning is learned most rapidly with a short delay between the conditioned and unconditioned stimuli, taste aversions are learned even if illness follows with a long delay after consumption, and these can be long lasting.[75] Goldfish have also shown appetitive Pavlovian conditioning in procedures in which the illumination of a lamp precedes the delivery of food.[76–78] Given such an arrangement, goldfish (as well as pigeons, rats, and many other organisms) direct behavior that normally occurs in the presence of the food (e.g., biting) to the stimulus.[79,80]

In a recent study that paired light and delayed shock with interstimulus delays of either 5 or 15 sec (i.e., a brief shock followed the onset of a light by 5 or 15 s), goldfish showed the ability to time a behavior by showing defensive swimming at the time of shock delivery.[25,81] An important finding was that the timing of the conditioned swimming appeared immediately in the first training trials, demonstrating that temporal learning has a reflexive character. Moreover, the timing of the behavior was *scalar*, showing a peak of activity coinciding with the shock delay and variance in the peak proportional to delay.[82] The demonstration of scalar timing is important because temporal control by environmental events is a key behavioral adaptation, important in many ecological situations.[83]

Pavlovian conditioning is implicated in a variety of task performances, including signaled avoidance.[36,84–86] Figure 15.1.1 shows a typical shuttle box used with fish.[87] The task arranges pairing of light and delayed shock (a Pavlovian preparation), but permits the organism to delete shock by crossing a barrier after the onset of the light (an operant contingency). Goldfish readily learn to avoid shock; the resulting avoidance performance is thought to involve both operant and Pavlovian learning processes. The operant component is the crossing behavior that prevents shock (i.e., negatively reinforced responding), and escape from the Pavlovian conditioned stimulus serves to reinforce the behavior.[36] (See Hineline[88] for a discussion of avoidance theory.)

Pharmacological, lesion, and electrophysiological studies have been conducted to understand the physiology of goldfish learning. Endorphins appear to impair avoidance learning by reducing fear responses in goldfish.[89] The NMDA (N-methyl-D-aspartate) antagonist dizocilpine (MK-801) impairs negatively reinforced escape learning in goldfish,[90,91] as does D(−)-2-amino-5-phosphonopentanoic acid injection to the telencephalon[92]; early learning within the first 6 to 12 trials of conditioning was phase sensitive to disruption by dizocilpine.[93] The role of the neuropeptide substance P has been found to mediate a memory-based enhancement when paired with a dopaminergic agonist.[94,95] Evidence for appetitive learning enhancement has been found with neuronal histamine administration.[96] Many other studies (e.g., Lee et al.[97]) have focused on cellular and developmental responses of goldfish neurons at various stages of life. Numerous ablation studies have shown that the telencephalon is critical for conditioning of fear responses in goldfish,[35,36,74,98–100] but the cerebellum is also implicated.[101]

Operant Conditioning

Given appropriate operant training, goldfish can learn to press a Plexiglas lever to earn food reinforcement, an appetitive-conditioning procedure.[31,102] Figure 15.1.2 shows a typical setup comprising a submerged lever, a stimulus projector behind the

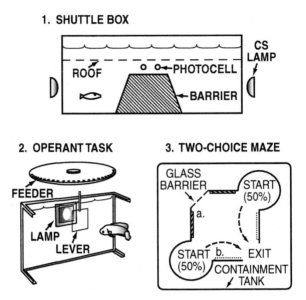

FIGURE 15.1 Apparatuses used to study learning and adaptive behavior in goldfish. (1) *Shuttle box* (side view). The shuttle box is used to study signaled-avoidance behavior[73] (see Hineline[88] for a theoretical review), as implemented by Bitterman.[214] In a typical procedure, illumination of the conditioned stimulus (CS) lamp signals a delayed shock that can be eliminated by swimming over the barrier to the other side of the box. Barrier crossing is recorded by photocells (between the barrier and a roof) that measure the response latency. (2) *Operant task* (perspective cutaway view). Operant behavior is studied by arranging food reinforcement for pressing a Plexiglas lever.[31] In a typical procedure, a stimulus lamp serves as a discriminative stimulus that signals occasions when lever presses produce food. This apparatus setup can be readily adapted to study classical conditioning.[113,215] (3) *Two-choice maze* (plan view). Problems in navigation are studied in submerged mazes. The two-choice maze shown here can be used to study spatial and nonspatial behavior.[117] A fish is trained by releasing it in either of the start boxes; the paths to a glass barrier and the maze exit (releasing the fish into a larger tank) are indicated by cues placed on the wall of the tank (a and b). The maze can be converted from a version that trains a nonspatial discrimination (shown here) to a version that trains a spatial discrimination by switching the locations of the two cues marked a and b.

lever, and a food dispenser above the lever; note that a similar apparatus is used in appetitive Pavlovian conditioning.[77] Talton et al.[31] arranged for food to be delivered according to a fixed-interval schedule whereby food is delivered for the first response following a fixed time since the last food delivery. In many species, including humans, the fixed-interval schedule produces a pause after the food delivery, followed by an accelerated rate of responding.[103] Operant lever pressing by goldfish also shows temporal control by the delayed food, with peak response rates occurring at the time of reinforcement,[31] though the timing of responses is not as precise as in the Pavlovian case.[25]

Operant choice has been studied in goldfish using ideal-free distribution of responding procedures in which several fish are presented with two spatially separated sources of food, f_1 and f_2, that differ in the rate of food presentation.[104–106] In such a situation, goldfish show a close approximate to the ideal-free distribution by dividing themselves between the food sources in proportion to the rate of food presentation at each source. For example, if 20% of the food delivered in a given hour is presented at f_1, then 20% of the fish will gather about the location of f_1, and the rest will gather at f_2. (For a discussion of the relationship between choice and timing, see Cerutti and Staddon,[107] Jozefowiez et al.,[108] and Staddon and Cerutti[109]; for a discussion of the relation between the ideal-free distribution and the matching law, see Baum and Kraft[110] and Mackintosh et al.[111])

A two-response appetitive-choice task has been used to study color perception in the goldfish by presenting two stimulus colors (e.g., blue and green), baiting only one color with food, and recording choice accuracy in the asymptotic performance.[112] A similar setup has been used to study shape discrimination with fish.[113] The effects of a variety of reinforcement manipulations show that memory in goldfish is not as extensive as that in rats and pigeons.[111,114]

Maze Learning

Spatial navigation has also been studied in goldfish, with the results showing that they can learn the spatial orientation of food patches in a tank[115] and that they can selectively use external landmarks (allocentric cues) to select the baited arm of a T-maze.[116] Technically speaking, maze learning is a form of operant behavior to the extent that running a maze is maintained by reinforcement, but it is of special interest because it involves learning to orient movements with respect to landmark cues.[111] A typical maze used to study spatial learning is illustrated in Figure 15.1.3.[37,38,116,117] In this procedure, a trial starts by placing the fish in one of the "start" areas and allowing the fish to swim out of the central area into a larger containment tank. Stimuli on the walls of the maze indicate the escape route and a "blind" exit containing glass. Given appropriate training, a fish will learn to swim in the direction of the escape route (i.e., escape from the central area is reinforcing).

López et al.[117] trained some goldfish in the nonspatial (or directly cued) version of the maze (as shown in Figure 15.1.3) and others in the spatial version of the maze (as in Figure 15.1.3, but with the positions of stimuli a and b reversed), and then tested the fish by removing some of the cues. The principal difference in test performance between groups was a large decrement in choice accuracy of the nonspatially trained fish when the cues around the exit were removed. These and other findings,[37] both qualitatively and quantitatively similar to those obtained with rodents and birds,[86] support the validity of a fish model of mammalian navigation behavior.

Many studies have examined the physiology of goldfish maze behavior.[37,118] Studies of telencephalic involvement find that ablation of the telencephalon impairs a previously learned spatial performance, but not a nonspatial performance; however, the spatial performance is readily reestablished following ablation.[37] Experiments with simple two-choice mazes suggest that the teleost telencephalon has a short-term memory role in learning the conditions present at the time of reinforcement.[65,119,120] Taken together, studies of telencephalic ablation seem to reveal several functions, including memory, arousal, and fear conditioning.[37,38,74,119]

Summary

The familiarity with goldfish and ready availability of behavioral assays has led to their use in various lines of investigation. For example, toxicological studies have found that acid stress (lowered ambient pH) interferes with maze learning in goldfish,[121] and that pesticides have detrimental effects on a variety of locomotor and social behaviors.[122-125] Goldfish have appeared in models of Parkinson's disease that include loss of noradrenaline and dopamine and show reduced ambulation.[126-128] Administration of apomorphine, a dopamine agonist, impairs spontaneous eye movements in the manner seen in Parkinson's disease and schizophrenia.[129] These and other examples of the complexity of learning and adaptive behavior in the goldfish ensure that it will be increasingly used alongside mammalian models of cognitive impairment.

ZEBRAFISH (DANIO RERIO)

Zebrafish are becoming a model of choice for studying the molecular basis of vertebrate neurodevelopment.[10,130,131] Their clear chorion allows continuous visualization of the process of development. Rapid development and accessibility to genetic analysis make the zebrafish an excellent model system for studies of neurodevelopment. The wide variety of genetic mutants available in zebrafish offers the promise of determining the molecular mechanisms of neurobehavioral function. Zebrafish studies have been critical in the identification of many genes affecting various aspects of neural development and function; a partial list is provided by Schier.[132] The potential of studying the zebrafish to aid in our understanding the genetics and physiology of learning and memory is gaining momentum[133]; many tasks are now able to tap behavioral processes previously only studied with rodents and goldfish.[9,134-140] In this section we highlight strategies of interest for models of cognitive impairment.

Reflexes and Habituation

A wide variety of excellent behavioral tests have appeared in the zebrafish literature, but the main focus has been on sensory-motor development (e.g., vision, swimming, and touch-elicited reflexes) in larvae or young fish.[141,142] The simplest behavior tested thus far is the tap-elicited startle reflex, which shows an increased latency due to early alcohol exposure.[143,144] The development of touch-elicited escape behavior is summarized by Granato et al.[141]: "Although the embryo is resting most of the time, touching the tail tip induces a fast and straight movement away from the stimulus source. In contrast, mechanical stimuli near the head of the embryo induce a fast escape response, where the embryo turns 180° along its horizontal body axis. At 96 hours the larva is freely swimming, changes swimming directions spontaneously, and is able to direct its swimming towards targets."

Exploratory behavior in novel environments has been used to assay anxiety in rodent models,[145-147] and analogous procedures have entered the zebrafish literature. Several experiments show that the fish first explores a stimulus predominantly using the right eye and subsequently approaches the stimulus favoring the left eye.[148] Figure 15.2.1 shows an apparatus employed by Miklosi and Andrew[24] to study lateralization

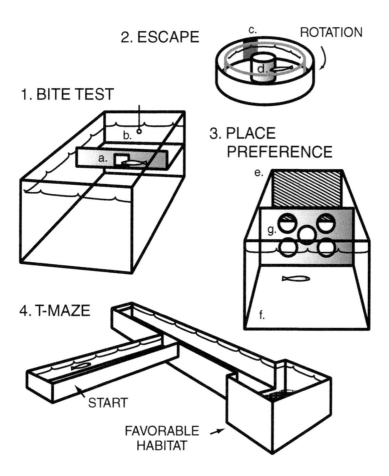

FIGURE 15.2 Four apparatuses used to study learning and adaptive procedure in zebrafish. (1) *Bite test*. In this procedure, a zebrafish is trained to enter the raised platform through the door (a) to explore a small submerged stimulus (b) such as a colored bead. Miklosi[24] employed this apparatus to study lateralization in the zebrafish and found habituation of biting and exploratory behavior elicited by a bead. (2) *Escape*. Darland and Dowling[9] and Li and Dowling[153] used elicited escape from a moving band to study visual function in zebrafish. In this apparatus, rotation of the dark band (c) surrounding the swim area elicits defensive hiding behavior behind a central pole (d). (3) *Place preference*. The place-preference procedure is used to assess affinity to conditioned stimuli. In a typical procedure, a space is divided into two distinctive halves (e and f) with a partition between them (g); the subject is exposed to an unconditioned stimulus in one-half of the space, and then later, with the partition opened, it is given a preference. For example, Darland and Dowling[9] found that zebrafish show a preference for a stimulus previously paired with cocaine. (4) *T-maze*. The T-maze can be used to study a variety of questions in learning and cognition, including discrimination[135] and spatial and nonspatial navigation, e.g., with goldfish.[116] The version shown here was employed by Darland and Dowling[9] with zebrafish in an experiment in which the primary datum was latency to reach the favorable habitat.

of visual exploration in the zebrafish. Subjects are first trained to visit a box suspended in their home tanks by entering a door (a) to eat. After they reliably enter the box, a colored bead (b) is lowered into the water, with the behavior of the fish recorded on video. Right-eye use and biting were highly probable the first time a stimulus was presented, and both declined in probability in two subsequent trials, demonstrating habituation.[149–152]

Zebrafish show a highly developed visually guided escape reflex known as the optokinetic response,[153] escaping a stimulus behind a place of concealment. This "concealment" reflex may be analogous to the "targeted response" concealment behavior described in mice by Blanchard et al.,[44] who showed that if mice are familiarized with a container containing a place of concealment, they flee directly to that place when threatened. Figure 15.2.2 shows an apparatus developed to study the visually guided escape reflex in zebrafish.[9,153] Fish are tested by rotating the outer cylinder of the apparatus that contains a vertical black band (c) and then observing the subject's orientation with respect to the band and a central cylinder (d) behind which it can hide. The test can be used to test visual function[154,155] and has been used to determine contrast sensitivity of zebrafish.[156]

Pavlovian Conditioning

Zebrafish have shown Pavlovian learning in several experiments. Figure 15.2.3 shows a "place-preference" task used by Darland and Dowling[9] to screen zebrafish for cocaine sensitivity (see also Swain et al.[157]). The apparatus consists of a tank divided into two distinctive chambers by a screen. During training, the screen is sealed and a zebrafish is exposed to cocaine in one of the chambers. In subsequent preference tests, the fish show an appetitive conditioning effect by approaching and staying in the chamber in which they had previously received cocaine (note, however, that the swimming toward the conditioned stimulus is likely to be an operant response).

Many studies with zebrafish have used shuttle-box procedures in which they learn to avoid an aversive conditioned stimulus.[136,137,158] Some of the earliest demonstrations of associative learning in zebrafish used a shock-deletion procedure to reinforce swimming away from a shock signal, as seen in Figure 15.1.1.[159] More recently, Pradel et al.[160] used a shuttle box and shock avoidance to study the role of cell adhesion molecules in memory consolidation.

Suboski et al.[20,21] demonstrated Pavlovian conditioning of fear by pairing morpholine and alarm substance (a chemical secreted by frightened or injured fish) and subsequently showing conditioned fear to morpholine alone.[21] The Pavlovian nature of their learning was later confirmed by showing that the conditioned alarm response could also be transferred between stimuli by sensory and second-order conditioning. The last finding in particular highlights the subtlety of learning possible in this unassuming, diminutive fish.

Operant Conditioning

Although it might be assumed that this predominance of aversive procedures exists because aversive procedures are more rapid than appetitive procedures, there are exceptions, as demonstrated by Williams et al.,[138] who trained fish to alternate between two feeding sites in an average of 14 trials. The task is essentially an appetitive version of the shuttle box shown in Figure 15.1.1, except that trials are initiated by the experimenter tapping on the center of the tank and 5 sec later dropping a small amount of food in one end of the tank (the location of food is alternated between trials); the dependent measure is the position of the fish immediately before the delivery of food. Carvan et al.[143] used this task to show dose-dependent detrimental effects of ethanol on learning and memory in zebrafish.

Mazes

Perhaps the earliest example of behavioral research with zebrafish is a maze-learning study in which approach to black or white stimuli was trained by eliciting an anode galvanotaxic reflex that caused approach to the target stimulus.[161] Colwill et al.[135] recently trained a color discrimination in zebrafish by placing different colors at the end of each arm of a T-maze (green vs. purple and red vs. blue) and feeding the fish only at one arm. These researchers unambiguously demonstrated discrimination of color by arranging discrimination reversals (i.e., a cross-over design) and experimenter-blind testing. In a similar T-maze apparatus shown in Figure 15.2.4, Darland and Dowling[9] reinforced choice of one arm by providing it with a goal box containing deep water, artificial grass, and marbles. (However, the dependent measure was the reduction in latency to reach the enriched arm, a result that could be due to habituation of fear in the novel maze apparatus.)

Arthur and Levin[134] have used the three-chamber maze shown in Figure 15.3. The start area is the middle "start chamber"; there are vertically sliding doors on either side of this central start area leading to left and right choice areas. At the outset of a trial the fish is placed in the start chamber and allowed to move about for a brief period. In the choice phase, the vertical sliding doors to the left- and right-choice chambers are opened, and the fish is allowed time to swim to one or the other; if it persists in the start chamber, a fish net is waved in the chamber (a threatening stimulus) until it makes a choice. After making a choice, both vertical sliding doors are closed. If the choice is correct (i.e., to the goal side), the fish is permitted to swim for a short period of time; if the choice is incorrect, the sliding partition is moved to the "restricting position" (see Figure 15.3) for a short period of time. This procedure is repeated for a fixed number of trials. Dependent measures in the three-chamber shuttle maze include latency to escape the start chamber and correct choices.[13,134] Initial tests of the maze[134] showed that zebrafish could be trained to turn in a particular direction (spatial learning) or to approach a particular color regardless of location (nonspatial learning). Based on the work of Levin et al.,[13] we have used the three-chamber maze to show that the delayed spatial-alternation behavior is a sensitive index of the persisting cognitive impairment caused by developmental exposure to chlorpyrifos. In a parallel line of investigation, Levin and Chen[162] found that acute nicotine administration causes a significant improvement in delayed spatial alternation at low doses but impairs performance at high doses. These results are shown in Figure 15.4. The biphasic effect of nicotine improvement of memory function at low doses and less improvement at higher doses

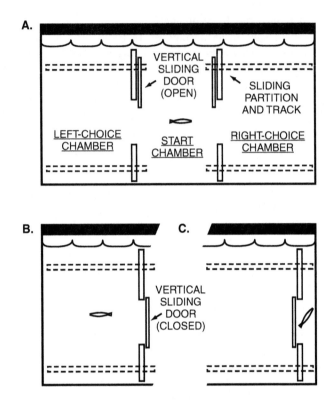

FIGURE 15.3 Three-chamber shuttle maze used to study learning and memory in zebra-fish.[13,134,216,217] As shown in the top diagram (A), trials begin with the fish in the start chamber. During a choice phase, both vertical sliding doors are opened. After a choice, the sliding doors are closed. If the fish chooses the correct chamber (as in B), it is allowed to swim freely for a short time, but incorrect choices are punished by sliding the partition to restrict the swimming of the fish (as in C).

is a common finding across a wide variety of species including rats, mice, monkeys, and humans.[163–165] The fact that the same effect was seen in zebrafish points to similarities of nicotinic effects on memory with mammalian species. This similarity can be advantageous, as molecular studies of neural function can be more easily studied in zebrafish than mammals.

Summary

A number of clever behavioral assays of zebrafish have appeared in the literature. Despite the small size of the fish, it is now clear that the zebrafish model of development can be used in studies of learning, memory, and cognition. There are both appetitive and aversive techniques, and they test a range of behavior from simple reflexes[143] and fear conditioning[158] to visual discrimination[135] and spatial orientation.[134]

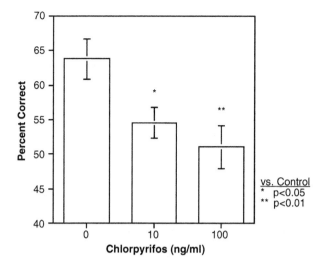

FIGURE 15.4 Persistent effect of early exposure to chlorpyrifos on delayed spatial alternation in the three-chamber shuttle maze.[13] Doses of 10 ng/ml ($p < 0.05$) and 100 ng/ml ($p < 0.01$) of chlorpyrifos during the first five days of development caused delayed spatial-alternation accuracy in adult zebrafish.

INVERTEBRATE MODELS

A wide variety of vertebrate and invertebrate species have been used as model systems to understand processes of learning and memory, each with their own advantages for studying the mechanisms of cognition.[166] Most famously, *Aplysia californica* have been used to understand the cellular and molecular processes of learning and memory.[51] *Hermissenda* have also been used. They show Pavlovian learning impairments in response to lead,[167] which may provide an interesting model of the cognitive impairments seen in children after lead exposure. The invertebrate species *drosophila* (fruit flies) and *C. elegans*, a nematode worm, have been widely used to help understand the molecular bases of cognition. They hold great promise as complementary models to the traditional rodent and primate mammalian models for understanding the basis of cognitive function and dysfunction.

Drosophila and *C. elegans* have been extensively used for studies of the molecular biology of development because they are relatively simple organisms in which the basics of the molecular biology of cellular communication can be discovered. *C. elegans* in particular is a very simple organism in which each cell has been identified and characterized in terms of its lineage, migration, and function. The connections of the *C. elegans* nervous system have been determined, and the activity of these connections in terms of sensory and motor function have been worked out. Neuroplasticity in terms of experience-based alterations in behavioral response, i.e., learning, has received recent attention. *C. elegans* and drosophila are excellent models to study the molecular bases of neurodevelopment and nervous system plasticity. These invertebrate models can be used to identify neuromolecular bases of learning and memory that can then be followed up with studies using vertebrate species such as zebrafish and other aquatic species (reviewed above) and to traditional mammalian species such as rodents and primates described in other chapters of this book.

C. ELEGANS

C. elegans, with a 302-neuron nervous system, shows a great variety of neurobehavioral plastic effects, as reviewed by Hobert.[168] The activity of individual neurons critical for memory can be visualized and directly related to the process of learning and memory.[169] *C. elegans* can easily be used for high-throughput studies that can screen great numbers of potential toxicants and therapeutic candidates. The results of such screens can provide useful starting points for further studies with vertebrate species, including mammalian models. However, because of the phylogenic distance between *C. elegans* and humans, *C. elegans* by themselves cannot be adequately used to predict chemical toxicity or therapeutic response in humans. *C. elegans* is best used as an initial whole-organism screen for triage of subsequent studies with higher organisms and to help identify molecular processes critical to neural plasticity.

The very simple forms of plasticity, habituation, and sensitization have been widely studied in *C. elegans*.[170,171] *C. elegans* are capable of specific habituation to an olfactory stimulus, which is separable from mere sensorimotor fatigue.[172] *C. elegans* respond to a light mechanical touch with an escape response that quickly habituates. The specific interneurons underlying this effect are being identified.[173] The habituation processes have been found to vary in a regular fashion according to the interstimulus intervals between 10 and 60 sec.[56,174] Memory for habituation in *C. elegans* lasts over one day. This memory process is disrupted by heat shock.[175]

C. elegans have also been used in more-complex studies of associative learning and memory.[176] *C. elegans* use taste, smell, temperature, and oxygen level to remember the location of food.[177] Association of the addition of sodium chloride with the lack of food can be learned by *C. elegans*.[178] They will avoid areas of low pH,[179] a behavior that can be used for association studies.

C. elegans show a behavior known as *isothermal tracking,* which is conditioning of movement along a temperature gradient to their cultivation temperature for food. Associative learning in *C. elegans* has been found to be context dependent,[180] indicating that this organism is capable of higher-level specific learning. Age-related impairment can be seen in isothermal-tracking associative learning. Processes of learning and memory can be manipulated separately. Cold stress before training has been found to impair learning and cold stress after the training to impair memory.[181]

C. elegans have been used to identify molecular factors in the basis of associative learning.[182] These can be differentiated from factors involved in nonassociative learning and sensorimotor performance factors. Genetic factors in learning can be readily determined with olfactory conditioning.[183] Different *C. elegans* strains have been identified that show selective deficits in learning and recall of a conditioned association of an odorant with a motor response.[184] Mutants with increased oxidative stress showed severely impaired learning, while mutants with reduced oxidative stress showed increased longevity and improved learning.[185] Olfactory imprinting depends on interneuron function and the expression of a G protein-coupled chemoreceptor family member encoded by the sra-11 gene.[186] The genetic factor hab-1 was found to be important for normal rates of habituation,[187] but it is not related to dishabituation and performance factors. In *C. elegans,* a genetic factor associated with Down's syndrome was related to cognitive impairment, and this effect was reversible.[188]

Neurochemical systems underlying cognitive function in *C. elegans* are in some ways similar to those in mammals. For example, glutamate neurons play a key role in habituation.[57,189] Adequate foraging behavior is dependent on correctly functioning NMDA glutamate receptors.[190] AMPA glutamate receptors have been shown to be important for olfactory conditioning.[191] Calcium channel signaling is key for adequate associative learning and memory in *C. elegans,* as determined by impaired isothermal tracking with knockout of the gene NCS-1 (neuron-specific calcium sensor-1) and improved conditioning with overexpression of this gene.[192]

DROSOPHILA

Methods for assessing cognitive function have been developed for drosophila, which can also play an important role as a model of cognitive impairment.[193] Mushroom bodies are critical structures for cognitive function in drosophila. Lesion of the mushroom bodies selectively impairs classical conditioning associating odor cues with an electrical shock without impairing other performance factors.[194] Genetic mutations that impair mushroom body development also selectively impair the same type of conditioning.[195] In the mushroom bodies of drosophila cAMP cascade is important for memory.[196]

Drosophila has been useful for detecting genetic factors involved in memory. Volado proteins located in the mushroom bodies have been found to be vital for olfactory memory.[197] Genetic factors for protein expression in the mushroom bodies such as rutabaga[198] and Leonardo[199] are critical for memory function in adult drosophila. The memory dysfunction in drosophila caused by deficits in the gene rutabaga in the mushroom bodies could be reinstated by transient reexpression in this area in adults, showing that the effect is not dependent on altered development.[200] Elevated Tau in the mushroom bodies of drosophila causes impairments in olfactory learning and memory without change in the sensorimotor responses needed for performance.[201] Tau elevations are also seen in humans with cognitive dysfunction. Drosophila can be conditioned to use an olfactory cue to avoid a shock. This learning is impaired by the dunce genetic mutation.[202] Disorders in the phosphorylation–dephosphorylation dynamic lie at the basis of the cognitive impairment in the dunce mutant.[203] Mutations affecting tyrosine kinase impair learning and memory.[204]

Drosophila shows aging-induced cognitive dysfunction, as do humans. Aging-induced memory loss is seen in drosophila with olfactory conditioning.[205] Drosophila is also useful in discovering molecular mechanisms of aging-induced memory impairment, including the involvement of cyclic AMP signaling.[206] The mechanistic relationships between drosophila and humans are still being worked out.[207] A simple habituation of proboscis extension for sugar has been shown to be sensitive to aging-related impairment between 3 and 35 days of age, much in the same fashion as associative learning.[208]

There are common mechanisms of memory for drosophila and mammals. Common factors underlying memory function in drosophila and mammals include control of microtubule formation by aPKC in glutamatergic synapses.[209] The transcription factor CREB (cyclic AMP response element binding protein) has been found in drosophila, mice, and humans to play a key role in learning and memory.[210] Drosophila with a loss of function due to mutation of pituitary adenylyl cyclase-activating polypeptide (PACAP) show memory impairments, as do mice.[211] This gene is active in the mushroom bodies of drosophila and hippocampus of mice. As with *C. elegans* and mammals, glutamate is important for cognitive function in drosophila. NMDA glutamate receptors are critical for learning and memory in drosophila as well as in mammals, including humans.[212] Attention can be measured in drosophila by determining the impact of a visual stimulus on flight pattern.[213] Developmental blockade of dopaminergic neurons impairs attention, as it does in mammals.

CONCLUSIONS

In the development and use of animal models of cognitive dysfunction, it is important to develop complementary models to exploit the unique advantages of the different species. Nonmammalian vertebrates such as fish provide the opportunity to directly observe neurodevelopmental processes and determine the impact of developmental permutations on learning and memory. Zebrafish in particular are valuable because of the availability of morpholine techniques to transiently suppress specific parts of genomic expression. Invertebrate models such as *C. elegans* and drosophila provide other advantages, particularly the elegant genetic manipulations available. The simple nervous systems in these models are useful in determining mechanisms of cognitive function. The development of new methods for high-throughput tests of cognitive function for fish can provide a means for rapid screening of potential toxic agents as well as promising therapeutic agents. It is equally important to develop specific tests of various aspects of cognitive function, including habituation, associative learning, memory, and attention as well as to be able to differentiate changes in sensorimotor function from cognition. Key in the use of nonmammalian models is the determination of which mechanisms of cognitive function are similar to mammals and which are different. Nonmammalian models can be used in concert with classic mammalian models to determine the neural bases of cognitive function and to aid in the discovery of toxicants and potential therapeutic agents.

ACKNOWLEDGMENTS

This research was supported by NIH ES10356.

REFERENCES

1. Beal, M.F., Experimental models of Parkinson's disease, *Nature Reviews Neuroscience*, 2, 325–332, 2001.

2. Arendash, G.W., Gordon, M.N., Diamond, D.M., Austin, L.A., Hatcher, J.M., Jantzen, P., Dicarlo, G., Wilcock, D., and Morgan, D., Behavioral assessment of Alzheimer's transgenic mice following long-term A beta vaccination: task specificity and correlations between A beta deposition and spatial memory, *DNA and Cell Biology*, 20, 737–744, 2001.
3. Bard, F., Cannon, C., Barbour, R., Burke, R.L., Games, D., Grajeda, H., Guido, T., Hu, K., Huang, J.P., Johnson-Wood, K., Khan, K., Kholodenko, D., Lee, M., Lieberburg, I., Motter, R., Nguyen, M., Soriano, F., Vasquez, N., Weiss, K., Welch, B., Seubert, P., Schenk, D., and Yednock, T., Peripherally administered antibodies against amyloid beta-peptide enter the central nervous system and reduce pathology in a mouse model of Alzheimer disease, *Nature Medicine*, 6, 916–919, 2000.
4. German, D.C. and Eisch, A.J., Mouse models of Alzheimer's disease: insight into treatment, *Reviews in the Neurosciences*, 15, 353–369, 2004.
5. Hsiao, K., Chapman, P., Nilsen, S., Eckman, C., Harigaya, Y., Younkin, S., Yang, F.S., and Cole, G., Correlative memory deficits, A beta elevation, and amyloid plaques in transgenic mice, *Science*, 274, 99–102, 1996.
6. Lipska, B.K. and Weinberger, D.R., To model a psychiatric disorder in animals: schizophrenia as a reality test, *Neuropsychopharmacology*, 23, 223–239, 2000.
7. Goldman-Rakic, P.S., Working memory dysfunction in schizophrenia, *Journal of Neuropsychiatry and Clinical Neuroscience*, 6, 348–357, 1994.
8. Castner, S.A., Goldman-Rakic, P.S., and Williams, G.V., Animal models of working memory: insight for targeting cognitive function in schizophrenia, *Psychopharmacology*, 174, 111–125, 2004.
9. Darland, T. and Dowling, J.E., Behavioral screening for cocaine sensitivity in mutagenized zebrafish, *Proceedings of the National Academy of Sciences of the United States of America*, 98, 11691–11696, 2001.
10. Guo, S., Linking genes to brain, behavior and neurological disease: what can we learn from zebrafish? *Genes, Brain and Behavior*, 3, 63–74, 2001.
11. Gerlai, R., Lahav, M., Guo, S., and Rosenthal, A., Drinks like a fish: zebrafish (*Danio rerio*) as a behavior genetic model to study alcohol effects, *Pharmacology, Biochemistry and Behavior*, 67, 773–782, 2000.
12. Linney, E., Upchurch, L., and Donerly, S., Zebrafish as a neurotoxicological model, *Neurotoxicology and Teratology*, 27, 175–175, 2004.
13. Levin, E.D., Crysthansis, E., Yacisin, K., and Linney, E., Chlorpyrifos exposure of developing zebrafish: effects on survival and long-term effects on response latency and spatial discrimination, *Neurotoxicology and Teratology*, 25, 51–57, 2003.
14. Reebs, S., *Fish Behavior in the Aquarium and in the Wild*, Comstock Pub. Assoc., Ithaca, NY, 2001.
15. Barlow, G.W., *The Cichlid Fishes: Nature's Grand Experiment in Evolution*, Perseus Books, Cambridge, 2000.
16. Tebbich, S., Bshary, R., and Grutter, A.S., Cleaner fish *Labroides dimidiatus* recognise familiar clients, *Animal Cognition*, 5, 139–145, 2002.
17. Reader, S.M., Kendal, J.R., and Laland, K.N., Social learning of foraging sites and escape routes in wild Trinidadian guppies, *Animal Behaviour*, 66, 729–739, 2003.
18. Santangelo, N. and Itzkowitz, M., Sex differences in the mate selection process of the monogamous, biparental convict cichlid, *Archocentrus nigrofasciatum*, *Behaviour*, 141, 1041–1059, 2004.
19. Payne, R.J.H., Gradually escalating fights and displays: the cumulative assessment model, *Animal Behaviour*, 56, 651–662, 1998.
20. Suboski, M.D., Acquisition and social communication of stimulus recognition by fish, *Behavioural Processes*, 16, 213–244, 1988.

21. Suboski, M.D., Bain, S., Carty, A.E., and McQuoid, L.M., Alarm reaction in acquisition and social transmission of simulated predator recognition by zebra danio fish, *Journal of Comparative Psychology*, 104, 101–112, 1990.
22. Braithwaite, V.A., Armstrong, J.D., McAdam, H.M., and Huntingford, F.A., Can juvenile Atlantic salmon use multiple cue systems in spatial learning? *Animal Behaviour*, 51, 1409–1415, 1996.
23. Miklosi, A., Andrew, R.J., and Savage, H., Behavioural lateralisation of the tetrapod type in the zebrafish (*Brachydanio rerio*), *Physiology and Behavior*, 63, 127–135, 1997.
24. Miklosi, A. and Andrew, R.J., Right eye use associated with decision to bite in zebrafish, *Behavioral Brain Research*, 105, 199–205, 1999.
25. Drew, M.R., Zupan, B., Cooke, A., Couvillon, P.A., and Balsam, P.D., Temporal control of conditioned responding in goldfish, *Journal of Experimental Psychology: Animal Behavior Processes*, 31, 31–39, 2005.
26. Reebs, S.G., Time-place learning in golden shiners (Pisces: Cyprinidae), *Behavioural Processes*, 36, 253–262, 1996.
27. Higa, J.J. and Simm, L.A., Interval timing in Siamese fighting fish (*Betta splendens*), *Behavioural Processes*, 67, 501–509, 2004.
28. Hollis, K.L., The role of learning in the aggressive and reproductive behavior of blue gouramis, *Trichogaster trichopterus*, *Environmental Biology of Fishes*, 54, 355–369, 1999.
29. Behrend, E.R. and Bitterman, M.E., Avoidance-conditioning in the fish: further studies of the CS-US interval, *American Journal of Psychology*, 77, 15–28, 1964.
30. Behrend, E.R. and Bitterman, M.E., Sidman avoidance in the fish, *Journal of the Experimental Analysis of Behavior*, 6, 47–52, 1963.
31. Talton, L.E., Higa, J.J., and Staddon, J.E.R., Interval schedule performance in the goldfish *Carassius auratus*, *Behavioural Processes*, 45, 193–206, 1999.
32. Bitterman, M.E., Phyletic differences in learning, *American Psychologist*, 20, 396–410, 1965.
33. Tinbergen, N., *The Study of Instinct*, Oxford University Press, Oxford, 1951.
34. Hollis, K.L. and Overmier, J.B., The function of the telost telencephalon in behavior: a reinforcer mediator, in *The Behavior of Fish and Other Aquatic Animals*, Academic Press, New York, 1978, pp. 1371–1395.
35. Wullimann, M.F. and Mueller, T., Teleostean and mammalian forebrains contrasted: evidence from genes to behavior, *Journal of Comparative Neurology*, 475, 143–162, 2004.
36. Overmier, J.B. and Papini, M.R., Factors modulating the effects of telost telencephalon ablation on retention, relearning, and extinction of avoidance behavior, *Behavioral Neuroscience*, 100, 190–199, 1986.
37. Salas, C., Broglio, C., Rodriguez, F., Lopez, J.C., Portavella, M., and Torres, B., Telencephalic ablation in goldfish impairs performance in a "spatial constancy" problem but not in a cued one, *Behavioral Brain Research*, 79, 193–200, 1996.
38. Broglio, C., Rodriguez, F., and Salas, C., Spatial cognition and its neural basis in teleost fishes, *Fish and Fisheries*, 4, 247–255, 2003.
39. Shin, S.W., Chung, N.I., Kim, J.S., Chon, T.S., Kwon, O.S., Lee, S.K., and Koh, S.C., Effect of diazinon on behavior of Japanese medaka (*Oryzias latipes*) and gene expression of tyrosine hydroxylase as a biomarker, *Journal of Environmental Science and Health — Part B: Pesticides, Food Contaminants, and Agricultural Wastes*, 36, 783–795, 2001.

40. Phillips, T.A., Summerfelt, R.C., and Atchison, G.J., Environmental, biological, and methodological factors affecting cholinesterase activity in walleye (*Stizostedion vitreum*), *Archives of Environmental Contamination and Toxicology*, 43, 75–80, 2002.

41. Brookshire, K.H. and Hognander, O.C., Conditioned fear in the fish, *Psychological Reports*, 22, 75–81, 1968.

42. Pinckney, G.A., Avoidance learning in fish as a function of prior fear conditioning, *Psychological Reports*, 20, 71–74, 1967.

43. Saglio, P. and Trijasse, S., Behavioral responses to atrazine and diuron in goldfish, *Archives of Environmental Contamination and Toxicology*, 35, 484–491, 1998.

44. Blanchard, R.J., Hebert, M.A., Ferrari, P., Palanza, P., Figueira, R., Blanchard, D.C., and Parmigiani, S., Defensive behaviors in wild and laboratory (Swiss) mice: the Mouse Defense Test Battery, *Physiology and Behavior*, 65, 201–209, 1998.

45. Blanchard, D.C., Griebel, G., and Blanchard, R.J., The Mouse Defense Test Battery: pharmacological and behavioral assays for anxiety and panic, *European Journal of Pharmacology*, 463, 97–116, 2003.

46. Laming, P.R. and McKinney, S.J., Habituation in goldfish (*Carassius auratus*) is impaired by increased interstimulus interval, interval variability, and telencephalic ablation, *Behavioural Neuroscience*, 106, 869–875, 1990.

47. Bolles, R.C., Species-specific defence reactions and avoidance learning, *Psychological Review*, 77, 32–48, 1970.

48. Marlin, N.A. and Miller, R.R., Associations to contextual stimuli as a determinant of long-term habituation, *Journal of Experimental Psychology: Animal Behavior Processes*, 7, 313–333, 1981.

49. Carew, T.J., Pinsker, H.M., and Kandel, E.R., Long-term habituation of a defensive withdrawal reflex in Aplysia, *Science*, 175, 451–454, 1972.

50. Falk, C.X., Wu, J.Y., Cohen, L.B., and Tang, A.C., Nonuniform expression of habituation in the activity of distinct classes of neurons in the Aplysia abdominal ganglion, *Journal of Neuroscience*, 13, 4072–4081, 1993.

51. Kandel, E.R. and Schwartz, J.H., Molecular biology of learning — modulation of transmitter release, *Science*, 218, 433–443, 1982.

52. Kandel, E.R., *Behavioral Biology of Aplysia: A Contribution to the Comparative Study of Opisthobranch Mollusks*, Freeman, San Francisco, 1979.

53. Marcus, E.A., Nolen, T.G., Rankin, C.H., and Carew, T.J., Behavioural dissociation of dishabituation, sensitization, and inhibition in Aplysia, *Science*, 241, 210–213, 1988.

54. Laming, P.R. and Ennis, P., Habituation of fright and arousal responses in the teleosts *Carassius auratus* and *Rutilus rutilus*, *Journal of Comparative and Physiological Psychology*, 96, 460–466, 1982.

55. Beck, C.D.O. and Rankin, C.H., Long-term habituation is produced by distributed training at long ISIs and not by massed training or short ISIs in *Caenorhabditis elegans*, *Animal Learning and Behavior*, 25, 446–457, 1997.

56. Rankin, C.H. and Broster, B.S., Factors affecting habituation and recovery from habituation in the nematode *Caenorhabditis elegans*, 106, 239–249, 1992.

57. Rose, J.K. and Rankin, C.H., Analyses of habituation in *Caenorhabditis elegans*, *Learning and Memory*, 8, 63–69, 2001.

58. Rankin, C.H., Gannon, T., and Wicks, S.R., Developmental analysis of habituation in the nematode *C. elegans*, *Developmental Psychobiology*, 36, 261–270, 2000.

59. Rankin, C.H., Context conditioning in habituation in the nematode *Caenorhabditis elegans*, *Behavioral Neuroscience*, 114, 496–505, 2000.

60. Staddon, J.E.R. and Higa, J.J., Multiple time scales in simple habituation, *Psychological Review*, 103, 720–733, 1996.
61. Staddon, J.E.R., On rate-sensitive habituation, *Adaptive Behavior*, 1, 421–436, 1993.
62. Thompson, R.F. and Spencer, W.A., Habituation: a model phenomenon for the study of neuronal substrates of behavior, *Psychological Review*, 73, 16–43, 1966.
63. Hinde, R.A., Behavioural habituation, in *Short-Term Changes in Neural Activity and Behaviour*, Horn, G. and Hinde, R.A., Eds., Cambridge University Press, London, 1970, pp. 3–40.
64. Overmier, J.B. and Curnow, P.G., Classical conditioning, pseudoconditioning, and sensitization in "normal" and forebrainless goldfish, *Journal of Comparative and Physiological Psychology*, 68, 193–198, 1969.
65. Savage, G.E., Temporal factors in avoidance learning in normal and forebrainless goldfish (*Carassius auratus*), *Nature*, 218, 1168–1169, 1968.
66. Rescorla, R.A., Behavioral studies of Pavlovian conditioning, *Annual Review of Neuroscience*, 11, 329–352, 1988.
67. Hawkins, R.D., Cohen, T.E., Greene, W., and Kandel, E.R., Relationships between dishabituation, sensitization, and inhibition of the gill- and siphon-withdrawal reflex in *Aplysia californica*: effects of response measure, test time, and training stimulus, *Behavioral Neuroscience*, 112, 24–38, 1998.
68. Frost, L., Kaplan, S.W., Cohen, T.E., Henzi, V., Kandel, E.R., and Hawkins, R.D., A simplified preparation for relating cellular events to behavior: contribution of LE and unidentified siphon sensory neurons to mediation and habituation of the Aplysia gill- and siphon-withdrawal reflex, *Journal of Neuroscience*, 17, 2900–2913, 1997.
69. Springer, A.D., Schoel, W.M., Klinger, P.D., and Agranoff, B.W., Anterograde and retrograde effects of electroconvulsive shock and of puromycin on memory formation in the goldfish, *Behavioral Biology*, 13, 467–481, 1975.
70. Bitterman, M.E., Classical conditioning in the goldfish as a function of the CS-UCS interval, *Journal of Comparative and Physiological Psychology*, 58, 359–366, 1964.
71. Gonzales, R.C., Longo, N., and Bitterman, M.E., Classical conditioning in the fish: exploratory studies of partial reinforcement, *Journal of Comparative and Physiological Psychology*, 54, 452–456, 1961.
72. Otis, L.S. and Cerf, J.A., Conditioned avoidance learning in two fish species, *Psychological Reports*, 12, 679–682, 1963.
73. Wodinsky, J., Behrend, E.R., and Bitterman, M.E., Avoidance conditioning in two species of fish, *Animal Behaviour*, 10, 76–78, 1962.
74. Portavella, M., Salas, C., Vargas, J.P., and Papini, M.R., Involvement of the telencephalon in spaced-trial avoidance learning in the goldfish (*Carassius auratus*), *Physiology and Behavior*, 80, 49–56, 2003.
75. Manteifel, Y.B. and Karelina, M.A., Conditioned food aversion in the goldfish, *Carassius auratus*, *Comparative Biochemistry and Physiology: A-Physiology*, 115, 31–35, 1996.
76. Brandon, S.E. and Bitterman, M.E., Analysis of autoshaping in goldfish, *Animal Learning and Behavior*, 7, 57–62, 1979.
77. Brandon, S.E., Satake, N., and Bitterman, M.E., Performance of goldfish trained on multiple schedules of response-independent reinforcement, *Journal of Comparative and Physiological Psychology*, 96, 467–475, 1982.
78. Papini, M.R. and Bitterman, M.E., The role of contingency in classical conditioning, *Psychological Review*, 97, 396–403, 1990.
79. Gamzu, E. and Williams, D.R., The source of keypecking in autoshaping, *Animal Learning and Behavior*, 3, 37–42, 1975.

80. Williams, D.R. and Williams, H., Auto-maintenance in the pigeon: sustained pecking despite nonreinforcement, *Journal of the Experimental Analysis of Behavior*, 12, 511–520, 1969.

81. Balsam, P., Drew, M., and Yang, C., Timing at the start of associative learning, *Learning and Motivation*, 33, 141–155, 2002.

82. Gibbon, J., Scalar expectancy theory and Weber's law in animal timing, *Psychological Review*, 84, 279–325, 1977.

83. Bateson, M., Interval timing and optimal foraging, in *Functional and Neural Mechanisms of Interval Timing*, Meck, W.H., Ed., CRC Press, Boca Raton, FL, 2003, pp. 113–141.

84. Woodward, W.T. and Bitterman, M.E., Pavlovian analysis of avoidance conditioning in the goldfish (*Carassius auratus*), *Journal of Comparative and Physiological Psychology*, 82, 123–129, 1973.

85. Zhuikov, A.Y., Couvillon, P.A., and Bitterman, M.E., Quantitative two-process analysis of avoidance conditioning in goldfish, *Journal of Experimental Psychology: Animal Behavior Processes*, 20, 32–43, 1994.

86. Shettleworth, S.J., *Cognition, Evolution and Behavior*, Oxford University Press, New York, 1998.

87. Woodward, W.T. and Bitterman, M.E., Classical conditioning of goldfish in the shuttlebox, *Behavior Research Methods and Instrumentation*, 3, 193–194, 1971.

88. Hineline, P.N., The several roles of stimuli in negative reinforcement, in *Advances in Analysis of Behaviour: Predictability, Correlation, and Contiguity*, Harzem, P. and Zeiler, M.D., Eds., Wiley, Chichester, England, 1981, pp. 203–246.

89. Olson, R.D., Kastin, A.J., Michell, G.F., Olson, G.A., Coy, D.H., and Montalbano, D.M., Effects of endorphin and enkephalin analogs on fear habituation in goldfish, *Pharmacology, Biochemistry and Behavior*, 9, 111–114, 1978.

90. Davis, R.E. and Klinger, P.D., NMDA receptor antagonist MK-801 blocks learning of conditioned stimulus-unconditioned stimulus contiguity but not fear of conditioned stimulus in goldfish (*Carassius auratus* L.), *Behavioral Neuroscience*, 108, 935–940, 1994.

91. Xu, X. and Davis, R.E., N-methyl-D-aspartate receptor antagonist MK-801 impairs learning but not memory fixation or expression of classical fear conditioning in goldfish (*Carassius auratus*), *Behavioral Neuroscience*, 106, 307–314, 1992.

92. Xu, X.J., Bazner, J., Qi, M., Johnson, E., and Freidhoff, R., The role of telencephalic NMDA receptors in avoidance learning in goldfish (*Carassius auratus*), *Behavioral Neuroscience*, 117, 548–554, 2003.

93. Xu, X., NMDA receptor antagonist MK-801 selectively impairs learning of the contiguity of the conditioned stimulus and unconditioned stimulus in goldfish, *Pharmacology, Biochemistry and Behavior*, 58, 491–496, 1997.

94. Mattioli, R., Santangelo, E.M., Costa, A.C., and Vasconcelos, L., Substance P facilitates memory in goldfish in an appetitively motivated learning task, *Behavioural Brain Research*, 85, 117–120, 1997.

95. Santangelo, E.M., Morato, S., and Mattioli, R., Facilitatory effect of substance P on learning and memory in the inhibitory avoidance test for goldfish, *Neuroscience Letters*, 303, 137–139, 2001.

96. Spieler, R.E., Nelson, C.A., Huston, J.P., and Mattioli, R., Post-trial administration of H1 histamine receptor blocker improves appetitive reversal learning and memory in goldfish, *Carassius auratus*, *Neuroscience Letters*, 277, 5–8, 1999.

97. Lee, R.K., Eaton, R.C., and Zottoli, S.J., Segmental arrangement of reticulospinal neurons in the goldfish hindbrain, *Journal of Comparative Neurology*, 329, 539–556, 1993.

98. Overmier, J.B. and Starkman, N., Transfer of control of avoidance behavior in normal and telencephalon ablated goldfish (*Carassius auratus*), *Physiology and Behavior*, 12, 605–608, 1974.

99. Portavella, M. and Vargas, J.P., Emotional and spatial learning in goldfish is dependent on different telencephalic pallial systems, *European Journal of Neuroscience*, 21, 2800–2806, 2005.

100. Portavella, M., Torres, B., Salas, C., and Papini, M.R., Lesions of the medial pallium, but not of the lateral pallium, disrupt spaced-trial avoidance learning in goldfish (*Carassius auratus*), *Neuroscience Letters*, 362, 75–78, 2004.

101. Yoshida, M., Okamura, I., and Uematsu, K., Involvement of the cerebellum in classical fear conditioning in goldfish, *Behavioural Brain Research*, 153, 143–148, 2004.

102. Lowes, G. and Bitterman, M.E., Reward and learning in the goldfish, *Science*, 157, 455–457, 1967.

103. Lejeune, H. and Wearden, J.H., The comparative psychology of fixed-interval responding: some quantitative analyses, *Learning and Motivation*, 22, 84–111, 1991.

104. Stenberg, M. and Persson, A., The effect of spatial food distribution and group size on foraging behaviour of a benthic fish, *Behavioural Processes*, 70, 41–50, 2005.

105. Magurran, A.E. and Pitcher, T.J., Foraging, timidity and shoal size in minnows and goldfish, *Behavioral Ecology and Sociobiology*, 12, 147–152, 1984.

106. Pitcher, T.J. and House, A.C., Foraging rules for group feeders: area copying depends upon food density in shoaling goldfish, *Ethology*, 76, 161–167, 1987.

107. Cerutti, D.T. and Staddon, J.E.R., Time and rate measures in choice transitions, *Journal of the Experimental Analysis of Behavior*, 81, 135–154, 2004.

108. Jozefowiez, J., Cerutti, D.T., and Staddon, J.E.R., Timing in choice experiments, *Journal of Experimental Psychology: Animal Behavior Processes*, 31, 213–225, 2005.

109. Staddon, J.E.R. and Cerutti, D.T., Operant conditioning, *Annual Review of Psychology*, 54, 115–144, 2003.

110. Baum, W.M. and Kraft, J.R., Group choice: competition, travel, and the ideal free distribution, *Journal of the Experimental Analysis of Behavior*, 69, 227–245, 1998.

111. Mackintosh, N.J., Lord, J., and Liddle, L., Visual and spatial probability learning in pigeons and goldfish, *Psychonomic Science*, 24, 221–223, 1971.

112. Muntz, W.R.A. and Cronly-Dillon, J.R., Colour discrimination in goldfish, *Animal Behaviour*, 14, 351–355, 1966.

113. Mackintosh, J. and Sutherland, N.S., Visual discrimination by the goldfish: the orientation of rectangles, *Animal Behaviour*, 11, 135–141, 1963.

114. Mackintosh, J. and Cauty, A., Spatial reversal learning in rats, pigeons, and goldfish, *Psychonomic Science*, 22, 281–282, 1971.

115. Pitcher, T.J. and Magurran, A.E., Shoal size, patch profitability, and information exchange in foraging goldfish, *Animal Behaviour*, 31, 546–555, 1983.

116. Rodriguez, F., Duran, E., Vargas, J.P., Torres, B., and Salas, C., Performance of goldfish trained in allocentric and egocentric maze procedures suggests the presence of a cognitive mapping system in fishes, *Animal Learning and Behavior*, 22, 409–420, 1994.

117. López, C., Broglio, C., Rodríguez, F., Thinus-Blanc, C., and Salas, C., Multiple spatial learning strategies in gold fish, *Animal Cognition*, 2, 109–120, 1999.

118. Overmier, J.B. and Hollis, K.L., Fish in the think tank: learning, memory and integrated behavior, in *Neurobiology of Comparative Cognition*, Kesner, R.P. and Olton, D.S., Eds., Lawrence Erlbaum Associates, Hillsdale, NJ, 1990, pp. 205–236.

119. Ohnishi, K., Effects of telencephalic ablation on short-term memory and attention in goldfish, *Behavioural Brain Research*, 86, 191–199, 1997.

120. Ohnishi, K., Telencephalic function implicated in food-reinforced colour discrimination learning in the goldfish, *Physiology and Behavior*, 46, 707–712, 1989.
121. Garg, R. and Garg, A., Operant learning of goldfish exposed to pH depression in water, *Journal of Environmental Biology*, 13, 1–6, 1992.
122. Saglio, P., Bretaud, S., Rivot, E., and Olsen, K.H., Chemobehavioral changes induced by short-term exposures to prochloraz, nicosulfuron, and carbofuran in goldfish, *Archives of Environmental Contamination and Toxicology*, 45, 515–524, 2003.
123. Saglio, P. and Trijasse, S., Behavioral responses to atrazine and diuron in goldfish, *Archives of Environmental Contamination and Toxicology*, 35, 484–491, 1998.
124. Bretaud, S., Saglio, P., Saligaut, C., and Auperin, B., Biochemical and behavioral effects of carbofuran in goldfish (*Carassius auratus*), *Environmental Toxicology and Chemistry*, 21, 175–181, 2002.
125. Bretaud, S., Saglio, P., and Toutant, J.P., Effects of carbofuran on brain acetylcholinesterase activity and swimming activity in *Carassius auratus* (Cyprinidae), *Cybium*, 25, 33–40, 2001.
126. Pollard, H.B., Dhariwal, K., Adeyemo, O.M., Markey, C.J., Caohuy, H., Levine, M., Markey, S., and Youdim, M.B.H., A parkinsonian syndrome induced in the goldfish by the neurotoxin MPTP, *FASEB Journal*, 6, 3108–3116, 1992.
127. Adeyemo, O.M., Youdim, M.B.H., Markey, S.P., Markey, C.J., and Pollard, H.B., L-deprenyl confers specific protection against MPTP-induced Parkinson's disease-like movement disorder in the goldfish, *European Journal of Pharmacology*, 240, 185–193, 1993.
128. Pollard, H.B., Kuijpers, G.A., Adeyemo, O.M., Youdim, M.B.H., and Goping, G., The MPTP-induced parkinsonian syndrome in the goldfish is associated with major cell destruction in the forebrain and subtle changes in the optic tectum, *Experimental Neurology*, 142, 170–178, 1996.
129. Dietrich, M., Hofmann, M.H., and Bleckmann, H., Effects of dopaminergic drugs and telencephalic ablation on eye movements in the goldfish, *Carassius auratus*, *Brain Research Bulletin*, 57, 393–395, 2002.
130. Fetcho, J.R. and Liu, K.S., Zebrafish as a model system for studying neuronal circuits and behavior, *Annals of the New York Academy of Sciences*, 860, 333–345, 1998.
131. Penberthy, W.T., Shafizadeh, E., and Lin, S., The zebrafish as a model for human disease, *Frontiers in Bioscience*, 7, D1439–D1453, 2002.
132. Schier, A.F., Genetics of neural development in zebrafish, *Current Opinion in Neurobiology*, 7, 119–126, 1997.
133. Anichtchik, O.V., Kaslin, J., Peitsaro, N., Scheinin, M., and Panula, P., Neurochemical and behavioural changes in zebrafish *Danio rerio* after systemic administration of 6-hydroxydopamine and 1-methyl-4-phenyl-1,2,3,6-tetrahydropyridine, *Journal of Neurochemistry*, 88, 443–453, 2004.
134. Arthur, D. and Levin, E.D., Spatial and non-spatial discrimination learning in zebrafish, *Animal Cognition*, 4, 125–131, 2001.
135. Colwill, R.M., Raymond, M.P., Ferreira, L., and Escudero, H., Visual discrimination learning in zebrafish (*Danio rerio*), *Behavioural Processes*, 70, 19–31, 2005.
136. Hall, D. and Suboski, M.D., Visual and olfactory stimuli in learned release of alarm reactions by zebra danio fish (*Brachydanio rerio*), *Neurobiology of Learning and Memory*, 63, 229–240, 1995.
137. Hall, D. and Suboski, M.D., Sensory preconditioning and second-order conditioning of alarm reactions in zebra danio fish (*Brachydanio rerio*), *Journal of Comparative Psychology*, 109, 76–84, 1995.

138. Williams, F.E., White, D., and Messer, W.S., A simple spatial alternation task for assessing memory function in zebrafish, *Behavioural Processes*, 58, 125–132, 2002.

139. Williams, F.E. and Messer, W.S., Jr., Memory function and muscarinic receptors in zebrafish, *Society for Neuroscience Abstracts*, 24, 182, 1998.

140. Mueller, T., Vernier, P., and Wullimann, M.F., The adult central nervous cholinergic system of a neurogenetic model animal, the zebrafish *Danio rerio*, *Brain Research*, 1011, 156–169, 2004.

141. Granato, M., van Eeden, F.J., Schach, U., Trowe, T., Brand, M., Furutani-Seiki, M., Haffter, P., Hammerschmidt, M., Heisenberg, C.P., Jiang, Y.J., Kane, D.A., Kelsh, R.N., Mullins, M.C., Odenthal, J., and Nusslein-Volhard, C., Genes controlling and mediating locomotion behavior of the zebrafish embryo and larva, *Development*, 123, 399–413, 1996.

142. Brustein, E., Saint-Amant, L., Buss, R.R., Chong, M., McDearmid, J.R., and Drapeau, P., Steps during the development of the zebrafish locomotor network, *Journal of Physiology: Paris*, 97, 77–86, 2003.

143. Carvan, M.J., III, Loucks, E., Weber, D.N., and Williams, F.E., Ethanol effects on the developing zebrafish: neurobehavior and skeletal morphogenesis, *Neurotoxicology and Teratology*, 26, 757–768, 2004.

144. Carvan, M.J., Mechanistic observations on the effects of neurotoxicants on zebrafish behavior, *Neurotoxicology and Teratology*, 25, 383, 2003.

145. Sagvolden, T., Russell, V.A., Aase, H., Johansen, E.B., and Farshbaf, M., Rodent models of attention-deficit/hyperactivity disorder, *Biological Psychiatry*, 57, 1239–1247, 2005.

146. Trinh, J.V., Nehrenberg, D.L., Jacobsen, J.P.R., Caron, M.G., and Wetsel, W.C., Differential psychostimulant-induced activation of neural circuits in dopamine transporter knockout and wild type mice, *Neuroscience*, 118, 297–310, 2003.

147. Jacobsen, J.P.R., Rodriguiz, R.M., Mork, A., and Wetsel, W.C., Monoaminergic dysregulation in glutathione-deficient mice: possible relevance to schizophrenia? *Neuroscience*, 132, 1055–1072, 2005.

148. Miklosi, A., Andrew, R.J., and Savage, H., Behavioral lateralization of the tetrapod type in the zebrafish (*Brachiodanio rerio*), *Physiology and Behavior*, 63, 127–135, 1998.

149. Barth, K.A., Miklosi, A., Watkins, J., Bianco, I.H., Wilson, S.W., and Andrew, R.J., fsi, Zebrafish show concordant reversal of laterality of viscera, neuroanatomy, and a subset of behavioral responses, *Current Biology*, 15, 844–850, 2005.

150. Miklosi, A., Andrew, R.J., and Savage, H., Behavioural lateralisation of the tetrapod type in the zebrafish (*Brachydanio rerio*), *Physiology and Behavior*, 63, 127–135, 1997.

151. Bisazza, A., Rogers, L.J., and Vallortigara, G., The origins of cerebral asymmetry: a review of evidence of behavioural and brain lateralization in fishes, reptiles and amphibians, *Neuroscience and Biobehavioral Reviews*, 22, 411–426, 1998.

152. Peitsaro, N., Kaslin, J., Anichtchik, O.V., and Panula, P., Modulation of the histaminergic system and behaviour by alpha-fluoromethylhistidine in zebrafish, *Journal of Neurochemistry*, 86, 432–441, 2003.

153. Li, L. and Dowling, J.E., A dominant form of inherited retinal degeneration caused by a non-photoreceptor cell-specific mutation, *Proceedings of the National Academy of Sciences of the United States of America*, 94, 11645–11650, 1997.

154. Neuhauss, S.C.F., Behavioral genetic approaches to visual system development and function in zebrafish, *Journal of Neurobiology*, 54, 148–160, 2003.

155. Li, L., Zebrafish mutants: behavioral genetic studies of visual system defects, *Developmental Dynamics*, 221, 365–372, 2001.

156. Rinner, O., Rick, J.M., and Neuhauss, S.C.F., Contrast sensitivity, spatial and temporal tuning of the larval zebrafish optokinetic response, *Investigative Ophthalmology and Visual Science*, 46, 137–142, 2005.

157. Swain, H.A., Sigstad, C., and Scalzo, F.M., Effects of dizocilpine (MK-801) on circling behavior, swimming activity, and place preference in zebrafish (*Danio rerio*), *Neurotoxicology and Teratology*, 26, 725–729, 2004.

158. Pradel, G., Schmidt, R., and Schachner, M., Involvement of L1.1 in memory consolidation after active avoidance conditioning in zebrafish, *Journal of Neurobiology*, 43, 389–403, 2000.

159. Gleason, P.E., Weber, P.G., and Weber, S.P., Effect of group size on avoidance learning in zebrafish: *Brachydanio rerio* (Pisces: Cyprinidae), *Animal Learning and Behavior*, 5, 213–216, 1977.

160. Pradel, G., Schachner, M., and Schmidt, R., Inhibition of memory consolidation by antibodies against cell adhesion molecules after active avoidance conditioning in zebrafish, *Journal of Neurobiology*, 39, 197–206, 1999.

161. Flanigan, W.F. and Caldwell, W.E., Galvanotaxic behavior and reinforcement of fish, *Brachydanio rerio*, *Genetic Psychology Monographs*, 84, 35–71, 1971.

162. Levin, E. and Chen, E., Nicotinic involvement in memory function in zebrafish, *Neurotoxicology and Teratology*, 26, 731–735, 2004.

163. Levin, E.D., Decker, M.W., and Butcher, L.L., *Neurotransmitter Interactions and Cognitive Function*, Berkhäuser, Boston, 1992.

164. Levin, E.D. and Simon, B.B., Nicotinic acetylcholine involvement in cognitive function in animals, *Psychopharmacology*, 138, 217–230, 1998.

165. Levin, E.D. and Slotkin, T.A., Developmental neurotoxicity of nicotine, in *Handbook of Developmental Neurotoxicology*, Slikker, W., Jr. and Chang, L.W., Eds., Academic Press, San Diego, 1998, pp. 587–615.

166. Miklos, G.L., Molecules and cognition: the latterday lessons of levels, language, and iac — evolutionary overview of brain structure and function in some vertebrates and invertebrates, *Journal of Neurobiology*, 24, 842–890, 1993.

167. Kuzirian, A.M., Epstein, H.T., Nelson, T.J., Rafferty, N.S., and Alkon, D.L., Lead, learning, and calexcitin in Hermissenda, *Biological Bulletin*, 195, 198–201, 1998.

168. Hobert, O., Behavioral plasticity in *C. elegans*: paradigms, circuits, genes, *Journal of Neurobiology*, 54, 203–223, 2003.

169. Samuel, A.D., Silva, R.A., and Murthy, V.N., Synaptic activity of the AFD neuron in *Caenorhabditis elegans* correlates with thermotactic memory, *Journal of Neuroscience*, 23, 373–376, 2003.

170. Rankin, C.H., Beck, C.D., and Chiba, C.M., *Caenorhabditis elegans*: a new model system for the study of learning and memory, *Behavioural Brain Research*, 37, 89–92, 1990.

171. Mah, K.B. and Rankin, C.H., An analysis of behavioral plasticity in male *Caenorhabditis elegans*, *Behavioral and Neural Biology*, 58, 211–221, 1992.

172. Bernhard, N. and van der Kooy, D., A behavioral and genetic dissection of two forms of olfactory plasticity in *Caenorhabditis elegans*: adaptation and habituation, *Learning and Memory*, 7, 199–212, 2000.

173. Kitamura, K.I., Amano, S., and Hosono, R., Contribution of neurons to habituation to mechanical stimulation in *Caenorhabditis elegans*, *Journal of Neurobiology*, 46, 29–40, 2001.

174. Broster, B.S. and Rankin, C.H., Effects of changing interstimulus interval during habituation in *Caenorhabditis elegans, Behavioral Neuroscience*, 108, 1019–1029, 1994.

175. Beck, C.D. and Rankin, C.H., Heat shock disrupts long-term memory consolidation in *Caenorhabditis elegans, Learning and Memory*, 2, 161–177, 1995.

176. Rankin, C.H., Invertebrate learning: what can't a worm learn? *Current Biology*, 14, R617–R618, 2004.

177. Rankin, C.H., Nematode memory: now, where was I? *Current Biology*, 15, R374–R375, 2005.

178. Saeki, S., Yamamoto, M., and Iino, Y., Plasticity of chemotaxis revealed by paired presentation of a chemoattractant and starvation in the nematode *Caenorhabditis elegans, Journal of Experimental Biology*, 204, 1757–1764, 2001.

179. Sambongi, Y., Takeda, K., Wakabayashi, T., Ueda, I., Wada, Y., and Futai, M., *Caenorhabditis elegans* senses protons through amphid chemosensory neurons: proton signals elicit avoidance behavior, *Neuroreport*, 11, 2229–2232, 2000.

180. Law, E., Nuttley, W.M., and van der Kooy, D., Contextual taste cues modulate olfactory learning in *C. elegans* by an occasion-setting mechanism, *Current Biology*, 14, 1303–1308, 2004.

181. Morrison, G.E. and van der Kooy, D., Cold shock before associative conditioning blocks memory retrieval, but cold shock after conditioning blocks memory retention in *Caenorhabditis elegans, Behavioral Neuroscience*, 111, 564–578, 1997.

182. Wen, J.Y., Kumar, N., Morrison, G., Rambaldini, G., Runciman, S., Rousseau, J., and van der Kooy, D., Mutations that prevent associative learning in *C. elegans, Behavioral Neuroscience*, 111, 354–368, 1997.

183. Morrison, G.E., Wen, J.Y., Runciman, S., and van der Kooy, D., Olfactory associative learning in *Caenorhabditis elegans* is impaired in lrn-1 and lrn-2 mutants, *Behavioral Neuroscience*, 113, 358–367, 1999.

184. Atkinson-Leadbeater, K., Nuttley, W.M., and van der Kooy, D., A genetic dissociation of learning and recall in *Caenorhabditis elegans, Behavioral Neuroscience*, 118, 1206–1213, 2004.

185. Murakami, S. and Murakami, H., The effects of aging and oxidative stress on learning behavior in *C. elegans, Neurobiology of Aging*, 26, 899–905, 2005.

186. Remy, J.J. and Hobert, O., An interneuronal chemoreceptor required for olfactory imprinting in *C. elegans, Science*, 309, 787–790, 2005.

187. Xu, X., Sassa, T., Kunoh, K., and Hosono, R., A mutant exhibiting abnormal habituation behavior in *Caenorhabditis elegans, Journal of Neurogenetics*, 16, 29–44, 2002.

188. Raich, W.B., Moorman, C., Lacefield, C.O., Lehrer, J., Bartsch, D., Plasterk, R.H., Kandel, E.R., and Hobert, O., Characterization of *Caenorhabditis elegans* homologs of the Down syndrome candidate gene DYRK1A, *Genetics*, 163, 571–580, 2003.

189. Rose, J.K., Kaun, K.R., and Rankin, C.H., A new group-training procedure for habituation demonstrates that presynaptic glutamate release contributes to long-term memory in *Caenorhabditis elegans, Learning and Memory*, 9, 130–137, 2002.

190. Brockie, P.J., Mellem, J.E., Hills, T., Madsen, D.M., and Maricq, A.V., The *C. elegans* glutamate receptor subunit NMR-1 is required for slow NMDA-activated currents that regulate reversal frequency during locomotion, *Neuron*, 31, 617–630, 2001.

191. Morrison, G.E. and van der Kooy, D., A mutation in the AMPA-type glutamate receptor, glr-1, blocks olfactory associative and nonassociative learning in *Caenorhabditis elegans, Behavioral Neuroscience*, 115, 640–649, 2001.

192. Gomez, M., De Castro, E., Guarin, E., Sasakura, H., Kuhara, A., Mori, I., Bartfai, T., Bargmann, C.I., and Nef, P., Ca^{2+} signaling via the neuronal calcium sensor-1 regulates associative learning and memory in *C. elegans* [see comment], *Neuron*, 30, 241–248, 2001.

193. Yoshihara, M., Ensminger, A.W., and Littleton, J.T., Neurobiology and the drosophila genome, *Functional and Integrative Genomics*, 1, 235–240, 2001.

194. de Belle, J.S. and Heisenberg, M., Associative odor learning in drosophila abolished by chemical ablation of mushroom bodies, *Science*, 263, 692–695, 1994.

195. de Belle, J.S. and Heisenberg, M., Expression of drosophila mushroom body mutations in alternative genetic backgrounds: a case study of the mushroom body miniature gene (mbm), *Proceedings of the National Academy of Sciences of the United States of America*, 93, 9875–9880, 1996.

196. Skoulakis, E.M., Kalderon, D., and Davis, R.L., Preferential expression in mushroom bodies of the catalytic subunit of protein kinase A and its role in learning and memory, *Neuron*, 11, 197–208, 1993.

197. Grotewiel, M.S., Beck, C.D., Wu, K.H., Zhu, X.R., and Davis, R.L., Integrin-mediated short-term memory in drosophila, *Nature*, 391, 455–460, 1998.

198. Mao, Z., Roman, G., Zong, L., and Davis, R.L., Pharmacogenetic rescue in time and space of the rutabaga memory impairment by using gene-switch, *Proceedings of the National Academy of Sciences of the United States of America*, 101, 198–203, 2004.

199. Philip, N., Acevedo, S.F., and Skoulakis, E.M., Conditional rescue of olfactory learning and memory defects in mutants of the 14-3-3 zeta gene Leonardo, *Journal of Neuroscience*, 21, 8417–8425, 2001.

200. McGuire, S.E., Le, P.T., Osborn, A.J., Matsumoto, K., and Davis, R.L., Spatiotemporal rescue of memory dysfunction in drosophila, *Science*, 302, 1765–1768, 2003.

201. Mershin, A., Pavlopoulos, E., Fitch, O., Braden, B.C., Nanopoulos, D.V., and Skoulakis, E.M., Learning and memory deficits upon TAU accumulation in drosophila mushroom body neurons, *Learning and Memory*, 11, 277–287, 2004.

202. Preat, T., Decreased odor avoidance after electric shock in drosophila mutants biases learning and memory tests, *Journal of Neuroscience*, 18, 8534–8538, 1998.

203. Devay, P., Pinter, M., Yalcin, A.S., and Friedrich, P., Altered autophosphorylation of adenosine $3',5'$-phosphate-dependent protein kinase in the dunce memory mutant of *Drosophila melanogaster*, *Neuroscience*, 18, 193–203, 1986.

204. Dura, J.M., Taillebourg, E., and Preat, T., The drosophila learning and memory gene linotte encodes a putative receptor tyrosine kinase homologous to the human RYK gene product, *FEBS Letters*, 370, 250–254, 1995.

205. Tamura, T., Chiang, A.S., Ito, N., Liu, H.P., Horiuchi, J., Tully, T., and Saitoe, M., Aging specifically impairs amnesiac-dependent memory in drosophila, *Neuron*, 40, 1003–1011, 2003.

206. Saitoe, M., Horiuchi, J., Tamura, T., and Ito, N., Drosophila as a novel animal model for studying the genetics of age-related memory impairment, *Reviews in the Neurosciences*, 16, 137–149, 2005.

207. Horiuchi, J. and Saitoe, M., Can flies shed light on our own age-related memory impairment? *Ageing Research Reviews*, 4, 83–101, 2005.

208. Fois, C., Medioni, J., and Le Bourg, E., Habituation of the proboscis extension response as a function of age in *Drosophila melanogaster*, *Gerontology*, 37, 187–192, 1991.

209. Ruiz-Canada, C., Ashley, J., Moeckel-Cole, S., Drier, E., Yin, J., and Budnik, V., New synaptic bouton formation is disrupted by misregulation of microtubule stability in aPKC mutants, *Neuron*, 42, 567–580, 2004.

210. Harum, K.H., Alemi, L., and Johnston, M.V., Cognitive impairment in Coffin-Lowry syndrome correlates with reduced RSK2 activation, *Neurology*, 56, 207–214, 2001.

211. Hashimoto, H., Shintani, N., and Baba, A., Higher brain functions of PACAP and a homologous drosophila memory gene amnesiac: insights from knockouts and mutants, *Biochemical and Biophysical Research Communications*, 297, 427–431, 2002.

212. Xia, S., Miyashita, T., Fu, T.F., Lin, W.Y., Wu, C.L., Pyzocha, L., Lin, I.R., Saitoe, M., Tully, T., and Chiang, A.S., NMDA receptors mediate olfactory learning and memory in drosophila, *Current Biology*, 15, 603–615, 2005.

213. Ye, Y., Xi, W., Peng, Y., Wang, Y., and Guo, A., Long-term but not short-term blockade of dopamine release in drosophila impairs orientation during flight in a visual attention paradigm, *European Journal of Neuroscience*, 20, 1001–1007, 2004.

214. Bitterman, M.E., The CS-US interval in classical and avoidance conditioning, in *Classical Conditioning: A Symposium*, Prokasy, W.F., Ed., Appleton-Century-Crofts, New York, 1965, pp. 1–19.

215. Roberts, C.M. and Loop, M.S., Goldfish color vision sensitivity is high under light-adapted conditions, *Journal of Comparative Physiology A: Sensory, Neural, and Behavioral Physiology*, 190, 993–999, 2004.

216. Levin, E.D. and Chen, E., Nicotinic involvement in memory function in zebrafish, *Neurotoxicology and Teratology*, 26, 731–735, 2004.

217. Levin, E., Limpuangthip, J., Rachakonda, T., and Peterson, M., Timing of nicotine effects on learning in zebrafish, *Psychopharmacology*, 184, 547–552, 2006.

16 Cognition Models and Drug Discovery

Michael W. Decker
Abbott Laboratories

CONTENTS

INTRODUCTION

Cognitive impairment is common in a variety of neurological and psychiatric conditions, including stroke, Alzheimer's disease (AD), Parkinson's disease, traumatic brain injury, and schizophrenia. Currently available medications are relatively ineffective in treating cognitive dysfunction. Thus, cognitive dysfunction represents an important unmet medical need, and developing treatments is a major area of research interest in the pharmaceutical industry.

Identification of potential pharmacological treatments for cognitive disorders is highly dependent on the availability of adequate animal models [1]. Animal models of cognitive function are used in the identification and validation of molecular targets and serve as screening tools to identify and evaluate specific compounds for their potential efficacy. Ideally, animal models of cognition could provide a prediction of the specific conditions for which compounds would be most useful and could be combined with toxicology and safety pharmacology models to estimate the therapeutic index. Given the duration and expense of clinical trials, particularly for progressive degenerative conditions such as Alzheimer's disease, it is becoming more important in drug discovery to maximize the potential for success through the development of high-quality animal models and their judicious use in selecting compounds for advancement into human studies.

The term "cognition model" requires some clarification. The term is frequently applied to the means by which cognitive processes are assessed in animals. Thus, the radial arm maze and the Morris water maze are often described as cognition models. However, perhaps a more precise use of the term "cognition model" would apply it exclusively to such measurements made in the context of some disease-relevant impairment of performance. Here, the term would not be applied to the radial maze or the water maze itself, but rather to the use of these tools in animals impaired, for example, by overexpression of β-amyloid in an attempt to mimic aspects of Alzheimer's disease. To avoid confusion, the use of the term "model" in this chapter is restricted to the latter case, and the terms "assay" or "measure" are used to describe the tasks and procedures used to assess cognitive function.

CONSIDERATIONS IN THE APPLICATION OF COGNITIVE ASSAYS IN DRUG DEVELOPMENT

Cognition assays have a variety of uses in the drug development process. The main focus of this chapter is on their use in identifying and characterizing potential cognition enhancers. However, before moving on to the discussion of this topic, it is worth briefly noting the value of cognition assays in side-effect profiling and target identification.

Cognitive assays can be used to assess potential adverse effects of compounds or drug targets being pursued for other indications. A number of CNS (central nervous system) drugs, including many antipsychotics, anticonvulsants, anxiolytics, and analgesics, can disrupt cognition, so it is prudent to evaluate this potential liability early in the discovery process for CNS drugs. Similarly, drugs that are not being pursued for CNS indications can also impair cognition. For example, β-blockers used for the treatment of hypertension can have cognitive side effects [2]. Thus, cognition assays can play an important role in assessment of CNS safety for new compounds.

Cognitive assays can also be used at an even earlier stage of drug discovery: target identification. Here, the assays might be used to characterize the effects of manipulations of potential molecular targets to evaluate the viability of these molecules as drug targets. Discovery of new targets by determining the phenotype of random genetic mutations created by administration of ENU (ethylnitrosourea) or of targeted knockouts of genes are two methods of broadly screening for potential drug targets [3–5]. Mice are a favorite species in which to conduct these studies and, of course, a range of cognitive behaviors can be assessed in mice. However, cognitive performance can also be measured in lower organisms, such as fruit flies and worms [6–8]. Information regarding the impact of manipulation of the target on cognitive function is an important aspect of the phenotype and can be used to assess both liabilities and therapeutic potential for a given target. For example, a receptor knockout that impairs cognitive performance suggests both that receptor antagonists might have undesirable cognitive effects and that agonists might be cognitive enhancers.

Whether developing a cognitive battery for phenotyping, side-effect profiling, or testing new compounds for cognition-enhancement efficacy, it is critical that the tests be reliable and their output reproducible. Throughput is typically viewed as an important consideration given that large numbers of compounds are often evaluated in a screening process. When identifying potential cognition enhancers, rapid screening assays can sometimes assist in early compound selection. For example, the ability of a muscarinic agonist or acetylcholinesterase inhibitor to prevent the disruption of inhibitory avoidance learning produced by a muscarinic antagonist provides verification of the desired pharmacological activity of the compound *in vivo*. This may provide initial guidance on doses or perhaps preliminary comparisons of therapeutic index within a class of compounds if combined with tests of side-effect liabilities. The use of a cognitive assay in this case is superior to the use of some other less relevant biological readout in that it increases the likelihood of selecting a compound that acts on systems important for cognition. However, this sort of *in vivo* bioassay for pharmacological activity and similar assays that can be used to screen compounds do not typically provide the rich characterization of effects required to make decisions on advancing a compound to clinical trials [9, 10]. To develop this more-advanced understanding of the pharmacology of a compound or to address questions of molecular target validation at early stages of the drug discovery process, it is critical to include assays that provide information that can be interpreted in the context of cognitive domains and neurological systems.

An assay battery should be constructed such that the behavioral profiles produced can distinguish between the cognitive effects of different manipulations. Thus, careful attention should be paid to the representation of distinct domains. Few behavioral assays can claim to measure with such specificity, but choosing diverse measures to provide a broad cognitive profile is possible. In interpreting data generated from cognitive assays, it is important to understand the cognitive domains and the neural systems involved in performing the tasks. This knowledge provides a fuller understanding of the nature of the drug effect and can sometimes provide information on a compound's mechanism of action. However, it should also be recognized that task conditions that vary across labs could influence the cognitive requirements of the task and thus the neural systems involved. The nature of the spatial cues in the environment in spatial learning tasks, for example, can vary appreciably across labs and is often not well described in publications. Clearly, this source of variability could have impact on the manner in which animals perform the task. Moreover, different neural systems are often engaged at different stages of task acquisition and could therefore influence effects of manipulations. To illustrate this point, performance of the five-choice serial reaction time task (5-CSRTT) of sustained attention elicits release of a number of neurotransmitters in the prefrontal cortex, including 5-HT, DA, NE, and ACh (acetylcholine), but different aspects of the task are involved in evoking the release of these neurotransmitters [11]. For example, sustained release of ACh is observed during task performance, whereas NE release is engaged more specifically by changes in contingencies [11]. When designing an assay battery, it is also important to include assays that have overlapping cognitive requirements but differ in motivational, sensory, and motor requirements to reduce the likelihood that effects on these noncognitive performance variables can account for observed effects [12, 13].

There are some special requirements for assays used in drug discovery that are related to the nature of the information needed to understand drug action. Advanced characterization of compounds typically requires answers to questions such as the relationship between duration of action and pharmacokinetics or the maintenance of efficacy after repeated dosing. To address duration of action issues, the measure should be made over a brief period of time so that the plasma concentration of the drug is likely to be maintained a relatively constant level during the period of measurement. Ideally, the same animal could be tested at different time points after administration of drug so that the duration can be assessed using a within-subject design. The ability to test animals repeatedly is also an advantage for the assessment of tolerance after repeated dosing, although in this case the delay between test sessions can be expanded to hours or days rather than the minutes or hours that would be required for duration-of-action studies. Moreover, to address these kinds of questions, assays of sufficient sensitivity are required. The sensitivity needed to determine that a compound "works" is not necessarily adequate to track perhaps subtle changes in efficacy across discrete time points in a duration-of-action study or during the course of repeated dosing over time in the assessment of tolerance liability.

DISEASE AND IMPAIRMENT MODELS IN DRUG DISCOVERY

An important consideration in the use of cognition models in drug discovery is the baseline performance that will be used to index cognitive enhancement. Clearly, baselines that are at or near the maximum possible level of performance do not provide a sufficient ceiling for observing enhancement, but the selection of a method for adjusting the baseline, i.e., impairing cognitive performance, can have important theoretical and interpretative implications.

Cognitive assays with normal animals are often used in the characterization of novel compounds. Typically, these tests are conducted under conditions that disrupt performance through increases in task difficulty. For example, an assay measuring visual attention might be made more difficult by employing stimuli of low salience or the inclusion of distracting stimuli. Similarly, a working-memory task can be made more difficult by increasing the number of items to be remembered or increasing the delay interval. It is important to recognize, however, that these parametric manipulations do not always simply make the task more difficult. Rather they can sometimes change the nature of the task and thereby change the neural systems engaged. For example, in the 5-CSRTT, noradrenergic lesions only impair performance when manipulations that activate this system are included, such as the inclusion of distractors or changes in the predictability of targets [11]. In addition, the inclusion of distractors can engage "top-down" executive functions that fundamentally change the neural processing involved in performing a task [14].

The use of normal animals to identify new compounds that improve memory or to validate molecular targets that have the potential for cognitive enhancement is probably most successful in identifying palliative or symptomatic treatments. For this approach, the mechanisms underlying the deficits are less important. Rather, the assumption is that cognition can be modulated similarly in normals and patients, albeit to different degrees given the ceiling effects encountered with normals. An analogy with cardiovascular disease is obvious here. Many blood pressure medications have been identified by tests in normal animals. Hypotensive effects of these compounds can be regarded as toxic in this context, but in someone suffering from hypertension, the effect is therapeutic. The limitation, of course, is that the underlying pathology is not necessarily addressed. At first glance, it would appear that a key difference between blood pressure and cognition is that both high and low blood pressure can be regarded as pathological, whereas pathological changes in cognitive function are always unidirectional. However, for at least some aspects of cognitive function, both increases and decreases in function could be pathological. For example, Sarter [15] has proposed that increased (but indiscriminate) attention can contribute to the distractibility characteristic of schizophrenia, whereas changes in the opposite direction characterize the impairments seen in AD. This view can be expanded to encompass the well-known "inverted-U" dose-response function observed with many cognition-enhancing drugs, where both increases and decreases in the underlying biological functions can disrupt performance. Here, it is important to understand just where on this continuum the targeted patient population is positioned. Conceivably, a treatment that is beneficial in patients may not be beneficial in normal people and could even be disruptive. Similarly, a compound that can improve cognitive processes in an abnormal brain might actually impair performance in preclinical assays in normal animals, resulting in a false negative outcome.

Testing normal animals under challenging conditions as an impairment model can also lead to false-positive outcomes. Parametric manipulations likely do not disrupt cognitive performance in the same way as the disease process of interest. Thus, it is possible that an approach that improves performance in normal animals under challenging conditions does so through actions on processes that are not involved in disrupting performance in a disease state. Of course, cognitive enhancement does not have to be mediated by correcting the process that is impaired in the disease state. A noncognitive example that illustrates this point is that a hearing-impaired person who learns to sign and read lips is not correcting an auditory deficit, but is overcoming hearing loss nonetheless. However, it is entirely possible that a compound that improves the performance of a normal brain will not do so in an abnormal brain, perhaps because a necessary substrate is missing. In an extreme example, a receptor agonist might improve performance in normal animals, or for that matter in normal people, but if the receptor is missing in a disease state or altered in a way that markedly changes the pharmacology of the compound, the compound is unlikely to be of value in the clinic. This is likely the basis for the mismatch between the good efficacy of cholinesterase inhibition in some preclinical models and the modest-to-poor efficacy in Alzheimer's disease. Inhibition of ACh metabolism is a less effective mechanism for increasing ACh levels in the AD brain than in the normal brain, given the marked destruction of cholinergic neurons in AD. Data from animal models are consistent with this interpretation. Cholinesterase inhibition can improve spatial learning in normal rats or in animals with modest disruption of cholinergic input to the hippocampus, but cholinesterase inhibition is ineffective in rats with more profound disruption of hippocampal cholinergic function [16, 17].

Treatments that appear promising in normal animals could also fail in the clinic because the impairment in the diseased state is more profound than that which can be modeled by increasing task difficulty in normal animals or because the treatment approach is not relevant to the disease. Perhaps a trivial human analogy will illustrate this latter point. One could impair reading performance in normal people by making the print very small. Under these conditions, their reading test scores might indicate impaired performance akin to that noted in people with a reading disability. The use of a magnifying glass would be a useful intervention in the normal population, but this would not translate into an effective therapy for reading disabilities. Thus, the means by which performance is impaired in a cognitive model must be carefully considered in interpreting the data.

An alternative to the use of cognitive assays in normal animals to assess drug effects is to manipulate neural function to mimic the changes that are observed in the disease process. These disease models then attempt to capture at least some part of the disease process that manifests itself in cognitive deficits. A common approach in modeling AD has been to disrupt the function of the cholinergic system, which is based on the observations that the most consistent neurochemical change in AD is a reduction of cholinergic markers in the forebrain and that muscarinic and nicotine cholinergic antagonists impair performance in cognitive tasks in both experimental animals and people [18–22]. Moreover, the profile of cognitive deficits produced by muscarinic blockade in normal volunteers bears some similarity to the profile of deficits observed in AD [23]. Thus, muscarinic antagonists like scopolamine have routinely been used in cognition models to induce deficits and assess the ability of test compounds to reverse these deficits. In the case of test compounds that are muscarinic agonists, this type of model is essentially a bioassay of relevant muscarinic agonist activity akin to that already discussed. However, when compounds without muscarinic agonist activity are evaluated, the model assesses something more than direct pharmacological interaction. Of course, one issue with the scopolamine-deficit model is that muscarinic receptors are blocked throughout the nervous system, whereas forebrain cholinergic function is particularly affected in AD. Moreover, in a scopolamine-deficit model, receptor function is disrupted but cholinergic neurons remain intact, which contrasts with the situation in AD, where the reverse is true.

ASSAY AND MODEL VALIDITY

Assays and models must have validity to be of use for drug discovery. Three types of validity are typically recognized for evaluating models and assays — face validity, predictive validity, and construct validity. Traditionally, cognitive assays and models have relied mainly on face validity; that is, they had the "look and feel" of a cognitive model or assay. More recently, however, better understanding of the biological substrates of cognitive function and the development of more sophisticated assays and models has allowed better assessment of construct validity. Thus, the assays and models can be assessed on the basis of how well they embody and conform to theoretical constructs. For example, the presence of a delay-dependent decay function in a working-memory task or a vigilance decrement over time in a sustained-attention task provide at least some reassurance that these tasks are measuring what they purport to measure. Moreover, comparing the effects of lesions or pharmacological manipulations on task performance in animals with clinical assessment of cognitive performance in patients with defined lesions or in subjects given drug challenges can also support the validity of an assay in animals. Advances in understanding the neuropsychological bases of cognitive function, emerging from the use of imaging techniques in both humans and experimental animals, provide another tool for establishing construct validity, although it should be noted that the validity of many of the cognitive assessment tools used in human research is not well established.

Some measure of face and construct validity is also available for preclinical disease models. For example, the age-related appearance of amyloid plaques and memory deficits in transgenic mice overexpressing human β-amyloid serves as partial validation of these animals as a model of AD. It is notable, however, that neurofibrillary tangles, one of the hallmarks of AD, are not observed in most mouse models of AD (with the notable exception of the "triple-transgenic" mouse from the University of California, Irvine [24]), nor is marked neurodegeneration observed in $A\beta$ transgenic mouse models. Thus, validation is always relative.

Despite the availability of some level of face and construct validity for assays and disease models used in cognition research, the critical feature for drug discovery is predictive validity. Here, the idea is to establish the degree to which the assay or disease model can be used to identify new treatments. Predictive validity in drug discovery, then, addresses a key question: how well do the effects of test compounds in the assays and models actually predict efficacy in humans? And to raise the bar yet a bit more, how well do they predict efficacious plasma levels and pharmacodynamic–pharmacokinetic relationships? Typically, in drug discovery research, assessing the effects of positive- and negative-control compounds provides this type of validation. Thus, an anxiety model might be validated by testing agents with clinically demonstrated anxiolytic activity as well as psychoactive drugs that do not have efficacy against anxiety in the clinic.

Unfortunately, the availability of efficacious treatments for cognition is much more limited than is the case for anxiety, depression, and positive symptoms of schizophrenia. Cholinesterase inhibitors are approved for use in AD, but their efficacy in the clinic is rather modest, as already noted. These compounds have efficacy in cognitive assays and disease models, including a transgenic mouse model of AD [25, 26], but interpretation of these results can be problematic. Does the observation of full efficacy of cholinesterase inhibition validate or invalidate a model? Demonstrating sensitivity of a model to cholinesterase inhibition provides validation at some level, but if the goal is to predict clinical efficacy, the observation of full efficacy could be regarded as a false positive result.

Even in conditions where highly efficacious compounds are available, lack of diversity among treatments can limit the kinds of predictions that can be made. In attention-deficit hyperactivity disorder, stimulants such as amphetamine and methylphenidate are clearly efficacious [27], but establishing predictive validity using only this class of compounds narrows the focus and increases the risk of developing models that can only identify stimulants.

The assessment of predictive validity is a dynamic process. The accumulation of preclinical and clinical data on the efficacy of new compounds allows for the continuous reevaluation of predictive validity. However, no single laboratory is capable of providing the full pharmacological characterization of the vast range of cognitive models, so advances in this area require the sharing of information. While the scientific literature is clearly an important vehicle for disseminating such data, it can be incomplete. Individual findings that may be viewed as unworthy of publication, e.g., negative results, can still contribute to the validation process. Accordingly, one recommendation emerging from the recent MATRICS (NIMH–Measurement and Treatment Research to Improve Cognition in Schizophrenia) initiative [28] was that a public database be established to serve as a repository of preclinical data on cognition models [29].

HOMOLOGY BETWEEN CLINICAL AND PRECLINICAL MEASURES

One important consideration in designing assays for preclinical evaluation of compounds is the degree to which they match up with the kinds of assessments that will be conducted in clinical trials. Generally, the correspondence is not particularly good. This is in part due to the inability to measure uniquely human aspects of cognitive function in experimental animals, but it is also the case that clinical measures are often general and more qualitative in nature. For example, efficacy is frequently assessed by a general clinical impression that essentially asks, "Did the patient get better or worse?" No similar assessment is made in animals, of course.

Still, there have been some important attempts to align preclinical and clinical measurements. One tack is to identify more-primitive cognitive processes that can be measured in both humans and experimental animals with only modest changes in parameters. The "preattentional" process of sensory gating can be measured by comparing evoked potential EEG (encephalogram) responses to pairs of auditory stimuli in both experimental animals and in humans. Similarly, sensory gating can be assessed across species using prepulse inhibition of the startle reflex [30, 31]. In both cases, pharmacological manipulations of these processes show good, but not perfect, translation from humans to rodents. For example, stimulants, such as amphetamine, impair gating assessed by EEG or prepulse inhibition across species. Eyeblink conditioning represents another function that can readily be assessed in both humans and experimental animals [32, 33]. Notably, impaired eyeblink conditioning is observed in AD, providing clinical relevance for this index of cognitive performance [34].

An alternative approach is to design animal analogs of measures used with humans. The five-choice serial reaction time test (5-CSRTT) was designed to approximate the continuous performance task frequently used to assess sustained attention in humans [11]. Similarly, Kesner and colleagues have designed several variations of the radial maze to demonstrate recency and primacy effects in rats, and they have demonstrated that small and large lesions of the forebrain cholinergic system in rats mimic the effects of early and advanced AD, respectively, on these measures [35–37].

Several limitations restrict the ability to predict clinical efficacy from preclinical cognitive assays and models. As already noted, many cognitive functions are unique to humans and cannot be addressed in experimental animals. Human verbal skills can also confound attempts to measure domains that are shared across species. For example, color can easily be used as a stimulus feature in matching-to-sample working memory tasks with monkeys, but humans can use verbal mediation to perform memory tasks with stimuli that can be named. Moreover, many preclinical tasks require extensive training that is obviated in human testing because the task can be explained verbally.

A related danger in using animal analogs of human cognitive measures is that these are often, at least initially, based on apparent similarities between the measures, i.e., face validity. It is critical that additional validation demonstrating the correspondence between the neurobiological substrates required for task performance be obtained before data collected in drug studies can be interpreted appropriately. Traditionally, pharmacological manipulations and lesion studies have been important means of providing this additional validation, although the latter were limited by the availability of corresponding lesion data from patients. Fortunately, with the development of noninvasive brain imaging, assessing the roles of specific brain structures in human cognition has become far less dependent on these "experiments of nature."

In pointing out the differences between preclinical and clinical measures, it is easy to complain about the sometimes vague and subjective clinical impression scales used to establish clinical efficacy. However, it is important to recognize that, despite their lack of precision, rating scales probably provide a better index of "meaningful" improvement. This is because they capture a broader functional picture and address the question: how is the patient doing in the real world? Improved performance in working memory or attention tasks may contribute to improved function, but they do not guarantee it. They are, in a sense, biomarkers that would seem to predict improved function but that have not been established conclusively as such. In this context, it is useful to consider an issue that Sarter and colleagues [38] have raised regarding cognition enhancement in AD but that can be readily extended to other disorders. They argue that it makes little sense to assume that drugs that improve cognition will have meaningful functional consequences unless they are coupled with behavioral intervention. In other words, drugs can improve learning capacity but cannot teach. Unless we keep this firmly in mind, functional outcomes with drugs that improve cognition preclinically will continue to disappoint.

REFERENCES

1. Decker, M.W., Animal models of cognition, *Crit. Rev. Neurobiol.*, 9, 321, 1995.
2. Müller, U., Mottweiler, E., and Bublak, P., Noradrenergic blockade and numeric working memory in humans, *J. Psychopharmacol.*, 19, 21, 2005.
3. Sayaha, D.M. et al., A genetic screen for novel behavioral mutations in mice, *Molecular Psychiatry*, 5, 369, 2000.
4. Tecott, L.H. and Nestler, E.J., Neurobehavioral assessment in the information age, *Nat. Neurosci.*, 7, 462, 2004.
5. Crabbe, J.C. and Morris, R.G.M., Festina lente: late-night thoughts on high-throughput screening of mouse behavior, *Nat. Neurosci.*, 7, 1175, 2004.

6. Atkinson-Leadbeater, K., Nuttley, W.M., and van der Kooy, D., A genetic dissociation of learning and recall in *Caenorhabditis elegans*, *Behavioral Neurosci.*, 118, 1206, 2004.

7. Dubnau, J. and Tully, T., Gene discovery in drosophila: new insights for learning and memory, *Ann. Rev. Neurosci.*, 21, 407, 1998.

8. Gomez, M. et al., Ca^{2+} signaling via the neuronal calcium sensor-1 regulates associative learning and memory in *C. elegans*, *Neuron*, 30, 241, 2001.

9. Sarter, M., Hagan, J., and Dudchenko, P., Behavioral screening for cognition enhancers: from indiscriminate to valid testing: part I, *Psychopharmacology*, 107, 144, 1992.

10. Sarter, M., Hagan, J., and Dudchenko, P., Behavioral screening for cognition enhancers: from indiscriminate to valid testing: part II, *Psychopharmacology*, 107, 461, 1992.

11. Robbins, T., The 5-choice serial reaction time task: behavioural pharmacology and functional neurochemistry, *Psychopharmacology*, 163, 362, 2002.

12. Wayner, M.J. and Sanberg, P.R., Neural mechanisms of behavior: performance vs. learning, *Brain Res. Bull.*, 23, 331, 1989.

13. McGaugh, J.L., Dissociating learning and performance: drug and hormone enhancement of memory storage, *Brain Res. Bull.*, 23, 339, 1989.

14. Sarter, M. et al., Unraveling the attentional functions of cortical cholinergic inputs: interactions between signal-driven and cognitive modulation of signal detection, *Brain Res. Rev.*, 48, 98, 2005.

15. Sarter, M., Neuronal mechanisms of the attentional dysfunctions in senile dementia and schizophrenia: two sides of the same coin? *Psychopharmacology*, 114, 539, 1994.

16. Decker, M.W., Bannon, A.W., and Curzon, P., Septal lesions as model for evaluating potential cognition enhancers, in *The Behavioral Neuroscience of the Septal Region*, Numan, R., Ed., Springer-Verlag, New York, 2000, p. 363.

17. Matsuoka, N. et al., Differential effects of physostigmine and pilocarpine on the spatial memory deficits produced by two septo-hippocampal deafferentations in rats, *Brain Res.*, 559, 233, 1991.

18. Decker, M. et al., Diversity of neuronal nicotinic acetylcholine receptors: lessons from behavior and implications for CNS therapeutics, *Life Sci.*, 56, 545, 1995.

19. Levin, E.D., Nicotinic systems and cognitive function, *Psychopharmacology*, 108, 417, 1992.

20. Bartus, R.T., On neurodegenerative diseases, models, and treatment strategies: lessons learned and lessons forgotten a generation following the cholinergic hypothesis, *Experimental Neurol.*, 163, 495, 2000.

21. Bartus, R.T. et al., The cholinergic hypothesis of geriatric memory dysfunction, *Science*, 217, 408, 1982.

22. Warburton, D.M. and Wesnes, K., Drugs as research tools in psychology: cholinergic drugs and information processing, *Neuropsychobiology*, 11, 121, 1984.

23. Christensen, H. et al., Cholinergic "blockade" as a model of the cognitive deficits in Alzheimer's disease, *Brain*, 115, 1681, 1992.

24. Billings, L.M. et al., Intraneuronal Aβ causes the onset of early Alzheimer's disease-related cognitive deficits in transgenic mice, *Neuron*, 45, 675, 2005.

25. Spowart-Manning, L. and van der Staay, F.J., Spatial discrimination deficits by excitotoxic lesions in the Morris water escape task, *Behavioural Brain Res.*, 156, 269, 2005.

26. Van Dam, D. et al., Symptomatic effect of donepezil, rivastigmine, galantamine and memantine on cognitive deficits in the APP23 model, *Psychopharmacology*, 180, 177, 2005.

27. Elia, J., Ambrosini, P.J., and Rapoport, J.L., Treatment of attention-deficit-hyperactivity disorder, *N. Engl. J. Med.*, 340, 780, 1999.

28. Green, M.F. and Nuechterlein, K.H., The MATRICS Initiative: developing a consensus cognitive battery for clinical trials, *Schizophrenia Research,* 72, 1, 2004.

29. Floresco, S.B. et al., Developing predictive animal models and establishing a preclinical trials network for assessing treatment effects on cognition in schizophrenia, *Schizophrenia Bulletin,* 31, 888, 2005.

30. Braff, D.L., Grillon, C., and Geyer, M.A., Gating and habituation of the startle reflex in schizophrenic patients, *Arch. Gen. Psychiatry,* 49, 206, 1992.

31. Geyer, M.A. et al., Startle response models of sensorimotor gating and habituation deficits in schizophrenia, *Brain Res. Bull.*, 25, 485, 1990.

32. Fortier, C.B. et al., Conditional discrimination learning in patients with bilateral medial temporal lobe amnesia, *Behavioral Neurosci.*, 117, 1181, 2003.

33. Weiss, C. et al., Impaired eyeblink conditioning and decreased hippocampal volume in PDAPP v717f mice, *Neurobiol. Dis.*, 11, 425, 2002.

34. Solomon, P.R. et al., Classical eyeblink conditioning in Alzheimer's disease patients, *Neurobiol. Aging,* 12, 283, 1992.

35. Kesner, R.P., Crutcher, K., and Beers, D.R., Serial position curves for item (spatial location) information: role of the dorsal hippocampal formation and medial septum, *Brain Res.*, 454, 219, 1988.

36. Kesner, R.P., Crutcher, K.A., and Measom, M.O., Medial septal and nucleus basalis magnocellularis lesions produce order memory deficits in rats with mimic symptomatology of Alzheimer's disease, *Neurobiol. Aging,* 7, 287, 1986.

37. Kesner, R.P. and Ragozzino, M.E., Structure and dynamics of multiple memory systems in Alzheimer's disease, in *Pharmacological Treatment of Alzheimer's Disease: Molecular and Neurobiological Foundations*, Brioni, J.D. and Decker, M.W., Eds., Wiley-Liss, New York, 1997, p. 3.

38. Sarter, M., Bruno, J.P., and Himmelheber, A.M., Cortical acetylcholine and attention: neuropharmacological and cognitive principles directing treatment strategies for cognitive disorders, in *Pharmacological Treatment of Alzheimer's Disease: Molecular and Neurobiological Foundations*, Brioni, J.D. and Decker, M.W., Eds., Wiley-Liss, New York, 1997, p. 105.

Index

Index

Anxiety disorders, 265
Anxiolytics, 53
AP5 (2-amino-5-phosphonopentanoate), 39, 258
Aplysia californica, 318, 327
Appetitive tasks, 26–27
Applications, 344–346, *see also* Models and
    modeling, applications and future
    developments
APV (2-amino-5-phosphonovalerate), 41
Aroclor 1016, 150–151, 155
Aroclor 1248, 150–151, 153, 155
Aroclor 1254, 152, 156
Arthur and Levin studies, 325
Arun, Aarti, 298
Ashe, Karen, 186, 192
Assays, 344–346, 348–350
Associative learning, 241–242, *242,* 329
Atkinson and Shriffin studies, 303
Atropine, 6, 8, 10
Attention
    cognition-enhancing drugs, 288–289
    cognitive testing, mutant mice, 233–241
    developmental lead exposure, 89–92
    ethanol effects, 57–58
    go/no-go testing, 239
    learning and memory, mutant mice, 233–241
    nAChRs, 211
Attention deficit hyperactivity disorder (ADHD)
    assay and model validity, 349
    cognition-enhancing drugs, 291
    go/no-go testing, 239–240
    MPTP exposure, 176
    orienting responses, 233, 235
    reinforcement, intermittent schedules, 74,
        76–77
Attention set-shifting, 9, 172, 174
Auditory-Verbal Learning Test, 159
Avoidance tests, 30–32, 243–246, *see also* Passive
    avoidance tests

B

BACE1 gene, 187
Background alleles, 213
Bamabuterol, 212
Banna studies, 101–135
Baranova studies, 301–312
Bartus studies, 6
Basolateral amygdaloid (BLA) nucleus
    conditioned taste aversion, 262
    fear conditioning, 267
    fear-potentiated startle, 264
Baum and Kraft studies, 321
BChE (butyrylcholinesterase), 211–212
Behavior and behavioral mechanisms

ethanol effects, 51–64
    methylmercury effects, 106–108, *107,* 128–134
    polyunsaturated fatty acids, 105
Behavior therapy, 113
Bellinger studies, 91
Benson and Geschwind studies, 304
Benvenga and Spaulding studies, 39
Benztropine, 7
Bernhoft studies, 149
BIBN-99, 207
Bienkowski studies, 54
Biperidine, 7
Blanchard studies, 324
Blokland studies, 8
Blood pressure medications, 347
Blurred vision, 11
Bolivar studies, 200
Bootstrapping, 316
Boston Naming Test, 104
Boston prospective study, 89–90
Bourgeois studies, 157
Bowman, Bushnell and, studies, 79
Bowman, Levin and, studies, 155
Bowman studies, 150, 152
Brightness discrimination, 105
Bruno, Sarter and, studies, 203
Buccafusco, Terry and, studies, 200, 206
Buccafusco and Terry studies, 200, 209, 212
Buccafusco studies, 1–2, 285–297
α-bungarotoxin (αBGT), 211
Bushnell and Bowman studies, 79
Bushnell and Rice studies, 149
Butelman studies, 39
Butyrylcholinesterase (BChE), 211–212
Bymaster studies, 203

C

Cadmium, 119, 121
Caenorhabditis elegans (flatworms and
    nematodes), 2, 327–329
Caffeine, 59
Calcium channels, 55
Caldarone studies, 210
California Verbal Learning Test (CVLT), 89, 104,
    160
Calmodulin kinase II, 191
Calon studies, 191
Cambridge Neuropsychological Testing
    Automated Battery (CANTAB), 83, 172
Cannon studies, 260
CANTAB, *see* Cambridge Neuropsychological
    Testing Automated Battery (CANTAB)
Carassius auratus (goldfish)
    basics, 2, 317, 322

Nonspatial transverse-pattern test, 250–252, *251*
Nonspatial visual discrimination, 77–78
Nontransgenic mice, 188, *see also* Mice
Noradrenergic pathways, 131
Norbinaltorphimine, 52
Norepinephrine, 240–241, 259
North Carolina cohort study, 160
Nose poking, 236–237, 239
NR2B subunit, 191
NR1 subunit, 225
NS-3(CG3703), 8
Nutrition, 105–106

O

Object-discrimination test, 247–248, *248*
Obsessive-compulsive disorder (OCD)
    fear conditioning, 265
    go/no-go testing, 240
    orienting responses, 233
Odors, 251–252, 266
Oflactory discrimination, 80–81
8-OH-DPAT, 52–53
Ohno-Shosaku studies, 202
O'Keefe and Nadel studies, 252, 257
Olfactory conditioning, 58–59
Olfactory memory, 329
Oligomers, 185
Olney and Farber studies, 42
Olton studies, 253
Ommaya studies, 304
Ondansetron, 53
One-way active avoidance, *243,* 245–246
Open field, 227
Operant behavior, neurotoxicants, 107
Operant conditioning
    Carassius auratus (goldfish), 319–321, *320*
    Danio rerio (zebrafish), 324–325
Operant procedures, working memory, 306
Opioids, 52
Orbach, Miller and, studies, 153
Orbital-frontal cortex, 131–132
Organic mercury, 103
Orienting responses, 233–235, *234*
Orr-Urtreger studies, 211
Ortho-substitutions
    basics, 149–163
    cognitive flexibility, 149–153, *151*
    inhibitory control, 160–163, *161*
    working memory, 153, *154–155,* 155–160
Oryzias latipes (Japanese medaka), 317
Oswego cohort study, 158–159, 163
Oye studies, 38, 43

P

Paletz, Newland and, studies, 108
Paletz studies, 101–135
Palop studies, 191
Parahippocampal region, 257
Para-substitutions, 148
Parkinsonism, modeling cognitive deficits
    attentional deficits, 171–175, *173*
    attention set-shifting ability, 172, 174, *174*
    basics, 169–177
    chronic low dose MPTP exposure, 173–176
    cognitive deficits, 169–170
    cued reaction time, 175
    executive function, 171–175
    focused attention, 175
    impulse control, 175
    modeling, 170–171
    motor readiness tasks, 175
    visuospatial attention shifting, 175
Parkinson's disease, 21
Passive avoidance tests
    antimuscarinic agents/amnestic agents, 12
    double mAChR KO mice, 208
    learning and memory, mutant mice, *243,*
        244–245
    memory task impairment, 6
    NMDA antagonists, 41
    polyunsaturated fatty acids, 105
    receptor deficiencies, 207
    scopolamine-reversal, 12–13
Paule studies, 49–64
Pausing, 125
Pavlovian conditioning
    Carassius auratus (goldfish), 318–319, *320*
    Danio rerio (zebrafish), *323,* 324
    fear-potentiated startle, 262
    NMDA antagonists, 41
PCB, *see* Polychlorinated biphenyls (PCBs),
        executive function
Penta-chlorine substitutions, 156
Pentobarbital
    drug challenges, 128
    drug discrimination, 52
    ethanol system interactions, 56
    gamma-aminobutyric acid, 53–54
    methylmercury effects, 134
Peripheral muscarinic blockade, 10
Perirhinal areas, 257
PET (positron emission tomography), 286
Petrie, Whishaw and, studies, 8
Pharmacological approaches, classical, 216
Pharmacologic models
    acetylcholine, 1–2
    ethanol, 49–64

importance, 1
muscarinic receptor antagonists, 5–14
nicotinic receptor antagonists, 21–32
N-methyl-D-aspartate system, 37–44
Pharmacology, aging macaques
    age-related differences, 289–291, *290*
    aging, 285–287
    Alzheimer's(-like) disease, 286–287
    cognition-enhancing drugs, 288–291, *290–291*
    computer-assisted DMTS, 287–288
    future directions, 297
    human aging data, 295, *296*
    macaque aging data, 291–293, *292–294*
Phencyclidine
    gamma-aminobutyric acid, 54
    NMDA antagonists, 39, 41
    NMDA system, human studies, 42
Phenotypic screening, preliminary
    anxiety, *226,* 226–227
    basics, 225
    neurophysiological screen, 227–228, *228–229,*
        230
    open field, 227
Phenylmercury, 103
Phillips and LeDoux studies, 267
Phillips studies, 216
Picciotto studies, 209–210
Pigeons, 54, 125
Pirenzepine, 6
Place fields, 257
Place preference task, 324
Plastic behavior, neurotoxicants, 106–108, *107*
Podhorna and Didriksen studies, 41
Poland cohort study, 92–93
Polychlorinated biphenyls (PCBs), executive
        function
    basics, 1, 147–148, 164
    classes, *148,* 148–149
    cognitive flexibility, 149–153, *151*
    inhibitory control, 160–163, *161*
    memory, 84
    ortho-substitutions, 149–163
    working memory, 153, *154–155,* 155–160
Polyunsaturated fatty acids (PUFAs), 104–105
Popke studies, 58
Positron emission tomography (PET), 286
Post Synaptic Density Marker-95 (PSD-95), 191
Posttraumatic stress disorder (PTSD), 265
Potentiation, 318
PR, *see* Progressive ratio (PR)
Pradel studies, 324
Preattentive processes, 230–232, *232*
Predictive validity, 9, 12
Prefrontal cortex, *see also* Cortex
    drug discovery, 345

go/no-go testing, 240
receptor deficiencies, 205
Pregnanolone, 56
Preliminary phenotypic screening, 225–228, 230
Premature responses, ethanol effects, 58
Prepulse inhibition (PPI), 230–232
Presynaptic markers, 191
Pretraining administration, 41
Primates, *see* Human studies; Nonhuman primates
Progressive ratio (PR), 62
Protofibrillar forms, 185
PSD-95 (Post Synaptic Density Marker-95), 191
PTSD, *see* Posttraumatic stress disorder (PTSD)
PUFA, *see* Polyunsaturated fatty acids (PUFAs)

Q

Quartermain studies, 38
Quaternary anticholinergics, 11

R

RA, *see* Repeated acquisition (RA); Retrograde
        amnesia (RA)
Raber studies, 187
Radial arm maze (RAM), *see also* Mazes
    drug discovery, 343–344
    experimental models comparison, 310
    learning and memory, mutant mice, 253–254
    memory, 6, 85, 187
    methylmercury effects, 108
    nicotinic receptor antagonists, 22, *23–24,* 24
    receptor deficiencies, 203–205
    scopolamine-reversal, 13
    working/short-term memory, 60–61, 306
RAGE (Receptor for Advanced Glycation
        Endproducts) protein, 187–188
RAM, *see* Radial arm maze (RAM)
Rasmussen, Newland and, studies, 122
Rate, reinforcement, 76
Rats, *see also* Rodents
    aging effects, TBI, 310
    anterograde amnesia, 307
    closed cortical impact, 303
    drug challenges, 128
    drug discrimination, 52, 54–55
    ethanol, 56, 58, 61–63
    fluid percussion, 302
    gamma-aminobutyric acid, 53
    go/no-go testing, 240
    inhibitory control, 163
    latent inhibition, 241
    learning, 80–81
    mAChRs, 208
    memory, 8–9, 83, 85, 88